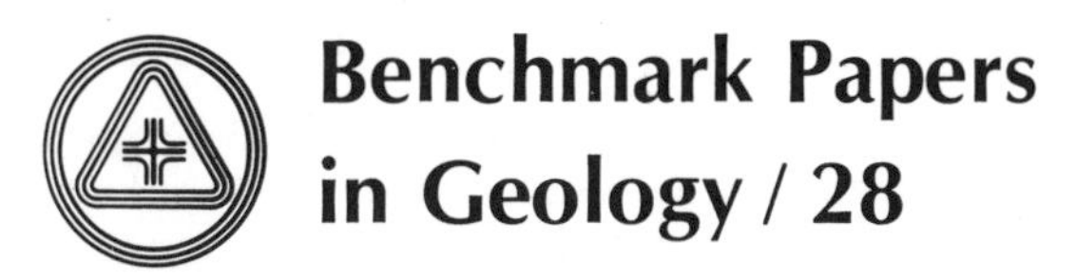

Benchmark Papers in Geology / 28

A BENCHMARK® Books Series

LANDFORMS AND GEOMORPHOLOGY
Concepts and History

Edited by

CUCHLAINE A. M. KING
The University of Nottingham

Dowden, Hutchinson & Ross, Inc.
STROUDSBURG, PENNSYLVANIA

Distributed by
HALSTED PRESS
A Division of John Wiley & Sons, Inc.

Benchmark Papers in Geology, Volume 28
Library of Congress Catalog Card Number: 76–3489
ISBN: 0–87933–192–5

78 77 76 1 2 3 4 5
Manufactured in the United States of America.

LIBRARY OF CONGRESS CATALOGING IN PUBLICATION DATA

Main entry under title:
Landforms and geomorphology
(Benchmark papers in geology/28)

Includes indexes and references.
1. Geomorphology—Addresses, essays, lectures. I. King, Cuchlaine A. M.
GB405.L36 551.4 76–3489
ISBN: 0–87933–192–5

Exclusive Distributor: **Halsted Press**
A Division of John Wiley & Sons, Inc.
ISBN 0–470–15054–8

ACKNOWLEDGMENTS AND PERMISSIONS

ACKNOWLEDGMENTS

BRITISH ASSOCIATION FOR THE ADVANCEMENT OF SCIENCE—*The Advancement of Science*
The Upland Plains of Britain: Their Origin and Geographical Significance

CENTRE FOR SOUTH-EAST ASIAN STUDIES. THE UNIVERSITY OF HULL—*Reprints of Publications by Staff Members*
Natural and Man-Made Erosion in the Humid Tropics of Australia, Malaysia, and Singapore

GEOLOGICAL SOCIETY OF AMERICA—*Bulletin of the Geological Society of America*
Dynamic Basis of Geomorphology
Erosional Development of Streams and Their Drainage Basins: Hydrophysical Application of Quantitative Morphology
Till Fabric

U.S. GEOLOGICAL SURVEY
U.S.G.S. Professional Paper 500A
The Concept of Entropy in Landscape Evolution
U.S.G.S. Professional Paper 500B
Geomorphology and General Systems Theory

UNIVERSITY OF UPPSALA, MINERALOGICAL AND GEOLOGICAL INSTITUTE—*Bulletin of the Geological Institute of Uppsala*
Studies of the Morphological Activity of Rivers as Illustrated by the River Fyris

PERMISSIONS

The following papers have been reprinted with the permission of the authors and copyright holders.

AKADEMIE DER WISSENSCHAFTEN IN GÖTTINGEN—*Nachrichten der Akademie der Wissenschaften in Göttingen, II. Mathematische-Physikalische Klasse*
Deductive Models of Slope Evolution

AMERICAN GEOPHYSICAL UNION—*Transactions, American Geophysical Union*
Yield of Sediment in Relation to Mean Annual Precipitation

AMERICAN JOURNAL OF SCIENCE (YALE UNIVERSITY)—*American Journal of Science*
Geomorphological Systems—Equilibrium and Dynamics
Interpretation of Erosional Topography in Humid Temperate Regions
Time, Space, and Causality in Geomorphology

Acknowledgments and Permissions

THE ASSOCIATION OF AMERICAN GEOGRAPHERS—*Annals of the Association of American Geographers*
The Geographic Cycle in Periglacial Regions as It Is Related to Climatic Geomorphology
A Technique of Morphological Mapping

B. T. BATSFORD LTD.—*The Cycle of Erosion in Different Climates*
Excerpts

CAMBRIDGE UNIVERSITY PRESS—*Great Britain Geographical Essays*
Relief

GEBRÜDER BORNTRAEGER—*Zeitschrift für Geomorphologie*
Principles in a Geomorphological Approach to Land Classification

GEOLOGICAL SOCIETY OF LONDON
Proceedings of the Geological Society of London
On Glaciers, and the Evidence of Their Having Once Existed in Scotland, Ireland, and England
Quarterly Journal of the Geological Society of London
On the River Courses of England and Wales
The Study of the World's Plainlands: A New Approach in Geomorphology

HARPER & ROW, PUBLISHERS, INC., and METHUEN & CO. LTD. for the BRITISH GEOMORPHOLOGICAL RESEARCH GROUP—*Spatial Analysis in Geomorphology*
Digital Simulation of Drainage Basin Development

INSTITUTE OF BRITISH GEOGRAPHERS—*Transactions of the Institute of British Geographers*
The Changing Sea Level

INSTITUTION OF CIVIL ENGINEERS—*Transactions of the Institution of Civil Engineers*
Field and Model Investigation into the Reasons for Siltation in the Mersey Estuary

INTERNATIONAL GLACIOLOGICAL SOCIETY—*The Journal of Glaciology*
The Mechanics of Glacier Flow

MACMILLAN & CO. LTD.
Climatic Geomorphology
Attempt at a Classification of Climate on a Physiographic Basis
Morphological Analysis of Landforms
Excerpt

MACMILLAN COMPANY OF AUSTRALIA PTY. LTD.—*Land Evaluations*
Review and Concepts of Land Classification

MACMILLAN JOURNALS LTD.—*Nature*
Catastrophic Storm Effects on the British Honduras Reefs and Cays

METHUEN & CO. LTD.—*Water, Earth, and Man*
Geomorphic Implications of Climatic Changes

THE ROYAL SCOTTISH GEOGRAPHICAL SOCIETY—*The Scottish Geographical Magazine*
Problems of Scottish Scenery

SOCIETY OF ECONOMIC PALEONTOLOGISTS AND MINERALOGISTS—*Journal of Sedimentary Petrology*
Brazos River Bar: A Study of the Significance of Grain Size Parameters

THE SWEDISH GEOGRAPHICAL SOCIETY—*Geografiska Annaler*
Recent Development of Mountain Slopes in the Kärkevagge and Surroundings, North Scandanavia

UNIVERSITY OF CHICAGO PRESS—*The Journal of Geology*
Magnitude and Frequency of Forces in Geomorphic Processes
Stream Lengths and Basin Areas in Topologically Random Stream Networks

UNIVERSITY OF TORONTO PRESS—*Permafrost in Canada. Its Influence on Northern Development*
Excerpt

MRS. E. M. WOOLDRIDGE and the late MRS. V. C. LINTON—*Structure, Surface, and Drainage in South-East England*
Excerpts

SERIES EDITOR'S PREFACE

The philosophy behind the "Benchmark Papers in Geology" is one of collection, sifting, and rediffusion. Scientific literature today is so vast, so dispersed, and, in the case of old papers, so inaccessible for readers not in the immediate neighborhood of major libraries that much valuable information has been ignored by default. It has become just so difficult, or so time consuming, to search out the key papers in any basic area of research that one can hardly blame a busy man for skimping on some of his "homework."

This series of volumes has been devised, therefore, to make a practical contribution to this critical problem. The geologist, perhaps even more than any other scientist, often suffers from twin difficulties—isolation from central library resources and immensely diffused sources of material. New colleges and industrial libraries simply cannot afford to purchase complete runs of all the world's earth science literature. Specialists simply cannot locate reprints or copies of all their principal reference materials. So it is that we are now making a concerted effort to gather into single volumes the critical material needed to reconstruct the background of any and every major topic of our discipline.

We are interpreting "geology" in its broadest sense: the fundamental science of the planet Earth, its materials, its history, and its dynamics. Because of training and experience in "earthy" materials, we also take in astrogeology, the corresponding aspect of the planetary sciences. Besides the classical core disciplines such as mineralogy, petrology, structure, geomorphology, paleontology, and stratigraphy, we embrace the newer fields of geophysics and geochemistry, applied also to oceanography, geochronology, and paleoecology. We recognize the work of the mining geologists, the petroleum geologists, the hydrologists, the engineering and environmental geologists. Each specialist needs his working library. We are endeavoring to make his task a little easier.

Each volume in the series contains an Introduction prepared by a specialist (the volume editor)—a "state of the art" opening or a summary of the object and content of the volume. The articles, usually some twenty to fifty reproduced either in their entirety or in significant extracts, are selected in an attempt to cover the field, from the key papers of the last century to fairly recent work. Where the original works are in foreign languages, we have endeavored to locate or commission translations. Geologists, because of their global subject, are often acutely aware of the oneness of our world. The selections cannot, therefore, be restricted to any one country, and whenever possible an attempt is made to scan the world literature.

To each article, or group of kindred articles, some sort of "highlight commentary" is usually supplied by the volume editor. This commentary should serve to bring that article into historical perspective and to emphasize its particular role in the growth of the field. References, or citations, wherever possible, will be reproduced in their entirety—for by this means the observant reader can assess the background material available to that particular author, or, if he wishes, he, too, can double check the earlier sources.

A "benchmark," in surveyor's terminology, is an established point on the ground, recorded on our maps. It is usually anything that is a vantage point, from a modest hill to a mountain peak. From the historical viewpoint, these benchmarks are the bricks of our scientific edifice.

RHODES W. FAIRBRIDGE

CONTENTS

PART II: MAJOR CONCEPTS IN GEOMORPHOLOGY

CONTENTS BY AUTHOR

INTRODUCTION

The science of geomorphology has developed alongside geology, and like most other sciences has grown in ever-widening circles, resulting in continuously expanding contacts with other fields of learning. This volume intentionally covers a very wide field, with the result that inevitably it must be highly selective. For this reason, rather more exerpts and condensations than full articles are reproduced. Part I deals with the history of the study of landforms, the field of geomorphology. From the early beginnings to the end of the eighteenth century constitutes the first phase of growth. The second and third phases occurred in the nineteenth century, and the fourth in the early twentieth century. During these phases the main concepts of geomorphology were formulated and discussed. Part I is devoted to an account of the main workers in the four phases and their major contributions and ideas. A consideration of the concepts themselves forms the beginning of Part II.

The concepts developed by the pioneers have been further studied and added to by new methods and ideas formulated in the last few decades, during which time there has been a great expansion in the number of workers, their techniques of study, and the relevance and significance of their results in the complex matter of present-day living in a modern technological age. The importance of and our dependence on the natural environment, its form, the processes that shape it, and the material of which it is composed, which make up the science of geomorphology, are continually increasing in the present way of life.

Early man was also dependent on his environment for his basic needs, and in more modern days, when leisure has allowed people more time to ponder the operations of nature, geomorphology grew out of a curiosity about the workings of the natural landscape. It is difficult to appreciate the imagination and originality needed to evolve ideas about geomorphology, when so much is now taken for granted. For example, it is now generally considered that fluvial and slope processes create valleys, but for a long time the relationship between a river and its valley was by no means clear.

One of the most difficult constraints from which the early European geologists and geomorphologists had to break loose was that imposed by the literal interpretation of the book of Genesis in the Bible. Bishop Ussher in 1650 proclaimed that the earth was created in 4004 B.C. In 1654, Dr. J. Lightfoot, vice-chancellor of Cambridge University, stated "Heaven and Earth, centre and circumstance were made in the same instance of time, and clouds full of water and mass were created by the Trinity on 26 October, 4004 B.C. at 9 o'clock in the morning." The time and date were based on a literal belief in the account of the Creation given in Genesis and the generations enumerated in the Old Testament. Under the circumstances of this period it must have required real courage, imagination, and strength of mind to break out of the straightjacket imposed by this chronology, especially in view of the power of ecclesiastical authority then.

When the time was ripe for a breakaway from these social constraints, the whole climate of thought was changing. Early ideas in many fields were being questioned, at the same time as new lands in different climatic zones were being discovered. Hardy, enquiring explorers were investigating the polor regions, the great deserts, the tropical forests, and the high mountains, bringing back accounts of a wide variety of phenomena and landscapes. In the promulgation of geomorphic principles, scientific ideas in other fields were used by those who first broke out of the biblical straightjacket. The principle of arguing by analogy was very important in making the necessary imaginative jumps. For the new ideas to be accepted, it was also essential that they be promulgated when the scientific climate was right for their reception. Otherwise they would fall on stony ground, wither away, and be forgotten for a long time. Genius is, therefore, not in itself enough, and genius has been said to be an ability to see analogy; but it must flourish in a period when lesser men are willing to change their minds. Scientists tend to be a conservative group, often withholding endorsement until the tide has really turned.

Some of the earliest ideas in geomorphology came to the surface about the eve of the French Revolution, when the religious dogmatism of the Middle Ages was being weakened or cast off, and an increasing interest was being shown in the mysteries of nature. The beginning of the Industrial Revolution was to provide a practical application for scientific knowledge. Thus toward the end of the eighteenth century the time was ripe for new ideas concerning the formation and character of the landscape. As they emerged, there was lively debate as old assumptions were slowly given up, to be replaced by logical conclusions reached from precise observation.

The four phases into which the historical section of this volume is divided cover the period during which geology became established as an academic discipline; geomorphology thus developed alongside it, as a branch of the main subject, which was initially defined in the broadest sense, literally as "earth science."

In the first phase, the period up to the beginning of the nineteenth century, geomorphology was dominated by arguments between the school led by Abraham Gottlob Werner in Germany and the British school under James Hutton in Edinburgh. Other notable names of the early period were de Saussure, a Swiss Alpinist, who was the first to have the opportunity to examine glaciers from a scientific point of view. Hutton's views were well ahead of the time in which he wrote, and even when exemplified and clarified by his friend, John Playfair, their collective ideas did not receive the attention they merited. Indeed it was not until the middle of the nineteenth century, at the beginning of the third phase, that Greenwood championed their cause with vigor.

Meanwhile, in the second phase geomorphic ideas were still constrained by biblical ties, with the Noachian deluge providing a useful explanation of many otherwise inexplicable phenomena, such as erratics and drift deposits of glacial origin. It was in phase 2 that this doctrine was contested, especially by Louis Agassiz and other supporters of the new theory of the Ice Age and the work of glaciers and ice sheets; the greater extent of those glaciers was coming to be recognized in many areas on the continent of Europe, in Britain, and in North America. A central figure of the second phase was Sir Charles Lyell, the first edition of whose influential *Principles of Geology* appeared in 1830. Dean Buckland provided a link with ecclesiastical ideas, and Conybeare also made important contributions.

In phase 3, the arguments shifted in Britain under the aegis of such great names as Jukes and Ramsey, who were arguing with Greenwood concerning the relative efficacy of marine and subaerial erosion. On the other side of the Atlantic, the western pioneers included some of the greatest geomorphologists of all time; G. K. Gilbert is outstanding, and his name is linked with other important workers, such as Powell and Dutton. These great men, who put forward some of the most basic and important concepts in geomorphology, set the scene for the fourth phase, the early twentieth century.

Phase 4 is dominated by the best known of all geomorphologists, William Morris Davis, who provides a useful link between America, Europe, and elsewhere, as he had followers, for example Sir Charles Cotton in New Zealand, and rivals, including Lester C. King and especially Walther Penck in Germany. The arguments between Davis and his friends, Albrecht and Walther Penck, father and son, provide much valuable material in the development of geomorphic ideas. The geographic and geological bases of geomorphology are made clear in their respective approaches to the study of landforms. Davis wrote more as a geographer and Penck as a geologist; it is, therefore, curious that at present geomorphology is taken as a branch of geology in the United States, while in Europe it comes largely under the aegis of geography. In the United States it tends to blend in with Quaternary stratigraphy; in Europe, including Britain, the two are more often kept apart.

The concepts of geomorphology were evolved during the four phases. Examples of the concepts themselves are given in Part II. The early philosophies of Werner and Hutton can be identified as catastrophism and uniformitarianism, respectively. The first is based on the literal biblical account of the Creation, in which time would not permit the slow development of landforms, if the earth were created in 4004 B.C. The biblical flood was a major catastrophe for which evidence was sought in the landscape. The "drift theory" was based on the discovery of certain deposits that were thought to provide the best evidence for this event. It could account for the erratics that were found so widely distributed. The flood idea thus became linked with the theory of the Ice Age in suggesting that the erratics were carried on floating icebergs, which were dispersed during the flood.

In contrast to the catastrophic concept stood the uniformitarian concept enunciated by James Hutton. He suggested an analogy between the circulation of the blood in the human body, a fact very familiar to him through his medical training, and the circulation of matter in the landscape through the cycle of weathering, erosion, transport, and deposition, to be revived from time to time by uplift. This concept was based on a recognition of unconformities in the geological strata of Scotland and an appreciation of the slow development of land forms through time. His famous dictum, "No vestige of a beginning—no prospect of an end," sums up his uniformitarian ideas and his pragmatic ability to take a very important step in philosophy.

As with so many concepts in geomorphology, modern ideas represent a compromise between the often too cut and dried ideas of the early workers, who tended to be rather dogmatic. It is now realized that neither strict uniformitarianism nor complete catastrophism are true; there ia an element of truth in both concepts. Some processes continue to act over prolonged periods, such as soil creep; others occur intermittently, for example rotational slumping. Nearly all vary in intensity over time, including river discharge and wave action. Some only take place at fairly prolonged intervals as isolated events, such as earthquakes, volcanic eruptions, glacier surges, floods, and hurricanes. Instead of uniformitarianism versus catastrophism, the concept of recurrence intervals in relation to the intensity or magnitude of phenomena has developed. This has now been established for many processes that vary in intensity, such as river discharge and wave height, both of which vary with a Poisson distribution, a skewed distribution in which the rarest events are those of maximum intensity and the common events are those of small dimensions. In a skewed pattern of this sort, the mean is toward the lower end of the range.

The early workers were impressed by the stability of the land, and much work was done to establish the effects of changes of base level related to fluctuations of sea level, Charles Darwin having brought special attention to them in his theory of coral reefs. Changes in sea level were thought to control the pattern of denudation, and hence to be recognizable in landforms. The writing of Suess in *Das Antlitz der Erde* (*The Face of the Earth*) did much to popularize this approach and extended it to stratigraphic cycles. The eustatic theory and its application by geomorphologists to denudation chronology owe much to the cyclic concept of W. M. Davis. These concepts were further developed during the first half of the twentieth century, and in the 1920s the International Geographical Union formed a commission that optimistically set out to correlate erosion surfaces around the world. This work was particularly stimulated by H. Baulig's contributions to French and British geomorphology in the late 1920s and 1930s. Many studies of denudation chronology of small areas continued to be undertaken through the 1940s and 1950s.

The problems inherent in approaching geomorphology through denudation chronology, which is a study with a time-bound base, became more obvious as new information indicated more and more the mobility of the earth's crust, both in the horizontal and vertical senses. The concept of mobility was at the basis of the work of W. Penck, often thought of as opposed to that of W. M. Davis. Penck sought in the landscape for clues that would enable him to deduce the crustal movements that were responsible for the operation of the earth-shaping processes, with particular reference to those forming and modifying slopes. Thus he attempted to deduce earth movements from morphology. The role of the earth's crustal mobility was noted early by G. K. Gilbert in his work on the warping of the shoreline features of Lake Bonneville, the larger precursor of the Great Salt Lake in Utah. Gilbert's work on Lake Bonneville was published in 1890. The effects of isostatic rebound in Scandinavia and the Canadian north, as a result of deglaciation, confirmed these findings;

more recently, since the beginning of the 1960s, the findings of plate tectonics have revealed the mobility of the earth's crust in the horizontal sense also, as sea-floor spreading and oceanic crust formation at the mid-ocean ridges indicate active movement, often on the order of several centimeters each year. Thus over the years the concept of simple eustasism against a background of crustal stability has increasingly given way to the realization that the earth's crust is everywhere mobile, and that this mobility is reflected by variable rates and on all scales, from the large plate tectonics to the small slope facets that reveal the effects of uplift or subsidence.

The extremely rapid change of land and sea level associated with the oscillations of large, continental-based ice sheets in the Pleistocene period has given rise to a new dimension and rate of change in the creation of landforms in the most recent of geological times. As a result, geomorphic processes currently observed in action are not typical of those operating throughout the major part of geological time. It is possibly no accidental coincidence that not only has the human race evolved in a period of almost unprecedented changes from the environmental point of view, but also is itself responsible for the exponential rising rate of activity of some recent landscape modifications that will be considered at the end of the book.

Another aspect of geomorphology is closely associated with the problems just mentioned. This is the concept of climatic control on landform development, which has been studied in depth especially by French and German geomorphologists during the present century. It is closely linked with the names of well-known French geomorphologists, including J. Tricart, A. Cailleux, P. Birot, as well as with Büdel in Germany and Fournier and Peltier in the United States. These climatic geomorphologists consider that the different climates, through their effect on soil, vegetation, runoff, and such parameters as precipitation effectiveness, determine the nature of the operation of processes and the resultant landforms. The newly recognized science of paleopedology is playing an increasingly significant role in the identification of former climatic controls and adds a new tool to denudation chronology. The situation is made more complex by the very rapid changes of climate, covering virtually the entire earth, associated with the waxing and waning of the large land-based ice sheets in North America and northern Europe, respectively the Laurentide and Scandinavian ice sheets. Thus very nearly all landscapes are both polycyclic, as a result of sea-level fluctuations and crustal motions, and polygenetic, as a result of climatic fluctuations, which were particularly extreme during the very changeable Quaternary period. The more work that is done, the greater is the complexity seen to be among the many processes and variables that affect the genesis and character of landforms.

Another aspect that has long been the foundation of an important theme in geomorphology is the control of rock type and structure on landforms. This aspect has not given rise to as unified an approach to the subject as has the climatic, eustatic, or mobilistic schools, but it has nevertheless produced some interesting and worthwhile studies. The large field of karst geomorphology is one. Rock type and structural control have been shown to exert a marked effect on many landforms, such as tors, mesas, buttes, and structurally controlled scarps. The relationship between drainage patterns and structure and rock type has been another fruitful field of study. There have been discussions concerning the relative effectiveness of climate and rock type in controlling landscape development. The dominance of the latter accounts for the relative rates of coastal erosion around much of the British shoreline. The slow erosion of the exposed, hard-rock west coast contrasts with the locally very rapid erosion on the sheltered eastern coast, where soft rocks and drift deposits form much of the shoreline. The west coast retains evidence of interglacial events, whereas parts of the east coast are eroding at rates of up to 2 to 3 m per year, and several kilometers have been lost in the last thousand years.

A new dimension has been given to geomorphic studies since the development of computers in the last two decades. This is based on the application of quantitative techniques in research at all stages of study. Many investigations begin with quantitative field measurements of form, material, and process, and continue with quantitative analysis of the resulting measurements, using increasingly complex statistical and numerical methods. Such methods allow a multivariate complex situation to be analyzed with increasing refinement, but there is a danger in these techniques in that the results tend to become divorced from reality and may be difficult to apply to the problems of the real landscape. Nevertheless, quantification is a very significant aspect of recent geomorphic work. The range of numerical techniques used is very great, extending from simple nonparametric statistics at one end to procedures at the other end of the scale that may involve complex multivariate, multidimensional techniques requiring a large computer. Most of the standard statistical techniques have been used with some success, and some special ones have been developed or adapted, such as vectorial data analysis and time series analysis.

Numerical methods have been applied increasingly to observations of processes in the field, as well as to the quantification of morphology and material description. The calculation of hypsometric integrals, mean slopes, and drumlin elongation exemplifies the former; the calculation of moment measure to describe sediment-size distribution and the recording of stone roundness exemplify the latter.

In common with other disciplines, geomorphology is now using the concept of models as a framework for many investigations. Models can be divided into a number of types, all of which have been used by geomorphologists. Some of the more important types include mathematical and/or physical models, hardware or scale models, and analog and simulation models. In the first category are several useful slope model studies, often based on iterative techniques, whereby an initial form is traced through a number of stages to a final form. The physical or deterministic models are based on the use of differential equations to model a dynamic situation. The study of sediment movement in a flowing medium and the flow of ice exemplify this method of study. Much of this work has been carried out by physicists, such as Bagnold (1941), who studied the movement of sand by wind and waves, and J. F. Nye, who has worked on the dynamic mechanism of ice flow.

Hardward models have been used to study the processes associated with rivers and waves. The models can be either general or specific; in the latter case a special area or problem can be modeled. This method is much used by engineers in their study of the problems of river control, coast defense or coastal reclamation, and improvement schemes. The use of scale model of an estuary, such as San Francisco Bay or Chesapeake Bay in the United States, or the Scheldt–Rhine delta region of Holland or the Wash, Mersey, or Humber in Britain, is a common way of establishing the effect of engineering works before they are actually constructed. Models of this type have also provided many valuable basic research results, for example on the movement of sand by waves under different conditions.

Models have the great advantage of both simplifying reality and allowing control of the many variables involved in all natural situations and processes. These advantages apply also to simulation models, which are often based on computer analysis. Models also allow simplification and control of variables, such as computer simulations of delta and spit formation. The types of models already mentioned all have these two fundamental advantages in common, and in this characteristic they differ from the analog model. The value of appreciating analogies has already been mentioned, and the ability to associate seemingly dissimilar features or processes has led to several important advances in geomorphology. The work of Lewis (1949) on the movement of a cirque glacier is a good example of reasoning by analogy; he related the flow of the ice in the cirque to the movement of a slump block in a rotational shearing landslide, such as occurs at the Warren near Folkestone in England. Both phenomena have in common an element of rotational movement.

The term "model" has been overworked in some senses. Nevertheless, it does provide a useful framework for individual studies of the type mentioned in the preceding paragraph. The term "model" has also been used to describe the framework of a whole discipline and to consider its relationship with adjoining disciplines. One problem of geomorphology is that it tends to lack such an overall framework into which individual studies may be fitted, so that their place in the whole study is apparent. The cycle scheme advanced and so vigorously canvased by W. M. Davis provided one such framework model into which geomorphology was fitted for a time, but from which it has subsequently emerged. However, no other suitable cohesive framework has been developed.

Davis's cycle scheme was based on the trilogy of *structure, process,* and *stage.* The latter control implies time, and Davis's scheme is essentially time-dependent. This control is not always relevant or useful, and modern schemes tend to stress the time independentness of some geomorphic situations. It might, therefore, prove valuable to change the trilogy of elements in the geomorphic sphere from structure, process, and stage to *process, material,* and *morphology.* The last element is the fundamental problem and goal of geomorphology, in that it must be explained in terms of the first two elements, the processes and material. In the trilogy of process, material, and morphology, only the first two are independent variables; morphology is dependent. In the trilogy of structure, process, and stage, all three variables are independent.

The contrast between the time-dependent geomorphology of W. M. Davis and the time-independent geomorphology of some recent writers, such as J. Hack, has been expressed well in the systems approach to geomorphology. This procedure has been championed by R. J. Chorley, who drew attention to the contrast between the time-dependent scheme of W. M. Davis and the alternative time-independent scheme. The Davisian scheme has certain elements of a closed-system approach; there is a certain amount of potential energy at the start of the cycle, which is used up as kinetic energy as the cycle progresses, leading to a final state of minimum energy. The system cannot be entirely closed, as the energy on which it operates is derived from outside forces in the form of air, flowing water, and sunlight.

The open-system approach is based on the concept of dynamic equilibrium; energy can pass through the system, which comes to a more or less stable form when the state of equilibrium has been reached. Because the controls are likely to vary through time, the equilibrium cannot be static; it must be dynamic, varying within a range imposed by the vagaries of the controlling processes. These processes are likely to be affected by the morphology and/or the material on which they are acting, so that feedback loops are set up; they can be either negative, which are self-adjusting, or positive, which are self-generating. The problems resulting from feedback loops add greatly to the complexity of geomorphic analysis.

The term "system" is now often heard in relation to the ecosystem, which brings in the organic element. The earth must be thought of not only as an inorganic system; the organic aspects are vitally important in many aspects of pure geomorphology as well as in so many other respects. The organic elements of the ecosystem include the microfauna and flora of the soil, together with the larger plants and animals that coinhabit the landscape; collectively these elements form the complex ecosystem, all elements of which are mutually interrelated and interdependent. A major part in controlling the processes that act on slopes is played by vegetation in association with climate and landform. It is this aspect that is most readily modified by man, either directly or indirectly, through changes in farming practices and the introduction or elimination of plant and animal species. Man is an important element of the ecosystem is becoming an ever-increasingly potent agent of geomorphic change. At the same time, man is becoming increasingly aware of his environment.

The increasing interest in and concern for the environment has led to the recent development of a new field of geomorphic study, the field of environmental or applied geomorphology, which is linked to the exploitation and preservation of the landscape. A type of study that comes within this field, for example, is that concerned with coastal protection and preservation. A sound knowledge of coastal processes is essential for a successful protection scheme. The problem must be considered as a whole. It is essential to delimit the whole coastal system in which the individual units mutually interact, and to treat the system in its entirety. It is useless to protect one section of coastline if the erosion as a result is only transferred farther downdrift along the coast. In all these and similar situations it is necessary to work with nature, so that its workings are fully understood and appreciated.

This volume attempts to point to the steps by which this knowledge has been gradually accumulated, as the workings of nature were slowly and laboriously probed by many generations of geomorphologists. The story is still by no means ended; much remains to be done before we have a complete knowledge of natural processes and the landforms to which they give rise. Answers to the many fascinating problems raised by geomorphology are being sought on an increasingly wide front; many different concepts are being explored and many different methods are being used to study these concepts.

REFERENCES

Bagnold, R. A. 1941. *The physics of blown sand and desert dunes.* Methuen, London.

Lewis, W. V. 1949. Glacial movement by rotational slipping. *Geog. Ann., 31,* 146–158.

Part I

HISTORY OF GEOMORPHOLOGY

Editor's Comments on Papers 1 Through 11

1 PLAYFAIR
Excerpts from *Illustrations of the Huttonian Theory of the Earth*

2 BUCKLAND
On the Excavation of Valleys by Diluvian Action, as Illustrated by a Succession of Valleys which Intersect the South Coast of Dorset and Devon

3 CONYBEARE
Excerpts from *Report on the Progress, Actual State, and Ulterior Prospects of Geological Science*

4 LYELL
Excerpts from *Principles of Geology*

5 AGASSIZ
On Glaciers, and the Evidence of Their Having Once Existed in Scotland, Ireland, and England

6 GILBERT
Excerpts from *Report on the Geology of the Henry Mountains*

7 RAMSEY
Excerpts from *On the River Courses of England and Wales*

8 GREENWOOD
Excerpts from *Rain and Rivers*

9 JUKES
Atmospheric vs. Marine Denudation

10 DAVIS
The Geographical Cycle

11 PENCK
Excerpt from *Morphological Analysis of Landforms*

Part I is devoted to a brief history of geomorphology, the four phases of which have already been mentioned. Each phase will be introduced separately.

PHASE 1

The first phase covers the seventeenth and eighteenth centuries, during which period the early ideas were formulated. The two main concepts of catastrophism and uniformitarianism were put forward. The first is associated with Abraham Gottlob Werner, who lived from 1749 to 1817, and the second with James Hutton, born in 1726, and his friend and collaborator, John Playfair, who died in 1819.

Werner was born into a mining family; he early showed an interest in rocks, an interest that was fostered by his father. Werner attended Freiberg Mining Academy in Saxony, and then the University of Leipzig, where he studied history, philosophy, and foreign languages before turning to mineralogy. Later he returned to lecture at Freiberg, where his influence became substantial.

Werner had a well-disciplined mind with which to study the problems raised by the rock strata that were his main concern. He recognized that each stratum had its own characteristic fossils. His thoughts led him to formulate a theory involving a terrestrial conception of the birth of the world in the bowl of a large ocean. In his opinion each rock layer marked a new advance of the waters, which laid down a new stratum on the previous one. He thought that the densest material settled out of a heavy aqueous solution first. This material, he suggested, formed a core of granite, separated out by a process of crystallization, with later rocks being laid down on this core. As the material gradually settled out, so the oceans gradually shrank until they were confined to their present area. He divided the rocks into five types in order of deposition. The first were the primitive granites with no fossils. The second were a transitional series, having some fossils. The third he called *Flötz;* they included limestone, ore-bearing rocks, coal, chalk, and basalt. The fourth were derivative materials, such as sand, clay, or pebbles; finally, the fifth were volcanic rocks.

Werner had a high reputation as a mineralogist, but he did relatively little on the study of process; his main contribution was the introduction of the first ordered worldwide scheme to account for the earth's rocks. He thus founded what became known as the neptunist school, which was later to argue vehemently with the opposing vulcanist school, whose supporters believed that not all rocks were of marine origin, with specific reference to igneous and volcanic rocks.

The study of process was the major contribution of James Hutton and John Playfair of Edinburgh in Scotland. They may be described as the founders of geomorphology as distinct from geology. Hutton was the first real fluvialist. He also put forward the first reasonable idea of geological evolution. He was a farmer in Berwickshire in southeast Scotland, near the English border, but had started work as a lawyer, later giving up this profession to study medicine and chemistry. He never practiced as a doctor although he qualified in medicine. Instead he turned to farming, from which pursuit his interest in rocks and minerals grew, for he recognized their importance in governing soil character from the farming point of view. Some of his chemical experiments were commercially successful. His academic interests gradually took precedence over farming and he moved into the inquiring and actively intellectual society of Edinburgh. The Royal Society of Edinburgh provided the main forum for discussion.

Hutton's first paper expressing his geological ideas,"Concerning the System of the Earth, Its Duration and Stability," was read in 1785. It was published in 1788 under the title "Theory of the Earth; or an investigation of the laws observable in the composition, dissolution and restoration of land upon the globe." It was not well received at first, probably because the ideas were too far advanced for the period. His major idea was the principle of uniformitarianism, although, it was not so designated at first, coupled with the recognition that catastrophic events were not essential to explain the character of the earth's surface and the rock strata. He considered naturally occurring processes sufficient to account for all phenomena; existing processes were regarded as a key to past processes.

Hutton also appreciated the slow change of landforms, and that present features are ephemeral, being only one stage of a long succession of stages, stretching both into the past and the future. He recognized the significance of unconformities as indicating a succession of revolutions, each unconformity marking one epoch. The sand and gravel of today's rivers were seen to be the sedimentary rocks of tomorrow. From his knowledge of chemistry he recognized sand as the breakdown of older rocks, thus correctly relating the relative ages of the two deposits. Limestone and chalk, he realized, were submarine deposits of largely fossiliferous material, and he appreciated the orderly deposition of particulate material on the seabed as the source of new sedimentary rocks. Thus he elaborated the idea of the cycle of erosion and sedimentation and the part played by rivers in attacking the land and carrying debris to the sea to form future sedimentary rocks. He also saw that the breakdown of rocks created soil, a necessary element for plant growth.

Hutton's great contribution was the recognition of the transience of landforms, of continual destruction leading to continual renewal. He realized that erosion would eventually wear down the earth, but the presence of marine shells on land indicated that upheavals periodically take place to lift up marine strata above sea level. The process includes a stage of consolidation and upheaval, as indicated by faults and folds and by high dips in the strata. He suggested volcanic action as a possible cause of the upheavals. Hutton was a vulcanist, identifying granite as igneous and having been at one stage in a fluid state. He was aware of the action of currently acting processes in wearing away mountains, and also of the slowness of the operation. He reached the now well-known conclusion that he could see "No vestige of a beginning—no prospect of an end."

A number of reasons led to the poor reception of his eminently sensible ideas. The time was not ripe for such advanced ideas, his style was poor and long-winded, his personality retiring, and the circulation of the publications of the Royal Society of Edinburgh was limited. He was, however, goaded through criticism into elaborating his ideas, and in 1795 produced a two-volume work, *Theory of the Earth with Proofs and Illustrations.* Hutton died before a planned third volume was published, and it was left to John Playfair to explain and promote his work.

John Playfair was professor of natural philosophy at Edinburgh University, and a good mathematician. He was fully familiar with Hutton's theory, which he really understood. He also had a good intellect and a good command of the English language. He could, therefore, put Hutton's ideas into good literary English, and he did this in a volume *Illustrations of the Huttonian Theory of the Earth,* first published in 1802. Playfair made many valuable additions of his own to Hutton's ideas, including the important principle of accordant stream junctions. His statement of this principle is given in Paper 1. Playfair's main point was that rivers were the main agents of terrestrial erosion, and that they had carved the valleys and plains, a view opposed to that of Werner and the neptunists.

Playfair's contribution to this aspect of Hutton's theory was his most important. He mentions many of the processes involved in the work of erosion, such as water vapor and ice, as well as rivers of all sizes. He recognized the importance of rivers as transporting agents, and also their work in causing attrition of both bedrock and transported debris by their movement from the hilltops to the coast; he recognized also the work of abrasion due to wave action as an important element of coastal erosion. To support his views he even attempted some quantitative assessment of a river's load, considering that a river in flood carries sediment amounting to 1/250 part of its volume. He observed that a river torrent can carry a load next to that of glaciers, and pointed out that the weight of the boulders is less in water, and that the applied force is proportional to the square of the velocity. He thus anticipated some of the work of G. K. Gilbert carried out nearly a century later.

A major advance was Playfair's recognition that each river basin was adjusted to the size of the stream in it, and that each tributary met another or the main stream at an accordant junction. He reached the conclusion that each river must adjust its own valley, and in so doing it tended to eliminate irregularities. He realized that it was pressing the bounds of coincidence too far to assume that valleys were developed independently of the rivers in them. This was a fundamental and novel concept, which even Charles Lyell could not fully accept, although now it seems so obvious and reasonable. Three important conclusions emerge: (1) rivers cut their own valleys: (2) each river slope is adjusted to an equilibrium among velocity, discharge, and load; and (3) the whole river system is integrated by the natural adjustment of its parts. This is the work of a mathematician with a very logical mind. Playfair recognized the role of river superimposition, and had very advanced views on river terraces and on the elimination of knick points in establishing a graded profile. It is evident from the second excerpt in Paper 1 that Playfair had really looked at the way in which rivers work in the field, and he appreciated that rate of change depends on the stage of development being greatest at the beginning when conditions are furthest from equilibrium.

Another point that Playfair emphasized, in countering Hutton's critics, was the importance of heat in the consolidation of rocks and in their subsequent elevation. He invoked such evidence as vein intrusions, pointing to the Whin Sill faults and unconformities. He pointed out that igneous intrusions demonstrate that not all igneous rocks are older than sedimentary ones.

Playfair disseminated his views through his involvement with the *Edinburgh Review,* which was first published in 1802, and which had a fairly wide circulation in Scotland and England. By the time Playfair died in 1819 the Huttonian theory was making only desultory progress, but Playfair had the wisdom to see that time was on their side, as indeed it has been.

PHASE 2: EARLY NINETEENTH CENTURY

Catastrophic ideas flourished during phase 2. Four authors have been selected to illustrate the state of knowledge of geomorphology during this period. One of the most influential was Sir Charles Lyell, who was born the year that Hutton died, in 1797, and who lived until 1875. The other representatives of this period, a time when biblical influence was still strong, are W. D. Conybeare, Dean Buckland, and Louis Agassiz.

Conybeare lived from 1787 to 1857 and was a contemporary of Dean Buckland, who was an influential geologist. The fourth scientist, Louis Agassiz, was born in 1807 and died in 1873. His claim to fame is concerned with the major part he played in introducing and popularizing the glacial theory to the geological community on the European continent as well as in Britain and North America.

Dean William Buckland (1784–1856) and his supporter, Adam Sedgwick (1785–1873), were both still under the influence of biblical ideas, and the Flood played a major part in their geomorphic thinking. Buckland was the son of a Devonshire clergyman. He collected fossils and studied at Oxford. He was ordained and became a Fellow at Oxford in 1809, being appointed reader in mineralogy in 1813. He visited Werner in Germany, where his leaning toward catastrophism gained support. In 1819 a special readership in geology was established for him at Oxford, because his lectures aroused so much interest. He had a great sense of humor and a good intellect, and was a devout churchman.

Buckland truly represents the geology of his day, and genuinely believed that there was no conflict between science and religion. He accepted the watery catastrophe of the Flood, finding evidence for it in nature, and his ideas on the Creation, in deliberate conflict with those of Hutton, were also on catastrophic lines. He followed Werner more closely, and in turn was followed by Sedgwick. He appreciated the hydrological cycle, but could not concede that rivers or even torrents had the power to create their own valleys. He was, however, inconsistent in his adherence to catastrophism, in that he considered that rocks could be formed by gradual processes. He tacitly stated that new rocks could be derived from older ones, following Hutton's ideas, but without giving Hutton any credit. The Noachian deluge was accounted responsible for the more recent erosion and deposition, destruction being its dominant role in his view. He ascribed the wide valleys and drift deposits to it, but he does not allow rivers to erode effectively. These views were put forward in a very popular lecture, which had an immense success, and which led to the formation of the diluvialists as a major geological group.

Paper 2 indicates that Buckland had a good field knowledge of geology, as far as the rocks themselves were concerned, but that his appreciation of geomorphic processes was not so good. In some instances he considers the relief the result of geological structure, although his detailed argument concerning the matching of strata across valleys indicates that he was aware that at least some valleys must be erosional features. He refers, however, to the "diluvial waters" as responsible for this obvious erosion of valleys and the stripping of chalk from the uplands. He also argues that material carried from the uplands and valleys may now be found in the sea. On the basis of the matching strata on either side of the channel, he correctly suggested that the English Channel is a submarine valley. The tide was considered to be major agent of submarine transport, and he considered this process to be responsible for the formation of the shingle structure of Chesil beach in Dorset.

Adam Sedgwick, who was the first Woodwardian professor of geology at Cambridge University, agreed very closely with Buckland's views, so that Sedgwick's original contribution is limited. William Conybeare was a fellow Oxford graduate of Buckland's. He based his ideas more on field evidence, being a rather more original thinker than Sedgwick. Conybeare recognized the need for several floods, and was intrigued by the course of the Thames River, which cuts through the Berkshire Downs rather than following the Oxford Vale to join the Ouse near Bedford; he suggested that the river started on a gently inclined plane before the vale was cut, but then he returned to the idea of the flood to account for the erosion. He argued that the Thames gravels were formed by a combination of river erosion and the flood. Paper 3 gives a brief account of his ideas of geology in 1832, by which time the diluvialists had displaced the neptunists; they reigned supreme, although not for very long.

The growth of the fluvialist school provided an increasingly strong opposition to the diluvialists. One influential member of the fluvialist school was George J. P. Scrope (1797-1876), who was a confirmed fluvialist and uniformitarian. He published a book on his geological work carried out in the Auvergne in France; this book was hailed by Charles Lyell as the best since Playfair's. Henry de la Beche represents a third school, the structuralists, which had some following at this time; they adopted a mixture of the views of the other schools, allowing an element of catastrophism in their ideas on the formation of valleys.

The next major advances were made by Charles Lyell, who went to Oxford and studied under Buckland. Lyell studied law at first, but gradually turned more and more to science. He lived in a period in which, by wide reading, he could keep abreast of several subjects. Lyell's best-known work was *Principles of Geology,* the first edition of which appeared in 1830. He set himself up as the "high priest of Uniformitarianism," according to Sir Archibald Geikie. Lyell took his examples from geology, botany, zoology, and human settings, and he also attacked religious prejudice. There is a large geomorphic content in the second volume of his *Principles,* which appeared in 1832.

Lyell's account of the progress in geology up to 1832 discusses the three major groups of geological strata. He based his discussion on the rocks exposed in Britain, but also briefly reviewed geological knowledge elsewhere in Europe and the world. His account contains relatively little geomorphology, although the excerpts included here reveal a contemporary interest in endogenetic processes; exogenetic processes do not receive so much attention in his account.

Lyell dealt with erosion and denudation, fluvial processes, tides, glacial eustacy, and climatic change. The work was popular and well received at once, and well reviewed by Scrope, who, however, noted defects that were later to loom larger. Among these were the emphasis on marine planation and a rather too rigid adherence to uniformitarianism. The popularity of the work may have been partly due to the preparation of the ground by slow changes of views and the less dogmatic atmosphere, and to the high standing of Lyell, who in 1831 became the first professor of geology at the new King's College of London University. Lyell, however, resigned the chair in 1833 as the professorship took too much time, and his eyes were giving trouble. The *Principles* continued to be published in various editions, a single-volume edition appearing in 1853, and two-volume editions in 1867–1868, 1872, and 1875. The excerpts given as Paper 4 are from the seventh edition of 1847. Each volume was revised and new material added, although much of the geomorphology remained the same.

The contributions reprinted indicate Lyell's enthusiasm for geology. He also clearly reveals his concern for the slow development of landforms over the long periods of geological time. The fact that not all processes operating at present can be observed is a point he makes forcefully, particularly with reference to endogenetic forces and processes, as well as submarine ones. The final excerpt on the action of running water also mentions the significance of frost action. Lyell was a very keen observer and appreciated many points concerning river load, discharge, and similar variables. Later he veered too much toward marine and structural influences, and he did not appreciate the capacity of land ice to move boulders and deposit drift; he invoked icebergs in the sea to account for drift deposits, and held extreme views concerning the capacity of icebergs to erode. It was left to others to open the glacial lock.

The glacial theory and its progress is another important aspect of the development of geomorphology during the period. It is exemplified by the work of Agassiz, whose contribution to the *Proceedings of the Geological Society of London* in 1840 is reprinted as Paper 5. The glacial key during the period 1826 to 1878 unlocked many previously closed doors by opening men's minds to the great effects that glaciation has exerted on the earth's surface. By 1787, Bernard Kuhn, a Swiss cleric, had suggested that glaciers had once been more extensive, following the early work of De Luc (1779) and De Saussure (1779–1796), both of whom had recognized the ability of glaciers to carry a heavy load. Hutton and Playfair had also recognized the power of glaciers and some of their effects on the landscape.

The main development of the glacial theory took place in the 1820s and 1830s. It was initially the result of the ideas of Ignace Venetz and Jean de Charpentier, and was spread largely through the efforts of Louis

Agassiz. Venetz was an engineer, and was responsible for trying to avert landslides in the Val de Bagnes in 1818. In 1821 he wrote a paper proposing a former greater extension of the Alpine glaciers, but it was not published for 10 years. In these papers, some of which were read to the Societé Helvetique des Sciences Naturelles, he enlarged his earlier views and now suggested boldly that the Swiss glaciers had once extended across the Swiss Plain to the Jura Mountains, depositing the many erratics that had long intrigued people.

Louis Agassiz had an international reputation as a zoologist before he became interested in glaciation in the early 1830s. He was converted from skepticism to being a confirmed glacialist on a field trip with de Charpentier, and he soon outdid his predecessors, from whom he became estranged through his vigorous and forceful intrusion into glacial affairs. Agassiz, with the help of Karl Schimper, developed the concept of the Ice Age, *die Eiszeit,* which was first outlined in the famous *Discours de Neuchatel,* written in one night in July 1837, and presented the next day to the Societé Helvetique at the Neuchatel meeting. Agassiz discussed erratics, polished rocks, striations, and moraines in Switzerland, showing that at one time the Swiss glaciers had been much larger and thicker. He even suggested wildly that all the northern hemisphere from the pole to the Mediterranean had been shrouded by ice, although his ideas of the cause of the Ice Age were very fanciful. He suggested that, as bodies go cold when a person dies, so the earth cools and becomes frigid as its various life episodes come to an end.

Agassiz did not start serious collection of evidence for his views until 1838, but by 1840 he had enough material to publish his *Etudes sur les glaciers,* which was dedicated to Venetz and de Charpentier. His forceful advocacy of glacial theories sparked off a major scientific argument. Hutton had been the first to suggest that the Swiss glaciers had once been more extensive, although neither he nor Playfair had suggested previous glaciers in Britain. Agassiz toured Britain in 1840, and it was in this year that he introduced his ideas to the Geological Society in London (see Paper 5). He did much to extend the ideas of glaciation among British geomorphologists, although he was probably not the first to advocate former glaciers outside the Alps.

R. Jameson of Edinburgh, although he never published work on glaciation, had suggested former glaciers in Scotland in his lectures. He had also published de Charpentier's papers in the Edinburgh *New Philosophical Journal,* of which he was editor. He also published the famous *Discours de Neuchatel.* Despite this access to the best papers on glaciation, British geomorphologists on the whole remained apathetic concerning glaciation.

Agassiz, however, became friends with Buckland, and found in him a convert. They toured Switzerland together examining evidence of glaciation, and jointly toured Scotland after the British Association meeting in Glasgow. During the tour, as discussed in Paper 5, they recognized the true nature of the Parallel Roads of Glen Roy as glacial lake beaches. Agassiz and Buckland also made many shrewd observations on glacial deposits and their structure, using the term "till" to describe the unstratified sediments of glacial origin. Agassiz and Buckland were concerned with the geomorphic aspects of glaciers, and discussed the character of moraines, striations, and glacial gravels. The initial response of the Geological Society of London to the paper was hostile. Buckland, however, in 1841 added new evidence of glaciation in Wales, and John Bell in 1849 contributed evidence from Ireland. Buckland temporarily converted Lyell to the glacial theory, although Lyell later reverted to his alternative iceberg theory. Indeed, Buckland himself did not altogether abandon the iceberg theory, which was associated with the Flood. The evidence of marine shells in glacial drift reaching up to quite high elevation, as on Moel Tryfan in North Wales, was long used as an argument in favor of the iceberg theory.

PHASE 3: LATE NINETEENTH CENTURY

The second half of the nineteenth century was a most active period of geomorphic development, especially in the American West, but ideas were also advancing in Britain, on the continent of Europe, and elsewhere.

The first glacial chronology was suggested by Ramsey in the 1850s; other notable geomorphologists in Britain at the time, who are represented in this volume, were J. B. Jukes and G. Greenwood. In America there were several notable figures; G. K. Gilbert is the best-known, although many others, including John Wesley Powell, C. E. Dutton, and Clarence King, are also of great stature. It was during the 1870s that the land ice theory began to gain ascendency over the glacial flood theory. One line of evidence that led to the long retention of the flood theory was the discovery of shelly drift at considerable heights on the mountains of North Wales, and even as late as 1881 Ramsey was advocating submergence to 600 m (2000 ft) in Wales.

There is an important connection between the glacial and fluvial theories in the recognition of the power of glaciers to erode hollows. This fact disposed of the limnological objection to fluvialism. The presence of lakes in valleys could not be explained by fluvial processes, but they could now be explained by glacial erosion, as noticed by Ramsey in 1859. The great power of glaciers to erode was recognized in the great depth and character of the Norwegian fjords and the Swiss lakes. In 1862, Ramsey presented a paper on rock basins as evidence of this aspect of glacial erosion.

Some of the most important contributions to geomorphology in the second half of the nineteenth century emerged from the exploration of the American West. The work of the western pioneers was a mixture of exploration of unknown country and of a new geomorphic environment. This had the advantage that preconceived ideas were minimal. The western exploration covered the period from 1790 to 1890.

John Wesley Powell is a good example of the combination of exploration with geological investigation that characterized this period. Powell was born in 1834 and lost an arm in the war of 1862. This did not deter him from being among the first group to travel by boat through the Grand Canyon, and eventually he became the director of the new Geological Survey. Most of his best scientific work was done on his earlier trips. These journeys took him notably to the Colorado River and the Uinta Mountains. His reports on these journeys were published in 1875 and 1876. The work on the Colorado was done in 1869 and 1871-1872. His major scientific contributions in these reports were (1) the principle of base level, (2) the nature and relative potency of the processes of erosion, and (3) the generic classification of landforms.

Powell had a great eye for country and was an excellent artist, making very fine sketches of the landforms he studied; he also wrote in a very good style. He appreciated that the vital element of the Colorado landscape was the readjustment of the rivers to the alteration of land level. If the land rose slowly, the antecedent river could continue to maintain its course by cutting down through the rising mountains, whether these were folded or in the form of a plateau. The river was seen to be main agent of erosion, often working mainly along lines associated with structural trends or faults. Powell realized that arid conditions produced their own special landforms, owing to a slow rate of chemical weathering due to the lack of thick vegetation, a condition that helps to reveal the structure. A characteristic feature of the landscape is the canyon cut by running water incised in a plateau or down the face of a mountain front.

One of Powell's major contributions was the recognition of base level; he considered the sea to be the ultimate base level, with hard rocks controlling temporary ones. He regarded the base-level concept as a theoretical ideal, inasmuch as he felt that examples of river beds completely adjusted to base level must be very rare. He suggested terminology for three types of genetically derived river systems: (1) *antecedent* systems established prior to faulting and folding, and maintaining their pattern across the structures as they develop, (2) *consequent* streams that were dependent upon the structure, and (3) *superimposed* systems where streams, initiated on strata now eroded away, at present flow discordantly across the older strata. In his work on the Uinta Mountains, he appreciated the very extensive amount of erosion achieved by the rivers, and introduced the concept of a fluvially eroded planation surface. He also saw that uplift was necessary to initiate erosion. This was later to become the central theme in the work of W. M. Davis.

Probably the greatest of the western geologists at this time, and a man of a stature rarely equaled anywhere else or at any time in the field of geomorphology, was Grove Karl Gilbert, who was born in 1843 and died in 1918. He laid the foundations on which much geomorphology now rests. He was born in Rochester, New York, and went to the local university, where he was an assistant instructor from 1863 to 1868. He then took up surveying, first with the Ohio Survey, and later in 1871 to 1874 with the Reconnaissance Survey of California, Nevada, Utah, Arizona, and New Mexico. The survey had to cover a very large area and was mainly concerned with geographical and topographical observations. Gilbert, therefore, did not have time for detailed geological and geomorphic investigation. His main contribution to these early surveys under Wheeler was the recognition of the importance of faulting in a series of horsts and graben. At this time he also collected material on the evidence for the ancient Lake Bonneville, the subject of one of his best-known later works.

Gilbert first met Powell in 1872 and joined his survey in 1874. During 1875–1876 he worked in the Henry Mountains, where he came to some of his most profound conclusions concerning structure and erosional processes. Gilbert's report on the Henry Mountains is one of his best works and excerpts are reproduced as Paper 6. In this report he explained the origin of laccolithic mountains, of which the Henry Mountains are a good example. He studied the effect of climate on subaerial erosion, and enunciated the law of diminishing slopes and the concepts of grade, lateral planation, the formation of pediments, and the interdependence and stability of drainage lines. Curiously enough, however, he never really used Powell's concept of base level. He worked in Utah in 1877–1879, when he became senior geologist in the Survey, a post that kept him in Washington until 1881. Gilbert published his monograph on Lake Bonneville in 1890. After this he was able to give up much of the administrative work, and could concentrate on serious writing.

Gilbert produced at this time, among other works, his monumental contribution, *Transport of Debris by Running Water.* Here he developed his deductions of laws concerning geomorphic processes from initial forms to adjusted features, thus relating form and process. His theory of grade provided a link between many processes and features and was associated with the concept of dynamic adjustment. He added much to the knowledge of rock breakdown and weathering, and also discussed the processes of fluvial action. Transport by streams he thought served a treble purpose: (1) transport of eroded material, (2) erosion of the river bed, and (3) reduction of the size of the bed load by attrition, thus increasing the suspension load of the river.

These three main physical processes of erosion he realized were helped by a number of interrelated factors. An increase of slope led to more rapid weathering, transport, and corrasion. An increase of discharge led to an increase in the river velocity and carrying capacity. He studied the flow of the river in detail, and recognized the importance of friction in absorbing the river's energy and also in acting as a level to raise particles from the bed by turbulence. He appreciated the difference between competence and capacity of a stream. If friction, load, and discharge are equal, the slope will be the main factor governing transportation. He related physical laws concerning friction to the character of the surface area of the river, showing that if the area does not increase markedly, an additional discharge will cause an increase in velocity, and hence raise the rate of transportation. He concluded that rivers can be in three states; in one the river has more energy than needed to carry its load in suspension, and in this state the river could corrade its bed. In the second and opposite state, the river is sluggish and depositing load. In the third and intermediate state, the carrying capacity just matches the amount of material to be carried. This equilibrium state is the *graded* state, which need not necessarily apply to the whole river. A steep slope is reduced by the effects of more rapid flow. Because of its greater discharge, the bed of the main stream is lowered at a greater rate, resulting in a gentler slope than in its tributary.

Gilbert also applied these principles to slopes. He developed the concept of the law of divides, following his view that grade also applies to slopes, whereby steeper declivities decline more rapidly than flatter ones. These arguments gave rise to the concepts of the laws of declivities and structures. The concept holds that slopes should get steeper nearer the divides, but the idea may not apply in all circumstances; for example, under dense vegetation this "law of structure" may not apply. In some badland areas, creep causes a flattening of slopes near the divides, rather than a steepening.

One of Gilbert's main aims and his major contribution was his enunciation of general principles and laws. He was also a master of logic and wrote on the scientific method, defining the role of deductive and inductive reasoning in geomorphic analysis. The American work of this period has the advantage of directness of approach; there were no doctrinaire explanations or philosphical arguments. Gilbert made a useful contribution on methodology, pointing out that facts must be used to attempt to overthrow theories. He was himself a great investigator, and he wrote that a great investigator must be rich in theories, so that the weak ones can be discarded. Gilbert thought in terms of equilibrium and ratios. He also considered that physical investigation must precede the formulation of theories. His work on running water, published in 1914, was far ahead of its time, and a very good quantitative study. In his work on Lake Bonneville, Gilbert put forward ideas useful also in studies of marine processes, emphasizing the distinction between marine and subaerial processes. He was also the first to draw attention to *pediments,* which are conspicuous features of the semiarid area where he did much of his fieldwork.

The work done by Powell, Gilbert, Dutton, and others in the American West had a profound effect on geomorphology both in the eastern United States and elsewhere, although only after a considerable time lag. During the period 1846 to 1875, while the American West was being explored, the main geomorphic enquiry in Britain and elsewhere in Europe was the argument between the relative effectiveness of marine and subaerial erosion. During this period geological surveys were established in many countries, and much basic geological mapping was carried out.

Lyell and his group supported the school that maintained the importance of marine erosion. They regarded themselves as uniformitarians, set against the school of structuralists and a small group of fluvialists. Andrew C. Ramsey, who was an advocate of marine erosion, was born in Glasgow in 1814. In 1841 he abandoned business for geology, when he became connected with the meeting of the British Association in Glasgow in that year. He became assistant geologist in the new Ordnance Survey, then under military jurisdiction. In 1845 he was appointed director of the English, Welsh, and Scottish areas.

In an essay "The Denudation of South Wales," Ramsey put forward two important concepts: (1) the visual reconstruction of former folded surfaces, and (2) marine planation. He thought that the sea could work on both a rising and a sinking land mass, or with a stationary sea level. He argued that the sea, instead of creating greater irregularity of the surface, as suggested by Lyell, could during transgression form an almost level platform or marine plain. Ramsey stressed *marine planation,* while Lyell equally erroneously stressed *marine dissection.* Paper 7 deals with the development of drainage lines over England and Wales. In this Ramsey rightly stresses the relationship between warping on the land surface and the direction of drainage; he also refers to the effects of transgressions, especially in the development of high-level erosion surfaces. He discussed notably the formation of scarps, referring to the Chalk and Oolite escarpments particularly, in the younger strata in the south and east parts of the country.

Ramsey's ideas on marine planation played an important role, even after his death. The evidence for marine planation that he saw included *accordant summits* and a general seaward slope formed by an encroaching sea, with subsequent subaerial dissection. At the time of their publication, his views were not more important than others currently under discussion, but they rose to prominence with the discussion concerning peneplanation, the cornerstone of W. M. Davis's cycle theory. Ramsey could never believe that rivers created their own valleys, and he thought that fluvioglacial drift was a relic of old sea-floor deposits. His ideas were not catastrophic, however, for he thought that the sea could encroach repeatedly, thus accounting for several erosion surfaces. He regarded the breaks in slope between them as old cliff lines. His ideas have survived into the twentieth century in relation to marine planation and some aspects of denudation chronology.

The importance of subaerial erosion by rivers and mass movement was better appreciated in America, which is a large country, much of it remote from the sea, especially where the western exploration was revealing new types of landscapes. Fluvial processes were not, however, entirely neglected in Britain, and de la Beche stressed the importance of rain and rivers in the removal and transport of soil in 1839. Jame Prestwich also advocated the competence of floods in denudation and transport. The number of those "fluvialists" was small, however, compared with the marine erosionists.

The most ardent supporter of the fluvialist cause was Colonel George Greenwood (1799–1875); he spared no effort in advocating the work of Hutton and Playfair, whose merits he fully appreciated. He was a rather solitary worker, aloof from the main geological circles, and as a result his work was not very popular, especially as it was presented in a rather dogmatic fashion; nor was he kind to his opponents. He was the first person to seriously criticize Lyell's ideas. Greenwood's arguments, however, have stood the test of time and are worthy of examination. He had no formal training in geology; after leaving Eton he went into the army and was a creditable soldier. He retired from the army in 1840 for medical reasons and settled in Hampshire; he learned his geology through personal observation. He had little respect for academic opinion. His argument was simple yet profound; he considered that all denudation of the land could be explained by the combination of rain and rivers. The rain carried down the loose soil across the slopes, collecting first in rills, and then in small streams, and finally joined the main river. The material finished up on the seabed.

Greenwood's major work, of which excerpts are given in Paper 8, was *Rain and Rivers,* published in 1853 and 1857. It combined common sense, confidence, and arrogance, as indicated by the subtitle, *Hutton and Playfair Against Lyell and All Comers.* In this work he stressed the importance of rain as an agent of denudation, both in causing rainwash on the slopes and maintaining the flow in rivers. He made the important point that it is not strictly accurate to claim that rivers denude the landscape, but that it is the combination of rivers and rain into one system. Both valleys and slopes form part of this integrated system. He argued that denudation was controlled by the steepness of the slope, and that when the gradient flattens material would collect in the valley. The increasing width of the valleys near the sea, he explained, was the result of sea level checking the river's ability to erode.

Greenwood appreciated the different shapes of valleys on chalk and clay, and that the pattern of valleys was due to the initiation of consequent streams flowing down an inclined plane, with later subsequent valley development along outcrops of soft clays as they became exposed. He also recognized river capture as being responsible for dry wind gaps, and he had a clear appreciation of both general and local base levels. He was the first geomorphologist since Playfair to realize the importance of rain, and he credited rainwash with the formation of dry valleys. He was rather intolerant of the views of others, and also had his own blind spots. Thus he did not believe in the Ice Age, and considered all drift to be shoreline deposits; he was an extreme uniformitarian.

Another supporter of the subaerial hypothesis was Joseph Beet Jukes (1811-1869), who had a short but important geological career. His main work, published in 1862, was on the rivers in Ireland and was done with the Geological Survey. In this influential work he argued that marine erosion could not form valleys, and he supported Greenwood's ideas; unlike Greenwood, he was almost apologetic in putting forward his ideas. He introduced the term *subsequent* in its modern usage in connection with the Wealden drainage. He considered that the sea creates plains at a high level, and that rain and rivers dissect them on emergence. Paper 9 puts his arguments succinctly and clearly, and shows how these views were gradually achieved. Jukes makes an interesting point concerning the effect of vegetation in controlling landforms when he refers to the importance of grass covering slopes, which shows that he appreciated the part vegetation plays in slowing down the processes of erosion. The views expressed by Greenwood and Jukes lead directly into the fourth phase of geomorphic development.

PHASE 4: EARLY TWENTIETH CENTURY

The fourth phase in the early years of this century was dominated by the best-known name in all geomorphology, William Morris Davis.

His views were not, however, universally accepted; indeed, strongly opposed views were developed in Germany during the latter part of this period, particularly under the influence of Walther Penck. During this period a number of influential geomorphologists were active in a large number of different countries. These include Albrecht Penck, Walther's father, the distinguished physical geographer of Berlin, Edward Suess of Vienna, the father of global geology, Emmanuel de Martonne in France, and du Toit, the great South African. Views contrary to those of Davis were also put forward in Germany by Siegfried Passarge and Alfred Hettner.

William Morris Davis was born in 1850 and lived until 1934, to become a notable figure of both centuries. He was born in Philadelphia, attended a public grammar school in Massachusetts, and then entered Harvard. He studied geology with Raphael Pumpelly, who was to be his most influential teacher. Davis was also influenced by Shaler, his professor of palaeontology, who subsequently persuaded him to join his staff at Harvard. After graduating, Davis obtained a post in the observatory at Cordoba in Argentina where he spent 3 years. He then undertook a world tour, which included a visit to England. At this stage he was a modest and retiring young man. Teaching was a tradition in his family, and in 1875 he joined Shaler's summer camp and started on his academic career in teaching; in 1877 he took up his teaching duties at Harvard, and married just over a year later. His early teaching years were not successful. He taught physical geography and meterology, for which he had little training, nor indeed had he been trained in teaching. He published his first article in 1880; thereafter his output increased very rapidly and continued at a very high level throughout his life.

The first full development of his major concept, the theory of the *geographical cycle,* was published in 1889 in his paper "The Rivers and Valleys of Pennsylvania." The definitive version did not appear until 10 years later. This article is reprinted as Paper 10. By the time this paper appeared his status at Harvard was fully assured. His work was characterized by the explanatory method of presentation, in which he very often used the inductive approach. In 1885, Davis became assistant professor of physical geography at Harvard, and he contributed many papers on meteorology at this time as his interests continued to widen. He was promoted to professor of physical geography in 1890, and during the subsequent decade devoted most of his time to geomorphology, which became his major field.

Davis wrote on a great many different aspects of the subject, including glacial geomorphology, and he did not neglect the purely geological field. He was greatly influenced by Gilbert during this period and thought highly of him. He developed his cycle concept while working on the

Northern Pacific Railroad Survey in Montana in 1883, and frankly acknowledged his debt to the earlier western exploration geologists. He appreciated that the geographical cycle as set out was a gross simplification of reality.

Davis's analysis of landscape in cyclic terms, as described in Paper 10, is based on the well-known trilogy of *structure, process,* and *stage,* and was a product in part of the thought modes of the period, in which historical development was to the fore, rather than functional associations. His cycle has a modern guise also, in that it is an excellent example of a theoretical model. To most of his contemporaries the model appeared eminently reasonable and satisfying, but there was some dissent, particularly from German geomorphologists. Once having expounded his cyclic theory, Davis lost no opportunity to spread the idea as widely as possible.

Davis's major contributions led to his recognition as one of the most eminent geomorphologists of the period. His views on peneplanation and other problems connected with this aspect of geomorphology are discussed more fully in G. F. Adams's Benchmark volume *Planation Surfaces* (Dowden, Hutchinson & Ross, Stroudsburg, Pa., in press).

With success in academic life, Davis became a more assured and successful teacher, and during the succeeding years he traveled widely, visiting Europe, where he met Albrecht Penck in 1899. In 1909 he spent a year in Germany, exchanging teaching positions with Penck. He visited Mexico in 1904 and 1906 and South Africa in 1905. In the first decade of the twentieth century Davis was working on various aspects of the cycle, applying the principles to different climatic regimes. During his visit to the University of Berlin in 1908 to 1909, he learned German and published his *Die erklärende Beschreibung der Landformen* in German, which is said to be clearer and more precise than his more dispersed English-language expositions. He also visited Britain during the period and studied glacial landforms in Wales.

Davis reached the zenith of his success in 1908 to 1912 as a geomorphologist with an international reputation, who lost no opportunity to advance his own case. He made his views more attractive by the very fine block diagrams with which he illustrated most of his geomorphic papers, but he used no quantitative methods. He did, however, make great use of detailed topographic maps and also visited the landforms in the field, where possible with people who knew the area well geologically. Everywhere he went he always sketched the view, having a high artistic talent for this purpose.

Davis resigned his Harvard chair in 1912 at the early age of 62 at the height of his eminence. He visited the Pacific coral islands in 1914. He remarried after his return, following the death of his first wife. His second wife, however, died in 1923, an event which was followed by several years of depression, during which he traveled again. He returned to America to live in Washington, rather than Cambridge, and continued to lecture, partly in the west. After his third marriage in 1930, he finally settled in California. Meanwhile, he had been working on the coral-reef problem and produced a major work in this field in 1928. This was essentially a systematic study, island by island, confirming the Darwinian submergence theory. He continued to work and publish during the 1930s. This period included work on semiarid landforms and karst features. He died in 1934 at the age of 83.

Davis's views were opposed by four main antagonists in Germany, Albrecht and Walther Penck, Siegfried Passarge, and Alfred Hettner. Davis was a personal friend of Albrecht Penck, although the friendship cooled during World War I. Passarge's work was mainly associated with desert landscapes in Africa and formed the basis of Davis's arid cycle. Passarge based his criticism of Davis's cycle on (1) its lack of reality, (2) its ignoring of climatic considerations, particularly climatic change, (3) its attempt to describe landscape in terms of one set of processes, ignoring local variations due to geology and broader ones due to climate, and (4) its leading to quick, superficial, and dangerous conclusions. Davis in turn wrote a critical review of Passarge's book *Beschreibende Landshaftskunde,* which he said was too empirical in its approach, merely consisting of a long list of possible landscapes without genetic discussion or rational explanation. The differences of opinion stem in part from Passarge's dislike of the deductive model, which was Davis's strongest approach. Passarge in his third volume did produce a detailed discussion of process and the resultant form, together with a genetic classification of landscape. Passarge ended his book with an attack on Davis's cycle of erosion.

Hettner was professor of geography at Heidelberg and published views in which he attacked Davis. These were mainly contained in a book published in 1921, *Die Oberflächenformen des Festlandes.* Hettner makes the point that Davis only deals with the effect of uplift. He also argues that to characterize landscapes by assumed age is an error and may merely reflect rock type. "Senility" can have many causes, and so loses its significance as a term, and with that the cycle idea collapses. Cases of genuine old age, Hettner argued, had not yet been proved. He considered Davis's approach to be purely geometrical as a result of its basically deductive method. Hettner stated ". . . only observation will tell us the different kinds of process. While Davis talks a lot about 'life,' his scheme lacks vitality, the landscape picture it gives has a moribund dismal emptiness." Hettner considered the whole approach too superficial, being founded neither on rock type nor on climatic influences, and wrote, " . . . as a whole it has been abortive, and studies founded upon it have produced many fallacies."

Davis's review of these strictures was indignant, and he put up a stout defense. He pointed out that "stage" is the third term of his triligy, and not "age" as stated by Hettner, and although it does depend on time it is not measured in years, but recognized in form. Davis also argued against the criticism that his scheme was too deductive, pointing out that all aspects were tested against actual field examples, with the exception of some deduced desert landforms. Davis did not, however, specifically counter some of Hettner's charges. First, he did not answer the argument that the cycle of erosion did not sufficiently admit of the influences of structure on landforms. Second, he did not answer the point that different initial conditions and processes are capable of leading to very similar end results, the principle of equifinality. Third, the point that some landscape features are typical of particular climates rather than particular stages in a specific climatic setting was not considered. Both Passarge and Hettner struck at the theoretical character of the model of the cycle of erosion, and it was this aspect that Davis thought was most valuable. Ideographic and nomothetic approaches were involved in the argument, insofar as unique description is opposed to generalizations leading to the formulation of laws.

Albrecht Penck was friendly with Davis in the early years of the century, and even adopted some of Davis's ideas, although there were always fundamental differences in their approaches to the study of landforms. Davis was in Berlin as a visiting professor from 1908 to 1909, where he expressed his views in German directly to German students of geomorphology. Davis also translated some of Albrecht Penck's views into English, and the mistakes that he made in his translation have led to the misconception of these ideas by English-speaking workers. The fundamental difference between the approaches of Albrecht Penck and Davis was between the inductive and deductive methods. The relations between the two cooled with the appearance in German of *Grundzuge der Physiographie* by Penck in 1911 and *Die erklärende Beschreibung der Landformen* by Davis in 1912.

Albrecht Penck came to be increasingly influenced by his son Walther. Walther Penck stressed the effects of active tectonics and climatic influence on landscape development. Albrecht Penck wrote on climatic geomorphology, which has for long remained a major concern of German and French geomorphologists. He is, however, probably best known for his work with E. Brückner, *Die Alpen in Eiszeitalter,* published in 1909. In that year Penck exchanged places with Davis and visited the United States. The rift between the two widened in 1920 when Davis published a bad review of the *Festband,* published in honor of Albrecht Penck's sixtieth birthday in 1918. The main criticisms were based on the fact that the work was not in the Davisian tradition. Davis was critical of what he called a geological rather than geographical approach. He also criticized the quality of production, ignoring the difficulties inherent in publishing after World War I. Davis particularly regretted the lack of good, clear block diagrams, of the type he himself was so adept at producing. By 1928 Albrecht Penck had abandoned the idea of the sequential development of landforms, partly out of loyalty to his son Walther.

Walther Penck was born in Vienna in 1888. He became a mountaineer and worked in the Andes, after studying geology at Heidelberg and Vienna. He went to the United States in 1908 with his father, and then obtained a professorship at Istanbul. He also worked in southern Germany and the eastern Alps. He died at the early age of 35. During his life he had frequently written to and received correspondence from Davis. However, he never accepted the cycle concept, partly because he was influenced by his father, for whom he had a deep affection. Walther Penck viewed landscapes from the point of view of a geologist. He was concerned with tectonic activity, and his views differed from Davis's as much as the Andes differ from the Appalachians.

Walther Penck's main concern in the study of landforms was to elucidate the nature of crustal movements. He sought the effects on these in a study of the development and character of slope profiles in particular. His book *Die Morphologische Analyse* (1922) was long ignored by English-speaking workers for it was not translated into English until 1953, under the title *Morphological Analysis of Landforms.* Paper 11 is the introduction to this work. In it, Penck makes clear that his approach is essentially geological rather than geographical. He makes the point that Davis is concerned with the genetic description of landforms from the geographical point of view, but that his own intention is to go further, and to use morphology to determine the nature of crustal movements and earth history. He is interested in the relationship between endogenetic forces acting within the earth and expressed as crustal movement, and exogenetic forces that cause denudation and erosion. Penck considered that the resultant landforms depend on the relative intensity of the two types of process, which mainly act in opposition to each other. In his consideration of exogenetic processes he drew a distinction between the weathering of material and its transport across the slopes and along the rivers. He used the term "denudation" to apply to the transportation aspect, which he differentiated from "erosion," which is the actual lowering of the solid rock of the river bed. He argued that the endogenetic processes should theoretically be able to be deduced by studying the exogenetic ones and the landforms themselves. These forms are the result of a combination of the effects of both endogenetic and exogenetic processes.

Penck includes in his introduction a criticism of Davis's cycle concept, pointing out that it assumes that uplift takes place before denudation begins. Penck's own views are important in that he discussed the effects of simultaneous uplift and denudation, which is much closer to reality and a major step forward. He is much concerned with the relationship between endogenetic and exogenetic processes, stating that they begin to act against each other from the moment uplift begins. To elucidate the probable effect of a variety of different combinations of uplift rates and intensity of denudation, Penck works by iterative methods, taking infinitesimally small steps to produce a continuous curve. This he calls the "differential method." He thus opens the way to the more recent work on mathematical models of slope development, which will be considered in Part II.

Owing to the obscurity of the original German text of Penck's book, and some mistranslations of his ideas by Davis, Penck's geomorphic contribution did not have the impact that it deserved; only fairly recently have his views been seriously reconsidered. Many geomorphologists of the 1920s and 1930s reverted to the Davisian tradition. One was Emmanuel de Martonne, an influential French geomorphologist, who published a well-known textbook on physical geography, *Traité de Géographie Physique,* in 1925 in three volumes. The second volume, *Le Relief du sol,* is devoted to geomorphology. This work follows closely in the Davisian tradition and includes an account of the cycle of erosion. De Martonne refers to the model of normal erosion, and he includes chapters on the effects of rock type and structure, as well as glacial, karst, coastal, and arid processes and forms. He follows Davis in the use of block diagrams and offers no quantification. De Martonne draws attention to other noted geomorphologists that were working at or shortly before his own time, including de la Noë and Emmanuel de Margerie, as well as earlier workers, such as Surrell in France and von Richthofen in Germany.

Another follower of Davis who must be mentioned was working on the other side of the globe in New Zealand. Sir Charles Cotton wrote a number of very widely used geomorphology textbooks, including the *Geomorphology of New Zealand* and *Climatic Accidents in Landscape-making.* He drew on the wealth of magnificent examples of different landforms that are so well exhibited in the landscape of New Zealand, illustrating his work with very fine sketches. In Britain in the 1930s the main supporter of the Davisian tradition was Wooldridge.

BIBLIOGRAPHY

Bagnold, R. A. 1941. *The physics of blown sand and desert dunes.* Methuen, London.

Baulig, H. 1928. Le Plateau Central de la France et sa bordure Mediterrannée, étude morphologique. Armand Colin, Paris.

——. 1952. Surface d'aplanissement. *Ann. Geog. 61,* 161-183, 245-262.
Buckland, W. 1841-1842. On the glacie-diluvial phaenomena in Snowdonia. *Proc. Geol. Soc. 3,* 579-584.
Büdel, J., 1957. Die deppelten Einebnungflächer in den feuchten Tropen. *Geomorph.* NS *1,* 201-208.
——. 1961. Die Morphogenese des Festlandes in Abhangigkeit von Klimazonen. *Naturwiss. 9,* 313-318.
——. 1969. Das System der klimagenetischen Geomorphologie. *Erdkunde 23,* 165-183.
Chorley, R. J., Dunn, A. J., and Beckinsale, R. P. 1964. *The history of the study of landforms.* Methuen, London.
Cotton, C. A. 1941. *Landscape, as developed by the processes of normal erosion.* Cambridge University Press, New York.
——. 1947. *Climatic accidents in landscape-making.* Whitcombe and Tombs, Wellington, New Zealand.
Davies, G. L. 1968. *The earth in decay.* Macdonald, London.
De Luc, J. A. 1779. *Lettres physiques et morales sur l'histoire de la terre et de l'homme.* Paris.
De Saussure, H. B. 1779-1796. *Voyages dans les Alpes,* 4 vol. Neuchâtel.
Dutton, C. E. 1880-1881. The physical geology of the Grand Canyon District. *U.S. Geol. Survey 2nd Ann. Rept.*
Forbes, J. D. 1843. *Travels through the Alps of Savoy.* Edinburgh.
Fournier, F. 1949. Les facteurs climatiques de l'érosion du sol. *Bull. Assoc. Geog. France 47* (166), 97-103.
Gilbert, G. K. 1890. *Lake Bonneville.* U.S. Geol. Survey. Monograph 1.
Hettner, A. 1898. Die Entwicklung der Geographie im 19 Jahrhundert. *Geog. Z. 4,* 305-320.
Hutton, J. 1795. *Theory of the earth with proofs and illustrations,* Vols. I and II. Edinburgh.
King, C. 1878. *Systematic geology.* U.S. Geol. Exploration of the Fortieth Parallel, Vol. 1.
Krumbein, W. C., and Graybill, F. A. 1965. *An introduction to statistical models in geology.* McGraw-Hill, New York.
Lewis, W. V. 1949. Glacial movement by rotational slipping. *Geog. Ann. 31,* 146-158.
——. 1954. Pressure release and glacial erosion. *Jour. Glaciol. 2,* 417-422.
—— (ed.). 1960. Norwegian cirque glaciers. *Roy. Geog. Soc. Res. Ser. 4,* 104 pp.
Lyell, C. 1830-1833. *Principles of geology.* London.
——. 1838. *Elements of geology.* London.
Martonne, E. de. 1925. *Traité de géographie physique,* Vol. 2, Le Relief du sol. 5 ed. Colin, Paris, 499-1037.
North, F. J. 1943. Centenary of the glacial theory. *Proc. Geol. Assoc. 46,* 1-28.
Passarge, S. 1924. *Vergleichende Landschaftkunde.* Berlin.
Penck, A., and Brückner, E. 1909. *Die Alpen in Eiszeitalter.* Leipzig.
Powell, J. W. 1875. *Exploration of the Colorado River of the West and Its Tributaries.* Smithsonian Institution, Washington, D.C.
Prestwich, J. 1886. *Geology: chemical, physical and stratigraphic,* 2 vols. Oxford.
Ramsey, A. C. 1846. The denudation of South Wales. *Mem. Geol. Surv. G. B.* Vol. 1. H.M.S.O., London, 297-335.
Scrope, G. J. P. 1829-1830. On the gradual excavation of the valleys in which the Meuse and Moselle and some other rivers flow. *Proc. Geol. Soc.* a, 170-171.

Sedgwick, A. 1825. On the origin of alluvial and diluvial formations. *Ann. Phil.* NS *9,* 214-257; *10,* 18-37.
Simons, M. 1962. The morphological analysis of landforms: a new review of the work of Walther Penck (1888-1923). *Trans. Inst. Brit. Geog. 31,* 1-14.
Suess, E. 1883-1885. *Das Antlitz der Erde,* Vol. I; 1888, Vol. II. Vienna.
——. 1906. *The face of the earth.* Translation by H. B. C. Sollas and W. J. Sollas. Oxford.
Sweeting, M. M. 1963. Erosion cycles and limestone caverns in the Ingleborough District. *Geog. Jour. 115,* 63-78.
——. 1972. *Karst landforms.* Macmillan, London.
Tricart, J., and Cailleux, A. 1972. *Introduction to climatic geomorphology.* English translation by C. J. K. de Jonge. Longman, London.
Tyndall, J. 1858. On some physical properties of ice. *Phil. Trans. Roy. Soc. 148,* 211-229.
Venetz, J. 1833. Mémoire sur les variations de la température dans les Alpes de la Suisse. *Mem. Soc. Helv. Sci. Nat. 12.*
Werner, A. G. 1787. *Kurze Klassifikation und Beschreibung der verchiedenen Gebirgsarten.* Dresden.
——. 1791. *Neue Theorie von der Entsehung der Gänge.* Freiberg.

1

Reprinted from *Illustrations of the Huttonian Theory of the Earth,* Cadell and Davis, 1802, pp. 102–105, 350–352

ILLUSTRATIONS OF THE HUTTONIAN THEORY OF THE EARTH

John Playfair

[*Editor's Note:* In the original, material precedes this excerpt.]

99. If we proceed in our ſurvey from the ſhores, inland, we meet at every ſtep with the fulleſt evidence of the ſame truths, and particularly in the nature and economy of rivers. Every river appears to conſiſt of a main trunk, fed from a variety of branches, each running in a valley proportioned to its ſize, and all of them together forming a ſyſtem of vallies, communicating with one another, and having ſuch a nice adjuſtment of their declivities, that none of them join the principal valley, either on too high or too low a level; a circumſtance which would be infinitely improbable, if each of theſe vallies were not the work of the ſtream that flows in it.

If indeed a river conſiſted of a ſingle ſtream, without branches, running in a ſtraight valley, it might be ſuppoſed that ſome great concuſſion,

cuſſion, or ſome powerful torrent, had opened at once the channel by which its waters are conducted to the ocean; but, when the uſual form of a river is conſidered, the trunk divided into many branches, which riſe at a great diſtance from one another, and theſe again ſubdivided into an infinity of ſmaller ramifications, it becomes ſtrongly impreſſed upon the mind, that all theſe channels have been cut by the waters themſelves; that they have been ſlowly dug out by the waſhing and eroſion of the land; and that it is by the repeated touches of the ſame inſtrument, that this curious aſſemblage of lines has been engraved ſo deeply on the ſurface of the globe.

[*Editor's Note:* Material has been omitted at this point.]

101. In the ſame manner, when a river undermines its banks, it often diſcovers depoſites of ſand and gravel, that have been made when it ran on a higher level than it does at preſent. In other inſtances, the ſame ſtrata are ſeen on both the banks, though the bed of the river is now ſunk deep between them, and perhaps holds as winding a courſe through the ſolid rock, as if it flowed along the ſurface; a proof that it muſt have begun to ſink its bed, when it ran through ſuch looſe materials as oppoſed but a very inconſiderable reſiſtance to its ſtream. A river, of which the courſe is both ſerpentine and deeply excavated in the rock, is among the phenomena, by which the ſlow waſte of the land, and alſo the cauſe of that waſte, are moſt directly pointed out.

102. It is, however, where rivers iſſue through narrow defiles among mountains, that the identity of the ſtrata on both ſides is moſt eaſily recognized, and remarked at the ſame time with the greateſt wonder. On obſerving the Patowmack, where it penetrates the ridge of the Allegany mountains, or the Irtiſh, as it iſſues from the defiles of Altai, there is no man, how-

ever

ever little addicted to geological ſpeculations, who does not immediately acknowledge, that the mountain was once continued quite acroſs the ſpace in which the river now flows; and, if he ventures to reaſon concerning the cauſe of ſo wonderful a change, he aſcribes it to ſome great convulſion of nature, which has torn the mountain aſunder, and opened a paſſage for the waters. It is only the philoſopher, who has deeply meditated on the effects which action long continued is able to produce, and on the ſimplicity of the means which nature employs in all her operations, who ſees in this nothing but the gradual working of a ſtream, that once flowed as high as the top of the ridge which it now ſo deeply interſects, and has cut its courſe through the rock, in the ſame way, and almoſt with the ſame inſtrument, by which the lapidary divides a block of marble or granite.

[*Editor's Note:* Material has been deleted at this point.]

NOTE XVI. § 100.

Rivers and Lakes.

314. Rivers are the cauſes of waſte moſt viſible to us, and moſt obviouſly capable of producing

producing great effects. It is not, however, in the greateft rivers, that the power to change and wear the furface of the land is moft clearly feen. It is at the heads of rivers, and in the feeders of the larger ftreams, where they defcend over the moft rapid flope, and are moft fubject to irregular or temporary increafe and diminution, that the caufes which tend to preferve, and thofe that tend to change the form of the earth's furface, are fartheft from balancing one another, and where, after every feafon, almoft after every flood, we perceive fome change produced, for which no compenfation can be made, and fomething removed which is never to be replaced. When we trace up rivers and their branches toward their fource, we come at laft to rivulets, that run only in time of rain, and that are dry at other feafons. It is there, fays Dr Hutton, that I would wifh to carry my reader, that he may be convinced, by his own obfervation, of this great fact, *that the rivers have, in general, hollowed out their valleys.* The changes of the valley of the main river are but flow ; the plain indeed is wafted in one place, but is repaired in another, and we do not perceive the place from whence the repairing matter has proceeded. That which the fpectator fees here, does not therefore immediately fuggeft to him what has been the ftate of things before the valley was hollowed

hollowed out. But it is otherwife in the valley of the rivulet; no perfon can examine it without feeing, that the rivulet carries away matter which cannot be repaired, except by wearing away fome part of the furface of the place upon which the rain that forms the ftream is gathered. The remains of a former ftate are here vifible; and we can, without any long chain of reafoning, compare what has been with what is at the prefent moment. It requires but little ftudy to replace the parts removed, and to fee nature at work, refolving the moft hard and folid maffes, by the continued influences of the fun and atmofphere *. We fee the beginning of that long journey, by which heavy bodies travel from the fummit of the land to the bottom of the ocean, and we remain convinced, that, *on our continents, there is no fpot on which a river may not formerly have run* †.

[*Editor's Note:* Material has been omitted at this point.]

---

* Theory of the Earth, vol. ii. p. 294.

† *Ibid.* p. 296.

2

Reprinted from *Trans. Geol. Soc. London,* Ser. 2, 1(1), 95–102 (1822)

On the Excavation of Valleys by diluvian Action, as illustrated by a succession of Valleys which intersect the South Coast of Dorset and Devon.

By the Rev. WILLIAM BUCKLAND, F.R.S. F.L.S. V.P.G.S.
AND PROFESSOR OF GEOLOGY AND MINERALOGY IN THE UNIVERSITY OF OXFORD.

[Read April 19, 1822.]

MY intention in the following communication is to consider the general causes to which valleys owe their origin, and particularly such as occur in horizontal and undisturbed strata within the limits of their escarpments. That portion of the coast of Dorset and Devon which lies on the east of Lyme and on the east of Sidmouth, affords some of the best examples of such valleys with which I am acquainted: I beg therefore to present to the Society two geological views of that coast, drawn at my request some years since by Hubert Cornish, Esq., which will tend also further to illustrate the description given of it in a preceding paper by Mr. De la Beche.

Many valleys may be ascribed to the elevation or depression of the strata composing the adjacent hills, by forces acting at very remote periods from within the body of the earth itself; and to similar forces we may principally refer the high inclination and contortions of strata that occur in the highest mountains, and sometimes also in minor hills: other valleys have been occasioned by the strata having been originally deposited at irregular levels, and others by some partial slip or dislocation of portions of the strata.

But at different periods of time intermediate between the deposition of the most ancient and the most recent formations, the irregularities of level arising from the preceding causes have been variously modified by the action of violent inundations hollowing out portions of the surface and removing the fragments to a distance: to such circumstances we must ascribe the water-worn pebbles of the old red sandstone, the red marl, and the plastic clay formations.

A cause similar to that last mentioned has wrought extensive changes on the surface (however variously modified by preceding catastrophes) at a period subsequent to the deposition and consolidation of the most recent of the regular strata; for rocks and strata of all ages bear on those portions of their surface which are not covered by other strata, the marks of aqueous excavation; and are strewed over with the mingled fragments of the most recent, as well as of the most ancient beds.

The diluvian waters to which these effects must be referred, (if we except the very limited and partial action of modern causes, such as of torrents in cutting ravines, of rivers in forming deltas, of the sea in eroding its cliffs, and of volcanos in ejecting and accumulating their exuviæ,) appear to have been the last agents that have operated in any extensive degree to change the form of the earth's surface.

When one or more sides of a valley are formed by abrupt escarpments, such as usually terminate the outgoings of our secondary strata, it is difficult to say to what extent the discontinuity of the strata, and the formation of the valley beyond the limits of the escarpment, are attributable to diluvian excavation; for we know not how far the strata originally extended beyond their present frontier, nor how much of the subjacent valley is referable to other causes than the most recent diluvial agency: for example, it is not possible to determine how far the escarpment, which may be seen in the annexed map to terminate the green sand-hills of Blackdown towards the vale of Taunton, might have extended northward beyond its present boundary towards the Quantock hills; nor how much of this vale is to be attributed to excavation by water, or to originally low position: and we are equally without data for judging to what distance the same green sand might have originally extended on the west of Haldon towards Dartmoor; or over any spaces exterior to its present escarpments.

But when a valley originates and has almost its whole extent within the escarpment of strata that are horizontal or nearly so, and which bear no mark of having been moved from their original position by elevation, depression, or disturbance of any kind; and when such valley is inclosed along its whole course by hills that afford an exact correspondence of opposite parts, it must be referred exclusively to the removal of the substance that once filled it; and the cause of that removal appears to have been a violent and transient inundation *.

* For a well-digested statement of the arguments in support of the theory of the formation of valleys by aqueous excavations, I beg to refer to the works of Dr. Richardson on the coast of Antrim, in the Philosophical Transactions; to Mr. Greenough's Geology; and Mr. Catcott's Treatise on the Deluge: Mr. Catcott, however, has carried his doctrine of denudation too far, and has applied it to explain phenomena that must be referred to other causes.

It is not easy to imagine how valleys of this last description could have been formed in any conceivable duration of years by the rivers that now flow through them, since all the component streams, and consequently the rivers themselves which are made up of their aggregate, owe their existence to the prior existence of the valleys through which they flow.

Of the same nature with those last described are the valleys which intersect the coast of Dorset and Devon. In passing along this coast (see the Map and Views, Plates XIII. and XIV.) we cross, nearly at right angles, a continual succession of hills and valleys, the southern extremities of which are abruptly terminated by the sea, the valleys gradually sloping into it, and the hills being abruptly truncated, and often overhanging the beach or undercliff, with a perpendicular precipice. The main direction of the greater number of these valleys is from north to south; that is, nearly in the direction of the dip of the strata in which they are excavated: the streams and rivers that flow through them are short and inconsiderable, and incompetent, even when flooded, to move any thing more weighty than mud and sand.

The greater number of these valleys, and of the hills that bound them, are within the limits of the north and north-west escarpment of the green sand formation; and in their continuation southward they cut down into the oolite, lias, or red marl, according as this or that formation constitutes the substratum over which the green sand originally extended. There is usually an exact correspondence in the structure of the hills inclosing each valley; so that, whatever stratum is found on one side, the same is discoverable on the other side upon the prolongation of its plane: whenever there is a want of correspondence in the strata on the opposite sides of a valley, it is referable to a change in the substrata upon which the excavating waters had to exert their force.

The section of the hills in this district usually presents an insulated cap of chalk, or a bed of angular and unrolled chalk-flints, reposing on a broader bed of green sand; and this again rests on a still broader base of oolite, lias, or red marl (see Plates XIII. and XIV.). With the exception of the very local depression of the chalk and subjacent green sand and red marl on the west of the Axe, at Beer Cliffs, the position of the strata is regular and very slightly inclined; nor have any subterraneous disturbances operated to an important degree to affect the form of the valleys.

If we examine the valleys that fall into the bay of Charmouth from Burton on the east to Exmouth on the west, viz. that of the Bredy, the Brit, the Char, the Axe, the Sid, and the Otter, we shall find them all to be valleys of diluvian excavation; their flanks are similarly constructed of parallel and respectively identical beds; and the commencements of them all originate within the area and on the south side of the escarpment of the green sand.

The valley of the Sid, as it is coloured in the annexed map, may from its shortness and simplicity be taken as an example of the rest; it originates in the green sand, but soon cuts down to the red marl, and continues upon it to the sea; in both these respects it agrees with the upper branches of the Otter, and with the valleys that fall from the west into that of the Axe.

But in those cases where the lias and oolite formations are interposed between the red marl and green sand, the base of the valley varies with the variation of the substratum; as will appear on comparing the opposite sides of the lower valleys of the Otter, the Axe, and the Char, with the variations of their substrata, as expressed in the map.

The valley of Lyme is of equal simplicity with that of Sidmouth, and differs only in that its lower strata are composed of lias instead of red marl: but the valleys of Chideock, Bridport, and Burton, being within the area of the oolite formation, have their lower slopes composed of oolite subjacent to the green sand; whilst that of Charmouth is of a mixed nature, having its western branches in green sand reposing on lias, and in some of its eastern ramifications intersecting also the oolite. In the same manner, the valley of the Axe has lias interposed between the green sand and red marl on its east flank, but none at all on its western side, below the town of Axminster. These apparent anomalies form no exception to the general principle, that the variation of the sides of the valleys is always consistent with that which is simply referable to the variation of the substrata, on which the denuding waters had to exert their force. It is moreover such as can be explained on no other theory than that of the strata having at one time been connected continuously across the now void spaces which constitute the valleys.

The following section, taken from a series of lias quarries on the two opposite sides of the valley of the Axe near Axminster, will show the degree of minuteness to which this correspondence extends:

		Ft.	In.
1. White lias...	Slaty and fissile, is used for flooring when split into slabs from two to three inches thick..	2	0
Clay.			
2. Burrs......	Rough blue building stone	0	10
Clay.			
3. Cockles.....	Flat and broad blue stone, containing shells and divided into two beds, each three inches thick, with a parting of clay; is used for building —Total	0	10

		Ft	In.
Clay.			
4. Anvils......	Blue building stone, forming a bed of irregular anvil-shaped blocks	1	0
Clay			
5. Graze Burrs.	Good blue building stone	0	10
Clay.			
6. Fire stone...	White building stone, used also for forming the arch-work of lime pits: it divides into two beds, each four inches thick, with a parting of clay.—Total	1	0
Clay.			
7. Half-foot bed.	Strong blue flagstone, the best for paving .	0	6
Clay.			
8. Foot stone .	Blue paving and building stone	0	10
Clay.			
9. Red-size....	White lias, inclining to grey, splitting into two or three thin slabs, and used for paving and building	0	6
Clay.			
10. Under bed. .	Blue building stone, used for paving, and the best bed of all for steps	0	8
Clay, varying from one to six feet.			
11. White rock.	White lias, rough and rubbly throughout;—not good for paving or building, but used largely to make lime, which is better than that of the other beds for plastering and in-door work: the thickness of this bed is variable; its average is	30	0

All the above strata are separated by thin beds of clay, varying from four inches to a foot, and exceeding the latter thickness in one case only, viz. between N°s 10 and 11: but the presence and relative position of each individual stratum of stone is constant; and the specific character and uses of each bed are of practical notoriety among the masons through the district round Axminster, in which there are many and distant quarries, to any one of which the above section is equally applicable; *e. g.* to the quarries of Fox hills on the south-east, of Waycroft on the north, and of Sisterwood, Battleford, Long Leigh, Small-ridge, Green-down, and Cox-wood, on the north-west of Axminster. There can be little doubt, therefore, that the component strata of all these quarries were originally connected in one continuous plane across the now void space which forms the valley of the Axe*.

The fact of excavation is evident from simple inspection of the manner in which the valleys intersect the coast, on the east of Sidmouth and the east of Lyme, as represented in the annexed views (Plate XIV.); and it requires but little effort, either of the eye or the imagination, to restore and fill up the lost portions of the strata, that form the flanks of the valleys of Salcomb, Dunscomb, and Branscomb, on the east of Sidmouth; or of Charmouth, Seatown, and Bridport, on the east of Lyme. By prolonging the corresponding extremities of the strata on the opposite flanks, we should entirely fill up the valleys, and only restore them to the state of continuity in which they were originally deposited.

An examination of the present extent and state of the remaining portions of the chalk formation within the district we are considering, will show to what degree the diluvian waters have probably interrupted its original continuity. The insulated mass of chalk, which at Beer Head composes the entire thickness of the cliff, rises gradually westward with a continual diminution and removal of its upper surface; till after becoming successively more and more thin, on the cliffs of Branscomb, Littlecomb, and Dunscomb, it finds in the latter its present extreme western boundary: beyond this boundary, on the top of Salcomb Hill and of all the highest table lands and insulated summits of the interior, from the ridges that encircle the vales of Sidmouth and Honiton to the highest summits of Blackdown, and even of the distant and insulated ridge of Haldon on the west of the valley of Exe, beds of angular and unrolled chalk flints (which can be identified by the numerous and characteristic organic remains which they contain) are of frequent occurrence; similar beds are found also on the green sand summits that encircle the valleys of Charmouth and Axminster; other large and insulated masses of chalk occur, along the coast from Lyme nearly to Axmouth, and in the interior, at Widworthy, Membury, White Stanton, and Chard; and these at distances varying from ten to thirty miles from the present termination of the chalk formation in Dorsetshire; though within the limits of the original escarpment of the green sand.

These facts concur to show that there was a time when the chalk covered all those spaces on which the angular chalk flints are at this time found: and

* The section of these quarries is also important as proving the alternation of white lias with blue: it is more common to find the white variety occupying only the lower regions of this formation; but in the case before us there is not only this alternation, but each individual slab of blue lias is inclosed within an outer crust or case of white lias; the colour of which passes by insensible gradations into blue.

that it probably formed a continuous or nearly continuous stratum, from its present termination in Dorsetshire, to Haldon on the west of Exeter *.

From the correspondence observed by Mr. Wm. Phillips between the strata of Dover and on the hills west of Calais †, and by Mr. De la Beche between the strata of the coast of Dorset and Devon, and those of Normandy ‡, it may be inferred (after making due allowance for the possible influence of those earlier causes, which in many instances have occasioned valleys) that the English Channel is a submarine valley, which owes its origin in a great measure to diluvian excavation, the opposite sides having as much correspondence as those of any valleys on the land. Mr. De la Beche, as I am informed, has already drawn the same inference from his own observations. According to Bouache, the depth of the straits of Dover is on an average less than 180 feet; and from thence westward to the chops of the channel the water gradually deepens to only 420 feet, a depth less than that of the majority of inland valleys which terminate in the bay of Charmouth; and as ordinary valleys usually increase in depth from the sides towards their centre, so also the submarine valley of the channel is deepest in the middle, and becomes more shallow towards either shore.

It seems probable that a large portion of the matter dislodged from the valleys of which we have been speaking by the diluvian waters that hollowed them out, has been drifted into the principal valley, the bed of the sea; and being subsequently carried eastward by the superior force of the flowing above that of the ebbing tide, and partially stopped in its further progress by the Isle of Portland, has formed that vast bed of pebbles known by the name of the Chesil Bank: the principal ingredients of this bank are rolled chalk flints and pebbles of chert; the softer materials that filled the valleys, such as chalk, sand, clay, and marl, having been floated off and drifted far into the ocean, by the violence of the diluvian waters.

* There is also reason to think that the plastic clay formation was nearly coextensive with the chalk; for on the central summits of Blackdown there are rounded pebbles of chalk flint, which resemble those found in the gravel beds of the plastic clay formation at Blackheath: and on the hills that encircle Sidmouth there are large blocks of a siliceous breccia, composed of chalk flints united by a strong siliceous cement, and differing from the Hertfordshire pudding-stone only in the circumstance of the imbedded flints being mostly angular, instead of rounded as in the stone of Hertfordshire: a variation which occurs in similar blocks of the same formation at Portisham, near Abbotsbury, and elsewhere.—The argument, however, arising from the presence of these blocks and pebbles is imperfect; as it is possible, though not probable, they may have been drifted to their actual place by the diluvian waters, *before* the excavation of the valleys.

† See Geological Transactions, Vol. v. pp. 47, &c.   ‡ See p. 89 of the present volume.

The quantity of diluvian gravel which remains lodged upon the slopes and in the lower regions of the valleys that intersect this coast, is very considerable; but it is not probable that many animal remains will be discovered in it, because the large proportion of clay with which it usually is mixed, renders it less fit for roads than the shattered chert strata of the adjacent hills, and consequently gravel pits are seldom worked in the diluvium. Enough, however, has been done to identify its animal remains with those of the diluvian gravel of other parts of England, by the discovery, a few years since, of several large tusks of Elephants and teeth of Rhinoceros in the valleys of Lyme and Charmouth.

On the highest parts of Blackdown, and on the insulated summits which surround the vale of Charmouth, I have found abundantly pebbles of fat quartz; which must have been drifted thither from some distant primitive or transition country, and carried to their actual place, before the present valleys were excavated and the steep escarpments formed, by which those high table lands are now on every side bounded. These cases are precisely of the same nature with those of the blocks of granite that lie on the mountains of the Jura and on the plains of the north of Germany and Russia, and with that of the quartzose pebbles found on the tops of the hills round Oxford and Henley; which latter I have described in the fifth volume of the Geological Transactions, as having been drifted thither from the Licky Hill and central parts of England, before the excavation of the present valley of the Thames.

3

Reprinted from *Rept. Brit. Assoc.*, 365, 370–371, 408–411 (1832)

REPORT ON THE PROGRESS, ACTUAL STATE, AND ULTERIOR PROSPECTS OF GEOLOGICAL SCIENCE

W. D. Conybeare

It cannot be necessary, before an assembly like the present, to expatiate on the interest of the science to which I have now to call your attention; a science which by investigating the traces indelibly impressed on the surface of our planet by the successive revolutions it has undergone, proposes to elucidate the history of these stupendous physical actions; and thus fully to develop what may be termed the archæology of the globe itself,—a science which associating itself to those branches of our knowledge which relate to organized nature, to zoology, and to botany, affords to each the important supplementary information of numerous species which have long vanished from the actual order of things;—thus unexpectedly extending our views of the various combinations of organic forms; and in many instances supplying links, otherwise wanting, in uniting the different terms of this series in a continuous and unbroken chain.

Nor, if from these higher views of scientific interest we advert to the more practical considerations of utilitarian importance, and applicability to those œconomical arts on which our national wealth and strength depend, can we think meanly of a science which guides us in the full development of our mineral resources; which, (to mention only a single instance,) in indicating the proper line in which researches for coal may offer the prospect of success, extends, facilitates, and œconomizes the supply of this article, the great element not only of domestic comfort, but of mechanical power.

[*Editor's Note:* Material has been omitted at this point.]

The progress of geology from the period at which it thus began to assume the systematic character of a regularly digested science, may be considered as having presented three marked stages, distinguished by three successive schools; each of these schools has selected for the more especial object of its attention a single member of the three great geological divisions in the series of formations, *i. e.* the primitive, secondary, and tertiary; and the succession of these schools has, by a singular coincidence, followed the same order with that of the formations to which they were devoted: it may also be observed that the leaders of each school have been distinguished geologists of three different nations,—Germany, England, and France. The first, or German school, is that of Werner: this directed its attention principally to the primitive and transition formations †, in which the distinctions of mineralogical character assume the greatest importance; and the imbedded minerals, from their variety, and relations to the rocks containing them, become the chief objects of the geologist's notice. The second, or English school, has distinguished itself by the ardent and successful zeal with which it has developed the whole of the secondary series of formations: in these the zoological features of the organic remains associated in the several strata, afford characters far more interesting in themselves and important in the conclusions to which they lead, than the mineral contents of the primitive series. This school generally recognises the masterly observations of Smith, first made public in 1799, as those which have principally contributed to its establishment; although the regular distribution of organic remains had before

† In the early works of one of the ablest British disciples of this school, whose meritorious labours undoubtedly contributed very largely to the diffusion of an ardour for geological inquiries in this island, there occurs a curious illustration of the exclusive attention to the older rocks. In the general view of geology contained in the Introduction to Professor Jamieson's *Account of the Hebrides*, 1800, after a sufficiently full detail of the various primitive formations, we find the whole secondary group dismissed in these few vague words: "They consist of limestone and argillite, with numerous petrifactions; also basalt, porphyry, pitchstone, greenstone, wacke, and the various coal strata."

been recognised in Italy by Steno, and in France by Rouelle; and although Werner in his lectures, and Saussure as above quoted, appear to have indicated generally, that the laws of this distribution bore a relation to the geological age of the formations containing them, yet a degree of vagueness hung over the whole subject, which precluded any extensive or useful application of this great principle, until the acute observations of Smith first brought it prominently forward in all the precision of exact detail as applied to a vast succession of formations, including the most important portion of the geological series; and as from his situation in life we must consider the discoveries of Smith as the extraordinary results of native and untaught sagacity of intellect, they must on this account be held to challenge a still warmer tribute of approbation, and may be regarded as strictly original in him, even where faint traces of anticipation may be found in Continental writings little likely to have fallen beneath his observation. The third school, or that of Tertiary Geology, owes its foundation to the admirable Memoir on the Basin of Paris, published by Cuvier and Brongniart, 1811. Although this school was certainly subsequent in point of date to that of Smith, yet those who had already directed their attention to such pursuits at this period, must well remember that the Wernerian school of primitive and mineralogical geology having previously obtained an undisputed and exclusive ascendancy in the minds of most of those who possessed any influential station in the scientific world, the observations of the individual alluded to had little chance of recommending themselves at first to public notice, and that in fact the knowledge of them appears to have been for ten years chiefly confined to a small circle in the neighbourhood of Bath,—until the high scientific distinction of Cuvier, and the striking and interesting nature of the facts developed in his brilliant Memoir, excited a marked sensation and commanded the general attention of men of science; for none such could peruse with indifference those masterly descriptions, which exhibited the environs of one of the great metropolitan cities of Europe as having been successively occupied by oceanic inundations and fresh-water lakes; which restored from the scattered fragments of their disjointed skeletons the forms of those animals, long extinct, whose flocks once grazed on the margins of those lakes; and which presented to our notice the case of beds of rock only a few inches in thickness, extending continuously over hundreds of square miles, and constantly distinguished by the same peculiar species of fossil shells.

[*Editor's Note:* Material has been omitted at this point.]

The next branch to which I would call your attention may,

I think, be termed the true dynamics of geology, with far more justice than that appellation has been applied to other branches of our science. I would so denominate the general consideration of the forces which appear to have been the agents in dislocating and elevating our strata; whether in the earlier geological disturbances, or in the actual phænomena of volcanoes and earthquakes. We have already seen how much Von Buch has contributed, of the most important generalizations, to our knowledge of actual volcanic phænomena; and Elie de Beaumont has been nearly the first to call attention to anything like generalized views with regard to geological elevations. But still it is but a very small portion indeed of the totality of these phænomena, which have yet been brought under our cognisance. Indeed, the more general science which includes this, our knowledge of the physical geography of the different regions of our planet, is still in its infancy; but, as it shall advance, is it too much to anticipate the most important conclusions, when we shall be able to speak of the periods of elevation of the Himmalaya, the Andes, and the American Rocky Mountains, with as much evidence as we now do of the Pyrenees and Alps? When we shall have a generalized view of the principal disturbances of the great continents, may we not hope to enter, with a prospect of satisfactory solution, on the great problem of the elevation of those continents, and the determination of their general forms, on the consideration of the forces which have produced these effects, and even on the dynamical investigation of the laws which those forces appear to have followed? There is one source of analogy which has always appeared to me as likely to throw illustration on this subject, and which I yet almost hesitate to allude to, lest I should incur the charge of indulging speculations altogether rash and visionary. However, I would premise the observation, that we must surely in no respect consider our planet as an isolated body in nature; it is one only of the general planetary system, and every fair presumption of analogy favours the supposition, that similar general causes have acted in all the members of that system. Now one of these members, our own satellite, is placed sufficiently near us to enable our telescopic observations to distinguish accurately the general outlines of its mountain chains, and other similar features of its physical geography. We have been able to discern even the eruption of volcanoes; and any one who has viewed its surface through a telescope, must be struck with the exact identity of the forms which he there contemplates with the maps and descriptions of the volcanic districts of our own globe. If we recall Von Buch's account, already referred to, of crateriform amphitheatres of many leagues in diameter, encircling central conical craters; of lines of these generally disposed in a linear direction; of such linear trains often radiating from a central focus of principal disturbance; we may almost fancy that this description was intended as an exact portrait of what we observe on the lunar surface. Is it, then, altogether unfounded to believe that by more carefully observing these phænomena, where we have a whole hemisphere of a planet at once open to our inspection,—by comparing the best of the early delineations of its telescopic appearance, with its exact actual forms, and watching diligently those forms, so that we may be able to detect any changes in them,—is it too much to hope that we may thus effectually extend our knowledge of the general laws of the volcanic forces, which should appear to be among the general planetary phænomena*?

The great branches of the comparative geology, and comparative palæontology (or study of fossil remains) of distant countries, much as they have recently advanced, have as yet even a still wider interval to pass over than that which they may have already accomplished, before they shall have obtained that degree of completeness which alone can qualify them to serve as sound bases in any geological theory.

First, as to comparative geology. The very introductory question is yet inadequately answered, Is there or is there not anything like such a general uniformity of type in the series of rock formations in distant countries, that we must conceive them to have resulted from general causes, of almost universal prevalence at the same geological æras? Now it is clear that this question, if intelligently proposed, does not require, for its affirmative solution, anything like an exact *identity* of formations in remote localities. It does not require any one to be able to take to Australia a detailed list of English strata, and to be able at once to lay his hands on the exact equivalents of our lias, oolites, and chalk. Such an idea would be almost to caricature the Wernerian dogma of universal formations. We are indeed unable to trace many of these formations, even through our own island, without observing such considerable modifications in their comparative types, in our northern and southern counties, as may sufficiently remind us that we are to look only for such *analogous* rather than *identical* results, as would naturally proceed

* The ancient selenographical maps of Hevelius, Ricciolus, and Cassini, are too defective in precision of outline to be of much use. Russel's lunar globe and Schrœter's detailed plates afford every desirable information; and Mr. Blunt has recently published in a cheaper form a very beautiful engraving on a single sheet, extremely accurate, and amply sufficient for the purpose.

from the contemporaneous action of similar causes in distant localities; in each of which many varying local circumstances must have affected those results; for two conditions obviously enter into this problem:—first, the contemporaneous prevalence and extent of similar geological causes; and secondly, how far these causes, even where active, may have been modified by varying local circumstances. Now, at present, our materials for answering these questions accurately are confined to Europe: of America indeed we have some information; and although this may as yet be considered too vague to be fully satisfactory, yet as far as it goes it is undoubtedly favourable to the presumption of even a greater degree of geological uniformity, than we should have been justified in anticipating *à priori*.

Humboldt indeed has remarked, that while on entering a new hemisphere we change all other familiar and accustomed objects,—while in the plains around we survey entirely new forms of vegetable and animal being, and in the heavens over our heads we gaze on new constellations,—in the rocks under our feet, alone, we recognise our old acquaintances. And with regard to the primordial rocks, there is undoubtedly much truth in this pointed remark. Granite, mica slate, and their contained minerals, present the most identical resemblance, whether we collect from Dauphiné, Norway, the Alleghanies, Egypt, India or Australia. But concerning the secondary series, our information is far less precise.

[*Editor's Note:* Material has been omitted at this point.]

4

Reprinted from *Principles of Geology,* 1830, pp. 174, 188-193

PRINCIPLES OF GEOLOGY

C. Lyell

[*Editor's Note:* In the original, material precedes this exerpt.]

How the facts may be explained by assuming a uniform series of changes.—The readiest way, perhaps, of persuading the reader that we may dispense with great and sudden revolutions in the geological order of events, is by showing him how a regular and uninterrupted series of changes in the animate and inanimate world, may give rise to such breaks in the sequence, and such unconformability of stratified rocks, as are usually thought to imply convulsions and catastrophes. It is scarcely necessary to state, that the assumed order of events must be in harmony with all the conclusions legitimately drawn by geologists from the structure of the earth, and must be equally in accordance with the changes observed by man to be now going on in the living as well as in the inorganic creation. It may be necessary in the present state of science to supply some part of the assumed course of nature hypothetically; but if so, this must be done without any violation of probability, and always consistently with the analogy of what is known both of the past and present economy of our system. Although the discussion of so comprehensive a subject must carry the beginner far beyond his depth, it will also, it is hoped, stimulate his curiosity, and prepare him to read some elementary treatises on geology with advantage, and teach him the bearing on that science of the changes now in progress on the earth. At the same time it may enable him the better to understand the intimate connexion between the second and third books of this work, the former of which is occupied with the changes of the inorganic, the latter with those of the organic creation.

In pursuance, then, of the plan above proposed, I shall consider in this chapter, first, what may be the course of fluctuation in the animate world; secondly, the mode in which contemporaneous subterranean movements affect the earth's crust; and, thirdly, the laws which regulate the deposition of sediment.

[*Editor's Note:* Material has been omitted at this point.]

Concluding remarks on the identity of the ancient and present system of terrestrial changes.—I shall now conclude the discussion of a question with which we have been occupied since the beginning of the fifth chapter; namely, whether there has been any interruption, from the remotest periods, of one uniform system of change in the animate and inanimate world. We were induced to enter into that inquiry by reflecting how much the progress of opinion in Geology had been influenced by the assumption that the analogy was slight in kind, and still more slight in degree, between the causes which produced the former revolutions of the globe, and those now in everyday operation. It appeared clear that the earlier geologists had not only a scanty acquaintance with existing changes, but were singularly unconscious of the amount of their ignorance. With the presumption naturally inspired by this unconsciousness, they had no hesitation in deciding at once that time could never enable the existing powers of nature to work out changes of great magnitude, still less such important revolutions as those which are brought to light by Geology. They, therefore, felt themselves at liberty to indulge their imaginations in guessing at what *might be*, rather than inquiring *what is ;* in other words, they employed themselves in conjecturing what might have been the course of nature at a remote period, rather than in the investigation of what was the course of nature in their own times.

It appeared to them more philosophical to speculate on the possibilities of the past, than patiently to explore the realities of the present; and having invented theories under the influence of such maxims, they were consistently unwilling to test their validity by the criterion of their accordance with the ordinary operations of nature. On the contrary, the claims of each new hypothesis to credibility appeared enhanced by the great contrast, in kind or intensity, of the causes referred to and those now in operation.

Never was there a dogma more calculated to foster indolence, and to blunt the keen edge of curiosity, than this assumption of the discordance between the ancient and existing causes of change. It produced a state of mind unfavourable in the highest degree to the candid reception of the evidence of those minute but incessant alterations which every part of the earth's surface is undergoing, and by which the condition of its living inhabitants is continually made to vary. The student, instead of being encouraged with the hope of interpreting the enigmas presented to him in the earth's structure,—instead of being prompted to undertake laborious inquiries into the natural history of the organic world, and the complicated effects of the igneous and aqueous causes now in operation, was taught to despond from the first. Geology, it was affirmed, could never rise to the rank of an exact science,—the greater number of phenomena must for ever remain inexplicable, or only be partially elucidated by ingenious conjectures. Even the mystery which invested the subject was said to constitute one of its principal charms, affording, as it did, full scope to the fancy to indulge in a boundless field of speculation.

The course directly opposed to this method of philosophizing consists in an earnest and patient inquiry, how far geological appearances are reconcileable with the effect of changes now in progress, or which may be in progress in regions inaccessible to us, and of which the reality is attested by volcanos and subterranean movements. It also endeavours to estimate the aggregate result of ordinary operations multiplied by time, and cherishes a sanguine hope that the resources to be derived from observation and experiment, or from the study of nature such as she now is, are very far from being exhausted. For this reason all theories are rejected which involve the assumption of sudden and violent catastrophes and revolutions of the whole earth, and its inhabitants,—theories which are restrained by no reference to existing analogies, and in which a desire is manifested to cut, rather than patiently to untie, the Gordian knot.

We have now, at least, the advantage of knowing, from experience, that an opposite method has always put geologists on the road that leads to truth,—suggesting views which, although imperfect at first, have been found capable of improvement, until at last adopted by universal consent; while the method of speculating on a former distinct state of things and causes, has led invariably to a multitude of contradictory systems, which have been overthrown one after the other,—have been found incapable of modification,— and which have often required to be precisely reversed.

The remainder of this work will be devoted to an investigation of the changes now going on in the crust of the earth and its inhabitants. The importance which the student will attach to such researches will mainly depend in the degree of confidence which he feels in the principles above expounded. If he firmly believes in the resemblance or identity of the ancient and present system of terrestrial changes, he will regard every fact collected respecting the causes in diurnal action as affording him a key to the interpretation of some mystery in the past. Events which have occurred at the most distant periods in the animate and inanimate world, will be acknowledged to throw light on each other, and the deficiency of our information respecting some of the most obscure parts of the present creation will be removed. For as, by studying the external configuration of the existing land and its inhabitants, we may restore in imagination the appearance of the ancient continents which have passed away, so may we obtain from the deposits of ancient seas and lakes an insight into the nature of the subaqueous processes now in operation, and of many forms of organic life, which, though now existing, are veiled from sight. Rocks, also, produced by subterranean fire in former ages, at great depths in the bowels of the earth, present us, when upraised by gradual movements, and exposed to the light of heaven, with an image of those changes which the deep-seated volcano may now occasion in the nether regions. Thus, although we are mere sojourners on the surface of the planet, chained to a mere point in space, enduring but for a moment of time, the human mind is not only enabled to number worlds beyond the unassisted ken of mortal eye, but to trace the events of indefinite ages before the creation of our race, and is not even withheld from penetrating into the dark secrets of the ocean, or the interior of the solid globe; free, like the spirit which the poet described as animating the universe.

——— ire per omnes
Terrasque, tractusque maris, cœlumque profundum.

BOOK II.

CHANGES IN THE INORGANIC WORLD.

AQUEOUS CAUSES.

CHAPTER XIV.

Division of the subject into changes of the organic and inorganic world — Inorganic causes of change divided into aqueous and igneous — Aqueous causes first considered — Destroying and transporting power of running water — Sinuosities of rivers — Two streams when united do not occupy a bed of double surface — Heavy matter removed by torrents and floods — Inundations in Scotland — Floods caused by landslips in the White Mountains — Bursting of a lake in Switzerland — Devastations caused by the Anio at Tivoli — Excavations in the lavas of Etna by Sicilian rivers — Gorge of the Simeto — Gradual recession of the cataracts of Niagara.

Division of the subject. — GEOLOGY was defined to be the science which investigates the former changes that have taken place in the organic, as well as in the inorganic kingdoms of nature. As vicissitudes in the inorganic world are most apparent, and as on them all fluctuations in the animate creation must in a great measure depend, they may claim our first consideration. The great agents of change in the organic world may be divided into two principal classes, the aqueous and the igneous. To the aqueous belong Rain, Rivers, Torrents, Springs, Currents, and Tides; to the igneous, Volcanos and Earthquakes. Both these classes are instruments of decay as well as of reproduction; but they may also be regarded as antagonist forces. For the aqueous agents are incessantly labouring to reduce the inequalities of the earth's surface to a level; while the igneous are equally active in restoring the unevenness of the external crust, partly by heaping up new matter in certain localities, and partly by depressing one portion, and forcing out another, of the earth's envelope.

It is difficult, in a scientific arrangement, to give an accurate view of the combined effects of so many forces in simultaneous operation; because, when we consider them separately, we cannot easily estimate either the extent of their efficacy, or the kind of results which they produce. We are in danger, therefore, when we attempt to examine the influence exerted singly by each, of overlooking the modifications which they produce on one another; and these are so complicated, that sometimes the igneous and aqueous forces co-operate to produce a joint effect, to which neither of them unaided by the other could give rise,—as when repeated earthquakes unite with running water to widen a valley; or when a thermal spring rises up from a great depth, and conveys the mineral ingredients with which it is impregnated from the interior of the earth to the surface. Sometimes the organic combine with the inorganic causes; as when a reef, composed of shells and corals, protects one line of coast from the destroying power of tides or currents, and turns them against some other point; or when drift timber, floated into a lake, fills a hollow to which the stream would not have had sufficient velocity to convey earthy sediment.

It is necessary, however, to divide our observations on these various causes, and to classify them systematically, endeavouring as much as possible to keep in view that the effects in nature are mixed, and not simple, as they may appear in an artificial arrangement.

In treating, in the first place, of the aqueous causes, we may consider them under two divisions; first, those which are connected with the circulation of water from the land to the sea, under which are included all the phenomena of rivers and springs; secondly, those which arise from the movements of water in lakes, seas, and the ocean, wherein are comprised the phenomena of tides and currents. In turning our attention to the former division, we find that the effects of rivers may be subdivided into those of a destroying and those of a renovating nature; in the destroying are included the erosion of rocks, and the transportation of matter to lower levels; in the renovating class, the formation of deltas by the influx of sediment, and the shallowing of seas.

Action of running water.—I shall begin, then, by describing the destroying and transporting power of running water, as exhibited by rain, torrents and rivers. It is well known that the lands elevated above the sea attract, in proportion to their volume and density, a larger quantity of that aqueous vapour which the heated atmosphere continually absorbs from the surface of lakes and the ocean. By these means, the higher regions become perpetual reservoirs of water, which descend and irrigate the lower valleys and plains. In consequence of this provision, almost all the water is first carried to the highest regions, and is then made to descend by steep declivities towards the sea; so that it acquires superior velocity, and removes a greater quantity of soil, than it would do if the rain had been distributed over the plains and mountains equally in proportion to their relative areas. Almost all the water is also made by these means to pass over the greatest distances which each region affords, before it can regain the sea. The rocks also, in the higher regions, are particularly exposed to atmospheric influences, to frost, rain, and vapour, and to great annual alternations of cold and heat, of moisture and desiccation.

Among the most powerful agents of decay may be mentioned that property of water which causes it to expand during congelation; its increase of bulk when converted into ice amounting to no less than one twentieth of the whole volume, so that when water has penetrated into the crevices of the most solid rocks, it rends them open on freezing with mechanical force.* For this reason, although, in cold climates the comparative quantity of rain which falls is very inferior, and although it descends more gradually than in tropical regions, yet the severity of frost, and the greater inequalities of temperature, compensate in some degree for this diminished source of degradation. The solvent power of water also is very great, and acts particularly on the calcareous and alkaline elements of stone, especially when it holds carbonic acid in solution, which is abundantly supplied to almost every large river by springs, and is collected by rain from the atmosphere. The oxygen of the atmosphere is also gradually absorbed by all animal and vegetable productions, and by almost all mineral masses exposed to the open air. It gradually destroys the equilibrium of the elements of rocks, and tends to reduce into powder, and to render fit for soils, even the hardest aggregates belonging to our globe.†

When earthy matter has once been intermixed with running water, a new mechanical power is obtained by the attrition of sand and pebbles, borne along with violence by a stream. Running water charged with foreign ingredients being thrown against a rock, excavates it by mechanical force, sapping and undermining till the superincumbent portion is at length precipitated into the stream. The materials cause a temporary obstruction ponding back the water until by its accumulated weight it sweeps down the barrier.

[*Editor's Note:* Material has been omitted at this point.]

* Quart. Journ. of Sci., &c. New Series, No. xiii. p 194.
† Sir H. Davy, Consolations in Travel, p. 271.

O

5

Reprinted from *Proc. Geol. Soc. London,* 3(72), 327-332 (1842)

ON GLACIERS AND THE EVIDENCE OF THEIR HAVING ONCE EXISTED IN SCOTLAND, IRELAND, AND ENGLAND

L. Agassiz

Nov. 4.—A paper was read on Glaciers, and the evidence of their having once existed in Scotland, Ireland, and England, by Professor Agassiz, of Neuchâtel.

M. Agassiz commences by observing, that the study of glaciers is not new, as Scheuchzer visited, and even drew, most of the glaciers of Switzerland; and as, at a later period, Gruner and De Saussure examined them in great detail, and left few of their phænomena uninvestigated. Hugi also, in his account of the Alps, and Scoresby, in his descriptions of the arctic regions, have communicated much valuable information respecting glaciers, but without giving rise to any important geological results. Venetz and De Charpentier first ascribed to the agency of glaciers, the transport of the erratic boulders of Switzerland, supposing that the Alps formerly attained a greater altitude than at present, and that the glaciers extended to the plains of Switzerland, and even to the Jura. This assumed greater height of the Alps M. Agassiz dissents from, as no geological phænomena compel him to admit it; and the arrangement of the boulders proves that the blocks were not pushed forward by the glaciers, as conjectured by M. de Charpentier. Moreover, the phænomena of erratic boulders extend over all the temperate and northern regions of Europe, Asia and America, and, consequently, could not depend upon so local an event as a greater altitude of the Alps. The consideration of these difficulties induced M. Agassiz to resume the study of glaciers; and after devoting the suitable portion of five successive summers to the study of their details, and all that has been written respecting their structure, he has arrived at the conviction, that the formation of glaciers did not only depend upon the actual configuration of the globe, but was also connected with the last great geological changes in its surface, and with the extinction of the great mammifers which are now found in the polar ice. He is also convinced that the glaciers did not advance from the Alps into the plains, but that they gradually withdrew towards the mountains from the plains which they once covered. In this belief, he says, he is supported by many considerations which escaped previous observers, depending chiefly on the form and relative position of the erratic blocks, and the commonly called diluvial gravel, the former being in Switzerland always angular, and resting on the latter, which consists of rounded materials. Considered in this point of view, glaciers assume an entirely new importance, for they introduce a long period of intense cold between the present epoch and that during which the animals existed, whose remains are buried in the usually called diluvial detritus.

Having established his theory as completely as he could, by repeated investigations of Switzerland and the adjacent portions of France and Germany, M. Agassiz became desirous of investigating a country in which glaciers no longer exist, but in which traces of them might be found. This opportunity he has recently enjoyed, by examining a considerable part of Scotland, the north of England, and the north, centre, west and south-west of Ireland; and he has arrived at the conclusion, that great masses of ice, and subsequently glaciers, existed in these portions of the United Kingdom at a period immediately preceding the present condition of the globe, founding his belief upon the characters of the superficial gravels and erratic blocks, and on the polished and striated appearance of the rocks *in situ*.

M. Agassiz does not suppose that his views respecting glaciers will at once meet with the general concurrence of geologists; and he admits that the study of the phænomena of glaciers in different latitudes, as well as at different altitudes, together with the examination of their different effects where in contact with the sea, will introduce many modifications in the consideration of analogous phænomena in countries where glaciers have disappeared; but he is prepared to discuss his theory within the limits of observed facts, conscious of having searched for truth solely to advance the interests of science.

To avoid useless discussion, he states, that in attributing to the action of glaciers a considerable portion of the results hitherto ascribed exclusively to that of water, he does not wish to maintain that everything hitherto assigned to the agency of water has been produced by glaciers; he only wishes that a distinction may be made in each locality between the effects of the different agents; and he adds, that long-continued practice has taught him to distinguish easily, in most cases, the effects produced by ice from those produced by water.

Proceeding to the consideration of facts, he says the distribution of blocks and gravel, as well as the polished and striated surfaces of rocks *in situ*, do not indicate the action of a mighty current flowing from north-west to south-east, as the blocks and masses of gravel everywhere diverge from the central chains of the country, following the course of the valleys. Thus in the valleys of Loch Lomond and Loch Long, they range from north to south; in those of Loch Fine and Loch Awe from north-west to south-east; of Loch Etine and Loch Leven from east to west; and in the valley of the Forth from north-west to south-east, radiating from the great mountain masses between Ben Nevis and Ben Lomond. Ben Nevis in the north of Scotland, and the Grampians in the south, are considered by the author to constitute the great centres of dispersion in that kingdom; and the mountains of Northumberland, Westmoreland, Cumberland, and Wales; as well as those of Ayrshire, Antrim, the west of

Ireland, and Wicklow, to be other points from which blocks and gravel have been dispersed, each district having its peculiar debris, traceable in many instances to the parent rock, at the head of the valleys. Hence, observes M. Agassiz, it is plain the cause of the transport must be sought for in the centre of the mountain ranges, and not from a point without the district. The Swedish blocks on the coast of England do not, he conceives, contradict this position, as he adopts the opinion that they may have been transported on floating ice.

In describing the phænomena presented by erratic blocks and gravel, M. Agassiz first insists upon the necessity of distinguishing between stratified gravel and mud containing fossils, which could not have been accumulated by true glaciers, although the materials may have often been derived from them; and unstratified masses, composed of blocks, pebbles, and clay. These stratified deposits he considers to be of posterior origin to the glacier epoch. The till of Scotland, or the great unstratified accumulation of mud and gravel, containing blocks of different size heaped together without order, and containing no organic remains but bones of Mammalia and insignificant fragments of shells, he is of opinion was also not produced by true glaciers, although intimately connected with the phænomena of ice. The polished and striated surfaces of the blocks leave no doubt on M. Agassiz's mind that these masses have been acted upon by ice in the same manner as the blocks which are observed under existing glaciers, and which are more or less rearranged by water derived from the melting of the glaciers.

Similar detritus fills the bottom of all the Alpine valleys, as that of the Rhone from its mouth to its junction with the Lake of Geneva, and the valley of Chamounix: it is found between the Hospice de Grimsel and the borders of the lower glacier of the Aar; thence to the neighbourhood of Goutharen in the valley of Oberhasli, at Im Grund, in the plains of Meiringen, and in Interlasken; also between Thun and Berne. At all these localities, M. Agassiz considers, the blocks were left, when the glaciers extended to them.

With respect to the valley of the Aar, M. Agassiz says it is easy to prove that the rounded pebbles of Alpine rocks spread along its whole course, were not transported to their present position by that river, because between the glacier from which it issues and Berne, the flowing of the stream is interrupted by the barrier of Kirchet, the Lake of Brienz, and the Lake of Thun; and because between these lakes its velocity is so small, that it transports only mud and very fine gravel, and that the pebbles over which the river flows below Thun do not issue from the lake. Supposing that the volume of the Aar was formerly greater, why, asks M. Agassiz, are not the lakes of Brienz and Thun filled in the same manner as the plain of Meiringen and the bottom of the valley which separates the two lakes? All difficulties, however, he is of opinion, vanish, if the pebbles be considered the detritus of retreating glaciers, and that the hollows occupied by the lakes of Brienz and Thun were filled with glaciers.

The existence of a glacier in this valley is not imagined by the author to explain the origin of the detritus, as its having existed is proved by the polish on the rocks *in situ*, from the glacier of the Aar to Meiringen, a distance of twenty English miles, at the height of 8000, 7000, and 6000 feet successively above the level of the sea; and even on the shores of the Lake of Thun. Similar phænomena have been noticed by M. Agassiz in Scotland, in the valleys of Loch Awe and Loch Leven, near Ballachalish, and in England in the neighbourhood of Kendal.

The author then proceeds to describe the moraines of Switzerland, or the accumulations of blocks and pebbles deposited longitudinally on the borders, and transversely in front, of glaciers, and successively abandoned by them in their retreat. The longitudinal moraines differ from glacier-detritus remodelled or spread out by water, in being disposed in ridges with a double talus, one flank of which is presented to the glacier, and the other to the side of the valley; and their continuity and parallelism at the same height easily distinguish them from the debris disposed along the bottoms of valleys by currents. They occur on the flanks of all glaciers, but they have been also observed by M. Agassiz where no glaciers exist, as in the valleys of the Rhone, the Arve, the Aar, &c.; likewise in Scotland, near Inverary, at Muc Airn, at the outlet of Loch Traig, at Strankaer, and on the borders of the bay of Beauley; in Ireland to the south-east of Dublin, and near Enniskillen; and in England in the valley of Kendal, as well as near Penrith and Shap.

The common origin of moraines, and of accumulations of rounded pebbles and of blocks, M. Agassiz says, cannot be doubted. The former are simple ridges formed on glaciers; the latter, materials rounded and polished under glaciers, or great masses of ice, and exposed by the melting of the ice, and re-disposed by the water thus produced.

The author next describes the differences in the internal arrangement of the various accumulations. In the stratified deposits the materials are comparatively much smaller than in glacier-detritus; the pebbles also are elongated, and fine gravel and mud ordinarily form the upper beds. On the contrary, in the detritus of glaciers large and small materials are associated without order, the largest blocks being often in the upper part; and where very large *angular* blocks occur, they rest on the surface. In moraines there is a further distinction, blocks of all dimensions and every form are intermingled; and this difference, he says, is easily understood, by recollecting that moraines are composed of the angular blocks which fall on the surface of the glacier, as well as of pebbles with rounded edges.

The striated and polished surfaces, so often observed on solid rocks *in situ*, are next described by M. Agassiz. Without denying absolutely the power of water to produce such effects, he says that he has sought for them in vain on the borders of rivers and lakes, and on sea-coasts; and that the effects produced by water are sinuous furrows proportioned to the hardness of the rocks; not even uniform polished surfaces, such as those presented by the rocks under discussion, and which are independent of the composition of the stone; moreover wherever the moveable materials which are pressed by the ice on rocks *in situ* are the hardest, there occur, independent of the polish, striæ more or less parallel, and in the general direction of the movement of the glaciers. Thus in the neighbourhood of glaciers are found those rounded bosses which Saussure distinguishes by the name of "roches moutonnés." These phænomena M. Agassiz has traced under the glacier of the Aar, and he has observed them in the valley of the Rhone, and of Chamounix; also in Scotland, on the banks of Loch Awe and Loch Leven; and he says they are very remarkable in the environs of Kendal.

The most striking points in the distribution of the striæ, are their diverging at the outlets of the valleys, and their being oblique, and never horizontal on the flanks, which they would be, were they due to the agency of water, or floating masses of ice. The cause of this obliquity the author assigns to the upward expansion of the ice, and the descending motion of the glacier.

The most remarkable striated rocks in the Alps are near Handeck, and near the cascade of Pissevache; and the best examples M. Agassiz has seen in Scotland, are those of Ballahulish, and in Ireland those of Virginia.

If the analogy of the facts which he has observed in Scotland, Ireland, and the north of England, with those in Switzerland, be correct, then it must be admitted, M. Agassiz says, that not only glaciers once existed in the British Islands, but that large sheets (*nappes*) of ice covered all the surface.

The author then details the proofs that glaciers did not descend from the mountain summits into the plains, but are the remaining portions of the sheets of ice which at one time covered the flat country. It is evident, he says, if the glaciers descended from high mountains, and extended forward into the plains, the largest moraines ought to be the most distant, and to be formed of the most rounded masses; whereas the actual condition of the detrital accumulations is the reverse, the distant materials being widely spread, and true moraines being found only in valleys connected with great chains of lofty mountains.

It must then be admitted, the author argues, that great sheets of ice, resembling those now existing in Greenland, once covered all the countries in which unstratified gravel is found; that this gravel was in general produced by the trituration of the sheets of ice upon the subjacent surface; that moraines, as before stated, are the effects of the retreat of glaciers; that the angular blocks found on the surface of the rounded materials were left in their present position at the melting of the ice; and that the disappearance of great bodies of ice produced enormous debacles and considerable currents, by which masses of ice were set afloat, and conveyed, in diverging directions, the blocks with which they were charged. He believes that the Norwegian blocks found on the coast of England have been correctly assigned by Mr. Lyell to a similar origin.

Another class of phænomena connected with glaciers, is the forming of lakes by the extension of glaciers from lateral valleys into a main valley; and M. Agassiz is of opinion, that the parallel roads of Glen Roy were formed by a lake which was produced in consequence of a lateral glacier projecting across the glen near Bridge Roy, and another across the valley of Glen Speane. Lakes thus formed naturally give rise to stratified deposits and parallel roads, or beds of detritus at different levels.

The connexion of very recent stratified deposits with glacier-detritus, M. Agassiz observes, is difficult to explain; but he conceives that the same causes which can bar up valleys and form lakes, like those of Brienz, Thun and Zurich, may have formed analogous barriers at the point of contact with the sea sufficiently extensive to have produced large salt-marshes to be inhabited by animals, the remains of which are found in the clays superimposed on the till of Scotland; and he adds, that the known arctic character of these fossils ought to have great weight with those who study the vast subject of glaciers.

In conclusion, the author remarks, that the question of glaciers forms part of many of the great problems of geology; that it accounts for the disappearance of the large mammifers inclosed in the polar ice, as well as for the disappearance of the organic beings of the so-called diluvian epoch; that in Switzerland it is associated with the elevation of the Alps, and the dispersion of the erratic blocks; and that it is so intimately mixed up with the subject of a general diminution of the terrestrial heat, that a more profound acquaintance with the facts, noticed in this paper, will probably modify the opinions entertained respecting it.

6

Reprinted from *Report on the Geology of the Henry Mountains,* Government Printing Office, Washington, D.C., 1880, pp. 115–117, 123–124, 140–142

REPORT ON THE GEOLOGY OF THE HENRY MOUNTAINS

G. K. Gilbert

[*Editor's Note:* In the original, material precedes this excerpt.]

II. SCULPTURE.

Erosion may be regarded from several points of view. It lays bare rocks which were before covered and concealed, and is thence called *denudation.* It reduces the surfaces of mountains, plateaus, and continents, and is thence called *degradation.* It carves new forms of land from those which before existed, and is thence called *land sculpture.* In the following pages it will be considered as land sculpture, and attention will be called to certain principles of erosion which are concerned in the production of topographic forms.

Sculpture and Declivity.

We have already seen that erosion is favored by declivity. Where the declivity is great the agents of erosion are powerful; where it is small they are weak; where there is no declivity they are powerless. Moreover it has been shown that their power increases with the declivity in more than simple ratio.

It is evident that if steep slopes are worn more rapidly than gentle, the tendency is to abolish all differences of slope and produce uniformity. The law of uniform slope thus opposes diversity of topography, and if not complemented by other laws, would reduce all drainage basins to plains. But in reality it is never free to work out its full results; for it demands a uniformity of conditions which nowhere exists. Only a water sheet of uniform depth, flowing over a surface of homogeneous material, would suffice; and every inequality of water depth or of rock texture produces a corresponding inequality of slope and diversity of form. The reliefs of the landscape exemplify other laws, and the law of uniform slopes is merely the conservative element which limits their results.

Sculpture and Structure; the Law of Structure.

We have already seen that erosion is influenced by rock character. Certain rocks, of which the hard are most conspicuous, oppose a stubborn resistance to erosive agencies; certain others, of which the soft are most conspicuous, oppose a feeble resistance. Erosion is most rapid where the resistance is least, and hence as the soft rocks are worn away the hard are left prominent. The differentiation continues until an equilibrium is reached through the law of declivities. When the ratio of erosive action as dependent on declivities becomes equal to the ratio of resistances as dependent on rock character, there is equality of action. In the structure of the earth's crust hard and soft rocks are grouped with infinite diversity of arrangement. They are in masses of all forms, and dimensions, and positions; and from these forms are carved an infinite variety of topographic reliefs.

In so far as the law of structure controls sculpture, hard masses stand as eminences and soft are carved in valleys.

The Law of Divides.

We have seen that the declivity over which water flows bears an inverse relation to the quantity of water. If we follow a stream from its mouth upward and pass successively the mouths of its tributaries, we find its volume gradually less and less and its grade steeper and steeper, until finally at its head we reach the steepest grade of all. If we draw the profile of the river on paper, we produce a curve concave upward and with the greatest curvature at the upper end. The same law applies to every tributary and even to the slopes over which the freshly fallen rain flows in a sheet before it is gathered into rills. The nearer the water-shed or divide the steeper the slope; the farther away the less the slope.

It is in accordance with this law that mountains are steepest at their crests. The profile of a mountain if taken along drainage lines is concave outward as represented in the diagram; and this is purely a matter of sculpture, the uplifts from which mountains are carved rarely if ever assuming this form.

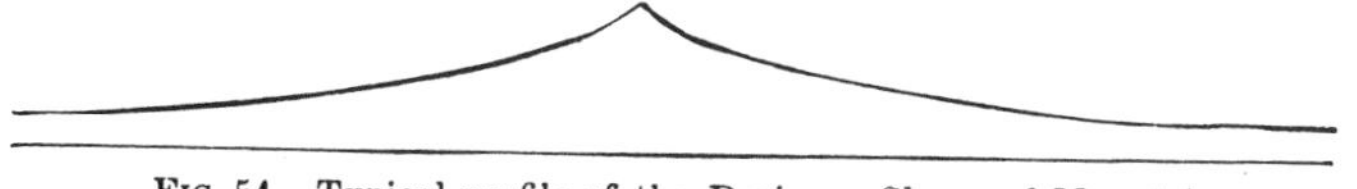

FIG. 54.—Typical profile of the Drainage Slopes of Mountains.

Under the *law of Structure* and the *law of Divides* combined, the features of the earth are carved. Declivities are steep in proportion as their material is hard; and they are steep in proportion as they are near divides.

The distribution of hard and soft rocks, or the geological structure, and the distribution of drainage lines and water-sheds, are coefficient conditions on which depends the sculpture of the land. In the sequel it will be shown that the distribution of drainage lines and water-sheds depends in part on that of hard and soft rocks.

In some places the first of the two conditions is the more important, in others the second. In the bed of a stream without tributaries the grade depends on the structure of the underlying rocks. In rock which is homogeneous and structureless all slopes depend on the distribution of divides and drainage lines.

The relative importance of the two conditions is especially affected by climate, and the influence of this factor is so great that it may claim rank as a third condition of sculpture.

[*Editor's Note:* Material has been omitted at this point.]

Equal Action and Interdependence.

The tendency to equality of action, or to the establishment of a dynamic equilibrium, has already been pointed out in the discussion of the principles of erosion and of sculpture, but one of its most important results has not been noticed.

Of the main conditions which determine the rate of erosion, namely, quantity of running water, vegetation, texture of rock, and declivity, only the last is reciprocally determined by rate of erosion. Declivity originates in upheaval, or in the displacements of the earth's crust by which mountains and continents are formed; but it receives its distribution in detail in accordance with the laws of erosion. Wherever by reason of change in any of the conditions the erosive agents come to have locally exceptional power, that power is steadily diminished by the reaction of rate of erosion upon declivity. Every slope is a member of a series, receiving the water and the waste of the slope above it, and discharging its own water and waste upon the slope below. If one member of the series is eroded with exceptional rapidity, two things immediately result: first, the member above has its level of discharge lowered, and its rate of erosion is thereby increased; and second, the member below, being clogged by an exceptional load of detritus, has its rate of erosion diminished. The acceleration above and the retardation below, diminish the declivity of the member in which the disturbance originated; and as the declivity is reduced the rate of erosion is likewise reduced.

But the effect does not stop here. The disturbance which has been transferred from one member of the series to the two which adjoin it, is by them transmitted to others, and does not cease until it has reached the confines of the drainage basin. For in each basin all lines of drainage unite in a main line, and a disturbance upon any line is communicated through it to the main line and thence to every tributary. And as any member of the system may influence all the others, so each member is influenced by every other. There is an interdependence throughout the system.

[*Editor's Note:* Material has been omitted at this point.]

Monoclinal Shifting.

In regions of inclined strata, the same process which gathers the waterways into the outcrops of the softer beds converts the outcrops of the harder into divides. As the degradation progresses the waterways and divides descend obliquely and retain the same relations to the beds. The waterways continuously select the soft because they resist erosion feebly, and the watersheds as continuously select the hard because they resist erosion strongly. If the inclination of the strata is gentle, each hard bed becomes the cap of a sloping table bounded by a cliff, and the erosion of the cliff is by sapping. The divide is at the brow of the cliff, and as successive fragments of the hard rock break away and roll down the

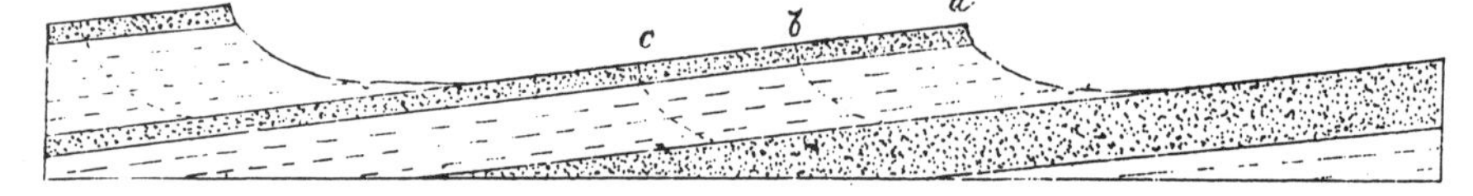

FIG. 69.—Ideal cross-section of inclined strata, to show the Shifting of Divides in Cliff Erosion. Successive positions of a divide are indicated at *a*, *b*, and *c*.

slope the divide is shifted. The process is illustrated in the Pink Cliffs of Southern Utah. They face to the south, and their escarpment is drained by streams flowing to the Colorado. The table which they limit inclines to the north and bears the head-waters of the Sevier. As the erosion of the cliffs steadily carries them back and restricts the table, the drainage area of the Colorado is increased and that of the "Great Basin", to which the Sevier River is tributary, is diminished.

Unequal and Equal Declivities.

In homogeneous material, and with equal quantities of water, the rate of erosion of two slopes depends upon their declivities. The steeper is degraded the faster. It is evident that when the two slopes are upon opposite sides of a divide the more rapid wearing of the steeper carries the divide toward the side of the gentler. The action ceases and the divide becomes stationary only when the profile of the divide has been rendered symmetric.

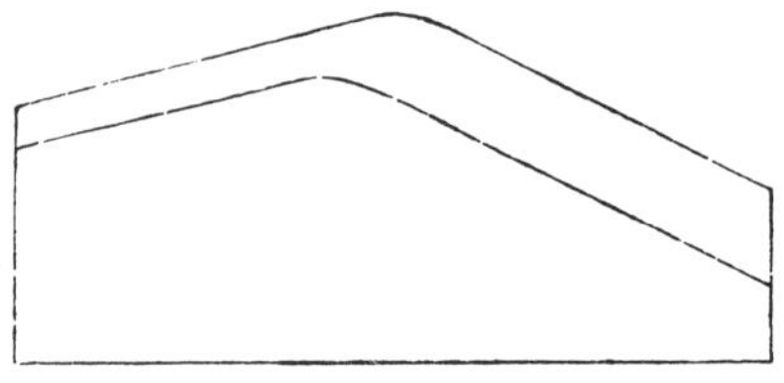

FIG. 70.—Cross-profile of a bad-land divide separating slopes of Unequal Declivity. Two stages of erosion are indicated, to illustrate the horizontal shifting of the divide.

It is to this law that bad-lands owe much of their beauty. They acquire their smooth curves under what I have called the "law of divides", but the symmetry of each ridge and each spur is due to the law of equal declivities. By the law of divides all the slopes upon one side of a ridge are made interdependent. By the law of equal declivities a relation is established between the slopes which adjoin the crest on opposite sides, and by this means the slopes of the whole ridge, from base to base, are rendered interdependent.

One result of the interdependence of slopes is that a bad-land ridge separating two waterways which have the same level, stands midway between them; while a ridge separating two waterways which have different levels, stands nearer to the one which is higher.

It results also that if one of the waterways is corraded more rapidly than the other the divide moves steadily toward the latter, and eventually, if the process continues, reaches it. When this occurs, the stream with the higher valley abandons the lower part of its course and joins its water to that of the lower stream. Thus from the shifting of divides there arises yet another method of the shifting of waterways, a method which it will be convenient to characterize as that of *abstraction.* A stream which for any reason is able to corrade its bottom more rapidly than do its neighbors, expands its valley at their expense, and eventually "abstracts" them. And conversely, a stream which for any reason is able to corrade its bottom less rapidly than its neighbors, has its valley contracted by their encroachments and is eventually "abstracted" by one or the other.

The diverse circumstances which may lead to these results need not be enumerated, but there is one case which is specially noteworthy on account of its relation to the principles of sculpture. Suppose that two streams which run parallel and near to each other corrade the same material and degrade their channels at the same rate. Their divide will run midway. But if in the course of time one of the streams encounters a peculiarly hard mass of rock while the other does not, its rate of corrasion above the obstruction will be checked. The unobstructed stream will outstrip it, will encroach upon its valley, and will at last abstract it; and the incipient corrasion of the hard mass will be stopped. Thus by abstraction as well as by monoclinal shifting, streams are eliminated from hard rocks.

Résumé.—There is a tendency to permanence on the part of drainage lines and divides, and they are not displaced without adequate cause. Hence every change which is known to occur demands and admits of an explanation.

(*a*) There are four ways in which abrupt changes are made. Streams are diverted from one drainage system to another, and the watersheds which separate the systems are rearranged,

(1) by *ponding*, due to the elevation or depression of portions of the land;
(2) by *planation*, or the extension of flood-plains by lateral corrasion;
(3) by *alluviation*, or in the process of building alluvial cones and deltas; and
(4) by *abstraction.*

(*b*) There are two ways in which gradual changes are effected:

(1) When the rock texture is variable, it modifies and controls by *monoclinal shifting* the distribution in detail of divides and waterways.
(2) When the rock texture is uniform, the positions of divides are adjusted in accordance with the principle of *equal declivities.*

The abrupt changes are of geographic import; the gradual, of topographic.

The methods which have been enumerated are not the only ones by which drainage systems are modified, but they are the chief. Very rarely streams are "ponded" and diverted to new courses through the damming of their valleys by glaciers or by volcanic *ejecta* or by land-slips. More frequently they are obstructed by the growing alluvial cones of stronger streams, but only the smallest streams will yield their "right of way" for such cause, and the results are insignificant.

[*Editor's Note:* Material has been omitted at this point.]

7

Reprinted from *Quart. Jour. Geol. Soc. London,* **28,** 148-150, 152 (1872)

On the River-courses *of* England *and* Wales.

By Prof. Andrew C. Ramsay, LL.D., F.R.S., V.P.G.S.

In the following paper I propose to show the origin of many of the principal rivers of England and Wales—that is to say, what are the special geological causes the operation of which led them to flow in the general directions they now take. I am not aware that any attempt has heretofore been made to do this on a large scale, though I have already done something on the subject with regard to the rivers of the Weald, in which line of argument I was afterwards followed by Mr. Foster and Mr. Topley.

I shall begin the subject by a rapid summary of certain physical changes that affected the English Secondary and Eocene strata, long before the Severn (leaving the mountains of Wales) took its present southern and south-western course along the eastern side of the palæozoic rocks that border that old land.

About the close of the Oolitic epoch the strata of these formations were raised above the sea, and remained a long time out of water; and during that period those atmospheric influences that produced the sediments of the great Purbeck and Wealden delta were slowly wearing away and lowering the land, and reducing it to the state of a broad undulating plain. At this time the Oolitic strata still abutted on the mountain country now forming Wales and parts of the adjacent counties. They also completely covered the Mendip Hills, and

Fig. 1.
Geological and River Map of England and Wales.

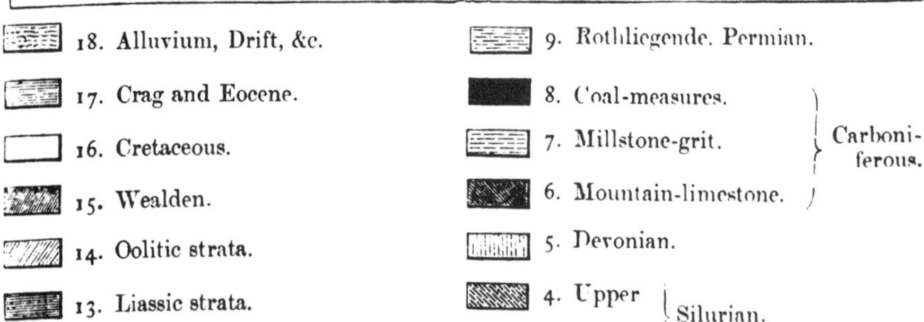

passed westwards as far as the hilly ground of Devonshire, running out between Wales and Devonshire through what is now the Bristol Channel. The whole of the middle of England was likewise covered by the same deposits, viz. the plains of Shropshire, Cheshire, Lancashire, and the adjoining areas; so that the Lias and Oolites passed out to what is now the Irish Sea, over and beyond the present estuaries of the Dee and the Mersey, between North Wales and the hilly ground of Lancashire, formed of previously disturbed Carboniferous rocks. In brief, most of the present mountainous and hilly lands of the mainland of Britain were mountainous and hilly then, and even higher than now, considering how much they must since have suffered by denudation.

At this period, south of the Derbyshire hills and through Shropshire and Cheshire, the Lower Secondary rocks lay somewhat flatly; while in the more southern and eastern areas they were tilted up to the west, so as to give them a low eastern dip. The general arrangement of the strata in the south would then be somewhat as shown in fig. 2 (p. 151).

The submersion of this low-lying area brought the deposition of the Wealden strata to a close, and the Cretaceous formations were deposited above the Wealden and Oolitic strata, so that a great unconformable overlap of Cretaceous strata took place across the successive outcrops of the Oolitic and older Secondary formations, as shown in fig. 3 (p. 151).

The same kind of overlapping of the Cretaceous on the Oolitic formations took place at the same time in the country north and south of the present estuary of the Humber, the proof of which is well seen in the unconformity of the Cretaceous rocks on the Oolites and Lias of Lincolnshire and Yorkshire.

At this time the mountains of Wales and other hilly regions formed of Palæozoic rocks must have been lower than they were during the Oolitic epochs, partly by the effect of long-continued waste, due to atmospheric causes, but probably even more because of gradual and greatly increased submergence during the time that the Chalk was being deposited. I omit any detailed mention of the phenomena connected with the deposition of the freshwater and marine Eocene strata, because at present this subject does not seem essential to my argument.

The Miocene period of old Europe was essentially a continental one. Important disturbances of strata brought this epoch to a close, at all events physically, in what is now the centre of Europe; and the formations formed in the great freshwater lakes that lay at the bases of the older Alps were, after consolidation, heaved up to form new mountains along the flanks of the ancient range; and all the length of the Jura, and far beyond to the north-west, was elevated by disturbances of the Jurassic, Cretaceous, and Miocene strata. The broad valley of the lowlands of Switzerland began then to be established, subsequently to be overspread by the large glaciers that deepened the valleys and scooped out the lakes.

One marked effect of this extremely important elevation, after Miocene times, of so much of the centre of Europe was, that the flat

[*Editor's Note:* Certain figures have been omitted at this point.]

or nearly flat-lying Secondary formations that now form great part of France and England (then united), were so far affected by this renewed upheaval of the Alps and Jura, that they were all tilted at low angles to the north-west. That circumstance gave the initial north-westerly direction to the flow of so many of the existing rivers of France, and led them to excavate the valleys in which they run, including the upper tributaries of the Loire and Seine, the Seine itself, the Marne, the Oise, and many more of smaller size; and my surmise is, that this same westerly and north-westerly tilting of the chalk of England formed a gentle slope towards the mountains of Wales, and the rivers of the middle and south of England at that time flowed westerly. This first induced the Severn to take a southern course, between the hilly land of Wales and Herefordshire and the long slope of Chalk rising to the east; and, aided by the tributary streams of Herefordshire, it cut a channel towards what afterwards became the Bristol Channel, and established the beginning of the escarpment of the chalk (*e*, fig. 4, p. 154), which has since gradually receded, chiefly by atmospheric waste, so far to the east. If this be so, then the origin of the valley of the Severn is of immediate post-Miocene date, and it is one of the oldest in England *.

The Avon, which is a tributary of the Severn, and joins it at Tewkesbury, is, at all events, partly of later date. It rises at the base of the escarpment of the Oolitic rocks east of Rugby, and gradually established its channel in the low grounds formed of Lower Lias and New Red Marl, as that escarpment retired eastward by virtue of that law of waste, that all inland escarpments retire opposite to the steep slope, and in the direction of the slope of the strata.

If the general slope of the surface of the chalk had been easterly instead of westerly at the post-Miocene date alluded to, then the initial course of the Severn would also have been easterly, like that of the Thames and the rivers that flow into the Wash and the Humber.

[*Editor's Note:* Material has been omitted at this point.]

* Many of the valleys of Wales may be much older.

8

Reprinted from *Rain and Rivers*, 2nd ed., Longmans, Green & Co. Ltd., London, 1857, pp. 53-54, 73-74, 102-104, 173-176

RAIN AND RIVERS*

G. Greenwood

[*Editor's Note:* In the original, material precedes this excerpt.]

How rivers run *through* ridges of hills even without ponding, as in the Weald.

We are not to imagine that it is in the Weald only that rivers run *through* ridges of hills. A glance at a physical map of England or the World, will show plenty of examples. Is Mr. Hopkins going to account for these " cross-fractures " by " mathematical methods ?"

A stream running *through* ridges, large or small, is the simple consequence of the differing hardness of the ground through which it runs. In all cases a stream cuts for itself a narrow channel, the depth of which is determined by the hardest part. For a stream cannot run down where its bed is soft, and up again where it is hard. But the wash of rain digs down where the ground is soft, and leaves hills or ridges where it is hard. And as a stream cuts through a hard stratum, the wash of rain is scooping out two lateral valleys *behind* it, that is, a valley behind each side of the gorge and ridge.

*Reproduced from microfilm of the original supplied by the History of Science Collections, University of Oklahoma Libraries.

Such are the valleys along which the western branches of the Arun and Wey flow. But in this *small* map only *large* streams are marked. The *débris* of these valleys is carried off by the lowering bed of the river. A ridge is then developed, and the river runs through a gorge in the ridge. In this sense the wash of rain may be said to form hills as well as valleys, since by abstracting intervening masses it develops them. And in *forming* a statue, what does the artist do more?

These gorges, cut through comparatively hard beds or ridges, form the exceptions to the widening of valleys as they descend. The great height of some of these ridges, though a *proof*, is a very inadequate *measure* of the enormous denudation which has gone on *behind* them, since the ridges themselves have suffered enormous denudation. The tops of all such ridges must originally have been lower than the hills from which the rivers rise. But whether all or any of them continue lower, and their comparative height one with the other, will depend simply on which is the hardest, that is, which is most quickly disintegrated and denuded. The hills from which the rivers rise *might* be denuded as low as the beds of the rivers in the gorges.

[*Editor's Note:* Material has been omitted at this point.]

Rivers have the power to cut narrow channels or *ravines*; but they have very little power of widening these. Disintegration and the wash of rain widen these ravines into broad valleys. While this is going on rivers convey to the sea what rain brings to them, which would otherwise pass out of the valleys more slowly by the wash of rain.

Rivers may cut deep channels, rain makes wide valleys.

But rivers no more *make* the sediment which they carry than railroads *manufacture* the wares of which their traffic consists. The sediment carried by rivers is brought to them by rain, from the entire surface of their *tributary ground*, or watershed. Erosion of banks is an exceptional calamity. But heavy rain never happens, but it loads the rivers with mud. Nay, it loads the along-shore waters of the sea with mud. These are matters of fact, not matters of opinion. We "need no ghost" nor good eyes to ascertain them; all we need good is the umbrella and macintosh, to enable us *to go out and see*.

The Atlantic round Madeira, or the Mediterranean along Spain, France, and Italy, or the

Egean round the Isles of Greece, forms no exception: vidimus flavum each of them; or rather I have seen the Atlantic *red* instead of yellow at Madeira during heavy rain. It is astonishing how soon the sea clears. Still more astonishing that not a vestige of dirt, mud, earth, or clay remains on or *near* the shore of the grand recipient of all these. Ocean not only instantly swallows all, but licks the platter clean.

Flat land may be denuded.

But in the same page Lyell is "overtaken by a violent thunderstorm," near San Lorenzo in Spain; and "I saw the whole surface, even the highest levels of some flat-topped hills streaming with mud, while on every declivity the devastation of torrents was terrific. The peculiarities of the physiognomy of the district were at once explained; and I was taught that on speculating on the *greater* effects which the direct action of rain may *once* have produced on the surface of certain parts of England, we need not revert to periods when the heat of the climate was *tropical.*"

[*Editor's Note:* Material has been omitted at this point.]

Mountains while rising may be decreasing in height.

It does not follow that while mountains are rising they are increasing in height. They may be decreasing in height. Suppose the Alps to have been rising six inches in a century for myriads of years, if their denudation has been seven inches in a century, they have been decreasing in height.

It is true that the *direct* action in waste and denudation of torrents and rivers is on lines only; and were it not for the atmospheric disintegration and the *lateral* wash of rain this their *direct* action would only cut ravines and channels to the sea, and the sides of all valleys, instead of sloping, would be *cliffs ;* that is, where a spring issues high up the mountain side, it will cut a deep ravine with precipitous *cliffs*, and the deeper it cuts, the more springs it will lay open. But what widens this ravine into a broad valley with gently sloping sides? The *lateral* wash of rain into the *longitudinal* valley. And what forms the broad valley, even *where there is no river at the bottom?* or within many miles? The *longitudinal* scooping power of the concentrated wash of rain, which in no respect differs from that of the torrent, except in its being a hundred-fold more powerful than the torrent. It is indeed intermittent; so is the real scooping force of the torrent, for torrents only really excavate *when swollen by rain.* A torrent swollen by rain to perhaps twenty times the volume of its usual spring water, and hurling fragments of rocks along of all sizes, is, in point of excavating and destructive power, as much more formidable than its usual self as a shotted gun is more formidable than an unshotted gun. We never *see* the clear torrent set its rocky ammunition in movement, though the shape of this ammunition tells us how often, and for what distances, it has been projected. But when the torrent is turbid with the wash of rain, we can *hear* its huge cannon balls rattling down, and grinding each other and their rocky bed and banks, till what has started from the mountain's brow as a huge rock arrives at the sea in the form of pebbles or of sand. For although, as the *flood of rain* subsides, *the flow of boulder stones* ceases, this is only for a time; each rain sets them on a stage on their journey, as, in lower levels and gentler gradients, will be seen of soil, and the more minute particles formed by disintegration and vegetable chemistry.

And it is not only that districts of rocks are amply supplied with this ammunition which districts of softer subsoil are without; but as rocky subsoils have little power to admit rain by absorption, it agglomerates in the ravines which it finds ready loaded to the muzzle with loose rocks and boulders. And thus the longitudinal cutting of ravines proceeds in rocky mountains perhaps as rapidly as in softer subsoils. But the widening of these longitudinal ravines into broad valleys by disintegration and the lateral wash of rain is a process *infinitely slow* in comparison to what takes place in districts of softer subsoils.

But as sure as dry land stands betwixt high heaven and the sea, the waters of heaven will wash it into the sea. And whether the dry land be of soft or of hard material simply makes the difference of the comparative time required for the operation. In soft and porous subsoils, as the tertiary and chalk, though nature does not make so much fuss about the affair, she proceeds equally surely.

[*Editor's Note:* Material has been omitted at this point.]

Rain, or a river, works at the whole length of the valley or river course at the same time. But the *chief* formation is *backward*, that is, from the sea to the hill. For, as the beds of rivers, if not horizontal, slope *down* to the sea, it is evident that the cutting down of the parts farthest from the sea must depend on the cutting down of the parts nearest the sea; that is, the parts farthest from the sea cannot be cut lower than the parts next the sea. So far the cutting of the parts farthest from the sea is dependent on the cutting of the parts next the sea. So that the *chief* longitudinal cutting of a valley or river-bed may be said to proceed from below upward as regards the valley, or *backward* as regards the stream of the river; that is, it pro-

The beds of rivers and valleys are formed chiefly *backward*, or *up* the valley.

ceeds from the sea to the hill, not from the hill to the sea.

No one will doubt that the Niagara gorge is cut *backward;* but take the case of a barrier without a lake behind it. Suppose a barrier of rock to run across any valley or river-bed; when the bed of the valley or river on the upper side of the barrier has been worn down to a horizontal level with this barrier, it can *not* go lower. The barrier is in this particular case what the sea is in all cases, a negative key to the level of the valley. Thus deep shalt thou go and no deeper. A river-bed cannot run down before it comes to the barrier and then up again over the barrier. But as the barrier is cut through, the bed of the valley or river will be deepened *backward*, or from below upward, or towards the hills. This principle is true of the *filling in* of valleys as well as of the *scooping* them *out.* For when the scooping out or deepening is stopped at the lower end by the sea or a barrier of rock, denudation of the inclined parts above still goes on, and makes that lower part horizontal. And the passage of the detritus and soil from the inclined upper parts of valleys is checked in the horizontal lower parts of valleys, and soil accumulates there. This is the origin of alluvial plains; and a river of any size, or any rapidity, may, at any distance from the sea, have *patches* of alluvial plain, where no lakes have ever been; that is, above every rapid or accidental barrier of hard ground. For as the barrier ponds the flood-water back on the horizontal valley above it, deposit will take place, that deposit will increase the ponding back *up* the valley, and, as long as overflow takes place, these patches will rise and *progress backward* up the valley. The only difference in the laws for the growth and gradient of these patches from those which regulate the growth and gradient of the plain at the level of the sea, is that they have no *increasing* cause for rising equivalent to the *forward* lengthening of the delta of the lower alluvial plain.

Rivers may have patches of alluvial plain anywhere,

and these patches *progress backward*, or *up* the valley.

These flat alluvial patches may be seen even in torrents, sometimes reaching from one cascade to the other; and though each cascade digs a hole deep in proportion to the height of the cascade, as the cascade recedes, that part of the hole or rather groove which is farthest from the cascade is filled with débris. It is easy to perceive that these patches must be liable to constant change. They must be perpetually shortened by the recession of the lower barrier, and lengthened by the recession of the upper one: also that the wearing down of a barrier or shallow from below will expose the alluvial patch above it to denudation; while the rise of an alluvial patch from below, over a rapid above it, may join two alluvial patches in one. These principles are eternally at work on all valleys, from the smallest to the largest.

With regard to the lower alluvial plain, the last yard protruded *at* the level of the sea, by preventing the flood-water spreading in the sea, that is, by ponding the water *back*, forces it *over* the yard *behind* it, and raises that yard *above* the level of the sea; and so every yard raises the yard *behind* it *backward* in succession not only to the highest point of the alluvial plain, but above it; that is, from the end of the delta at the sea, the river builds its alluvial plain at a certain gradient (say that given by Ellet, eight inches in a mile), and its banks, say at a gradient of three inches in a mile, till these gradients strike the unalluvial bed, which it progresses *up;* and whether the whole river is artificially embanked, or whether it has patches embanked, in either case, the banks must eternally be raised with the eternal rising of this *natural* gradient.

[*Editor's Note:* Material has been omitted at this point.]

9

Reprinted from *Geol. Mag.*, 3, 232–235 (1866)

ATMOSPHERIC VS. MARINE DENUDATION

J. Beete Jukes

Sir,—Having publicly advocated atmospheric action as the power by which the present "form of the ground" has been produced, I would wish to say a word or two on the clever articles you are publishing by Mr. D. Mackintosh, in which the sea is treated as the chief agent.

I am glad to see that Mr. Mackintosh does not allude to the action of internal force as having any *direct* effect on the external features of the ground. So long as we were hoodwinked by the *hocus-pocus* of "grand convulsions," and believed it possible for mountain chains to jump out of the interior of the earth like so many "jacks in the box," no advance in real knowledge was possible.

It may be taken for granted, then, that all the external features of the ground (except of course volcanic cones and craters) are the direct result of external agencies (Presidential Address to Section C. British Association, Cambridge, 1862). It may also be taken for granted that as all lands have risen out of the ocean, marine denudation has done something towards the production of their present form, and that during the time they have stood as dry land atmospheric agencies have also done something towards it. The problem is to apportion to the marine and atmospheric agencies the amount of work each has performed.

In reading Mr. Mackintosh's articles I recognise ideas which a few years ago I held as stoutly as he does now, and I believe therefore that he is following the same path which I did towards a fuller appreciation of the precise operation of these natural agencies. I think I was hardly aware of the change which had taken place in my own convictions, as the result of constant observation in the field, till I hit upon the solution of a problem that had long puzzled me, namely, the precise mode of production of the river valleys of the South of Ireland. (See Quart. Jour. Geol. Soc., London, vol. xviii.)

This solution requires that the rivers should never have ceased to run through the ravines, by which they traverse isolated hills between their sources and the sea, during the denudation of the plains by which those hills are surrounded. Had the sea ever marched across the country, and worn it down to form those plains, it must have obliterated all the old features which formerly existed over the areas where the plains are now; and the subsequently formed rivers would never have regained their channels in those ravines, which would have been left as shallow passes through the hills at a greater or less height above the plains. The plains have many broad openings to the sea, without the intervention of any hills, and these would have been the natural outlets of the rivers, if the "form of the ground" had been made for them by the sea.

The reason why the rivers choose to run through the hills by deep ravines, instead of by much easier routes which are now open to them, is that when they began to run these hills did not exist. The hills were then buried, as it were, in much higher ground, by which they were surrounded, and over which the rivers originally ran. The rivers choosing, of course, the lowest ground they could find in their course to the sea, happened here and there to cross the parts where these hills subsequently became disclosed by the waste and erosion of the rock which surrounded them. The rivers, however, having once cut channels for themselves, have ever since kept those channels open, and it is through those channels that the waste of the interior has been carried off. Although, then, the interior was worn down into a plain, while the hill ground resisted that action and was left standing as a hill, the river channel, through that hill, was always cut lower than any part of the plain, for it was only in consequence of the deepening of that channel that the waste could be carried off and the erosion of the surface of the plain continued.

In Ireland the rock that was thus wasted in the interior was Carboniferous Limestone, the ground that stood as a hill was Old Red Sandstone or some other siliceous rock.

The calcareous rock was acted on both by mechanical erosion and chemical solution, the siliceous rock only by mechanical erosion. The siliceous rock therefore resisted the atmospheric action far more than the calcareous rock did, but it would not have thus resisted the sea, which would have cut into Old Red Sandstone just as easily as into Carboniferous Limestone.

This alone is an argument in favour of atmospheric action, but the great argument is the continued running of the rivers during the denudation. Rivers only run over the land, therefore the denudation took place upon the land.

This conclusion, to which I found myself unconsciously and almost reluctantly brought, acted on me like a sudden revelation. It connected together and explained to me all that had been mysterious in the "form of ground" in Wales and England, and other parts of the world, during my observations of the last thirty years, including many of the localities mentioned by Mr. Mackintosh. I saw how it could be applied to the Weald, as my colleagues Professor Ramsay and Dr. Foster and Mr. Topley have since applied it; and, in fact, that its application was universal.

There are, doubtless, several difficulties to be got over in many cases. Some of those instanced by Mr. Mackintosh are easily removable; the rest will yield to patient investigation, if only we do not assume that there is nothing to be investigated.

In the meantime I confidently rely on two conclusions, which in our islands are specially applicable to Palæozoic districts, but apply *mutatis mutandis* to rocks of all ages. These are—

1st. The sea has removed vast masses of rock, and left undulating surfaces, the highest points of which ultimately become the summits of mountains.

2nd. When those undulating surfaces are raised high into the air they are attacked by the atmospheric agencies, and hills, valleys, and plains gradually carved out of the rock-mass below their particular features depending on original varieties in the nature of that mass, and variations in the action of the atmospheric agencies. The latter depend largely on variations of temperature, by which water is made to assume the different forms of vapour, water, snow, and ice.

It must be recollected that the forms of our Palæozoic grounds are of very ancient date, anterior to the period of the New Red Sandstone, and that the great denudation of the Older Palæozoic Rocks took place even before the deposition of the Old Red Sandstone. The time, then, during which the atmospheric agencies have been modelling the minor features is inconceivably great. The recent temporary depression beneath the waters of the glacial sea did little or nothing in the way of denudation, the principal effect then, being the transport of blocks, or the washing about of materials, already loose on the surface.

Much instruction as to the amount of atmospheric action may be gained by comparing volcanic cones with each other. I observed in Java that small volcanic cones of recent origin had their sides quite smooth and even, while others of older date, as was shown by the young trees growing on them, began to show gullies widening and deepening on all sides. The flanks of the great volcanic mountains were a mere series of deep glens, separated by sharp knife-edged crests, radiating like the spokes of a half-shut umbrella, as described by Dr. Junghuhn. (See Lyell's Elements, 6th ed., p. 620.)

Still older volcanos, as those in the South Pacific, described by Dana, have merely narrow vertical walls, radiating from the central mass, between flat-bottomed valleys, which gradually contracting towards the interior where at the head of each may be seen a little rill of water leaping from crag to crag, still going on with the work it has performed, and to which it seems at first so utterly inadequate.

It has sometimes occurred to me to ask how long grass has existed? and especially those grasses which make our matted turf? Conclusions as to the rate of atmospheric erosion drawn from our turf-covered downs would be apt to lead us astray if applied to hills not so covered. In many parts of Australia, for instance, where you come to ride over a hill that looks quite green in the distance you find you can see the ground between the roots of the grass, very much as if you were riding through young wheat. The rain, when it does come down in a torrent, must exert much more effect on such ground than where there is matted turf.

Supposing no grass at all to exist, the rate of erosion will be still more rapid, as on the recent volcanic cones mentioned above, or as may be seen on a new railway embankment or cutting where one or two years' storms produce perfect models of mountain glens and ravines in miniature.

J. Beete Jukes.

Dublin, April 6th, 1866.

10

Reprinted from *Physiographic Essays,* Ginn and Co., Boston, 1909, pp. 254–256

THE GEOGRAPHICAL CYCLE

W. M. Davis

[*Editor's Note:* In the original, material precedes this excerpt.]

The Ideal Geographical Cycle. The sequence in the developmental changes of land forms is, in its own way, as systematic as the sequence of changes found in the more evident development of organic forms. Indeed, it is chiefly for this reason that the study of the origin of land forms — or geomorphogeny, as some call it — becomes a practical aid, helpful to the geographer at every turn. This will be made clearer by the specific consideration of an ideal case, and here a graphic form of expression will be found of assistance.

In Fig. 4 the base line $\alpha\omega$ represents the passage of time, while verticals above the base line measure altitude above sea-level. At the epoch 1 let a region of whatever structure and form be uplifted, B representing the average altitude of its higher parts and A that of its lower parts, AB thus measuring its average initial relief. The surface rocks are attacked by the weather. Rain falls on the weathered surface and washes some of the loosened waste down the initial slopes to the trough lines, where two converging slopes meet; there the streams are formed, flowing in directions consequent upon the descent of the trough lines. The machinery of the destructive processes is thus put in motion, and the destructive development of the region is begun. The larger rivers, whose channels initially had an altitude

A, quickly deepen their valleys, and at the epoch 2 have reduced their main channels to a moderate altitude represented by *C*. The higher parts of the interstream uplands, acted on only by the weather without the concentration of water in streams, waste away much more slowly, and at epoch 2 are reduced in height only to *D*. The relief of the surface has thus been increased from *AB* to *CD*. The main rivers then deepen their channels very slowly for the rest of their lives, as shown by the curve *CEGJ*, and the wasting of the uplands, much dissected by branch streams, comes to be more rapid than the deepening of the main valleys, as shown by comparing the curves *DFHK* and *CEGJ*. The period 3–4 is the time of the most rapid consumption of the uplands, and thus stands in strong contrast with the period 1–2, when there was the most rapid deepening of the main valleys. In the earlier period the relief was rapidly increasing in value,

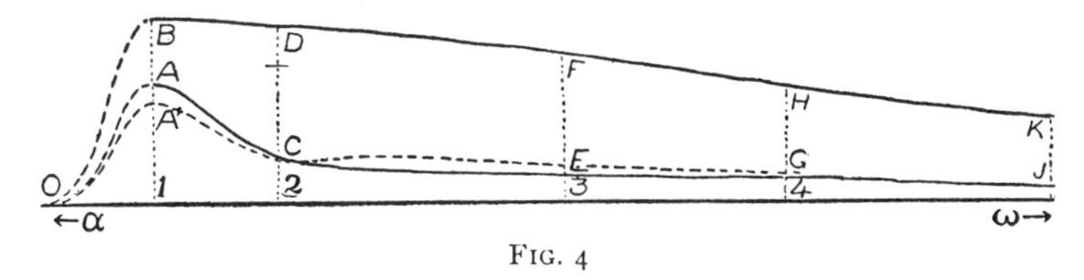

FIG. 4

as steep-sided valleys were cut beneath the initial troughs. Through the period 2–3 the maximum value of relief is reached, and the variety of form is greatly increased by the headward growth of side valleys. During the period 3–4 relief is decreasing faster than at any other time, and the slope of the valley sides is becoming much gentler than before; but these changes advance much more slowly than those of the first period. From epoch 4 onward the remaining relief is gradually reduced to smaller and smaller measures, and the slopes become fainter and fainter, so that some time after the latest stage of the diagram the region is only a rolling lowland, whatever may have been its original height. So slowly do the later changes advance that the reduction of the reduced *JK* to half of its value might well require as much time as all that which has already elapsed; and from the gentle slopes that would then remain, the further removal of waste must indeed be exceedingly slow.

The frequency of torrential floods and of landslides in young and in mature mountains, in contrast to the quiescence of the sluggish streams and the slow movement of the soil on lowlands of denudation, suffices to show that rate of denudation is a matter of strictly geographical as well as of geological interest.

It follows from this brief analysis that a geographical cycle may be subdivided into parts of unequal duration, each one of which will be characterized by the degree and variety of the relief, and by the rate of change, as well as by the amount of change that has been accomplished since the initiation of the cycle. There will be a brief youth of rapidly increasing relief, a maturity of strongest relief and greatest variety of form, a transition period of most rapidly yet slowly decreasing relief, and an indefinitely long old age of faint relief, in which further changes are exceedingly slow. There are, of course, no breaks between these subdivisions or stages; each one merges into its successor, yet each one is in the main distinctly characterized by features found at no other time.

[*Editor's Note:* Material has been omitted at this point.]

11

Reprinted from *Morphological Analysis of Landforms,* H. Czech and K. C. Boswell, trans., Macmillan & Co. Ltd., London and Basingstoke, 1953, pp, 1-18, 300-302

MORPHOLOGICAL ANALYSIS OF LANDFORMS

W. Penck

CHAPTER I

INTRODUCTION

1. NATURE OF THE PROBLEM

Study of the morphology of the earth's surface has developed as a borderland science linking geology and geography. The reason for this is the knowledge that land forms owe their shape to the processes of destruction which engrave their marks on the solid structure of the earth's crust; and that the properties of this crust decide the details of the sculpturing, as well as the arrangement in space of the individual forms. Its immediate purpose being to explain the origin of the multifarious land forms which appear at the surface of the earth, morphology very soon had a specially close connection with geography; and today, it is considered an integral part of geography proper, the study of the earth's surface, and treated as such[1]. Thus work on morphological problems has been overwhelmingly, though not exclusively, in the hands of geographers; and there has been scarcely any attempt to go beyond the aims prescribed by geography.

The material of morphology, however, contains within itself problems which reach far beyond the limits set by geography, and are neither exhausted nor solved by a genetically based description of the surface forms of the planet. *The significance of these problems lies in the realm of geology*; and their solution seems reserved for general geology, especially physical geology. *The problem is that of crustal movement.*

To see the matter more clearly, we must examine the character of the main and fundamental question of morphological science. However keenly geology and geography may be interested in the solution of this question as to the origin and development of land forms, the problem is neither specifically geological nor geographical, but is of a *physical nature*[2]. This results from the relationship and interdependence of those forces, or sets of forces, which produce the surface configuration of our planet. The destructive processes sculpturing the land, all of them together included in the concept *'exogenetic' forces*, cannot become effective until the earth's crust offers them surfaces of attack, until it is exposed to them. Parts of the crust covered by the sea are as much protected from the sculpturing forces as are those parts of the dry land which are being not destroyed but built up. Here, as there, material is not being removed from the earth's crust, but is being piled up on it in the form of rock strata; no unevennesses, no definite relief, are being created, but any unevenness of the crustal surface is being levelled up as by freshly falling snow. The activity of the destructive forces is limited to those parts of the earth which rise above these zones of deposition. Thus the indispensable prerequisite for attack by exogenetic processes of destruction is the activity of those *endogenetic* forces, originating within the planet, which are responsible for raising individual portions of the crust above sea level, and which, on dry land, raise individual blocks above their surroundings, in general create *upstanding parts*, thus giving rise to the altitudinal form of the earth's surface. Leaving aside endogenetically-caused volcanic accumulation (which attains morphological importance only to a modest extent, is limited to a few localities, and moreover has its further fate determined by that of its substratum), the endogenetic processes consist of *movements of the earth's crust*. In view of our present geological experience, their existence requires no further proof. The fact that we know of no piece of land which has not been once or several times submerged below sea level, is by itself striking evidence. Adequate proof that it is a matter of movement of the solid land and not fluctuations of sea level is this: the displacements of level which are of morphological significance, i.e. those which determine present altitudinal relationships on the earth, have never been of a corresponding amount everywhere on the earth's surface, now or at any other period[3].

The earth's surface is not only a limiting surface between different media, nor merely a surface of section giving the desired information about the structure of the earth's crust; *it is a limiting surface between different forces working in opposition to one another*. Both produce displacements of the rock material: the endogenetic forces displace it by raising parts of the earth's crust above their surroundings, or sinking them below these—at present it does not matter whether the direction of the forces is vertical or otherwise, so long as they lead to vertical displacement of level at the planetary surface; the exogenetic forces displace it by transporting solid material along the earth's surface. In the latter case, the transference usually takes place from higher to lower parts, the motive power always being the force of gravity. The endogenetic transference of material, on the other hand, is independent of the force of gravity. It is manifestly against gravity that magma reaches the crust and even the surface. *On this contrast depends the characteristic of mutual opposition between endogenetic and exogenetic processes.* The conflict goes on at the surface of the crust and it finds its visible expression in the tendency—long recognised—for exogenetic transference of material to lower the projecting parts of the land and to fill in depressions, in short to level, to remove any unevennesses which the endogenetic processes have created and are still producing on the crustal surface.

Thus the earth's surface is a field of reaction between opposing forces, and the effectiveness of the one depends upon the preceding activity of the other. On all surfaces so conditioned as to be the scene of interaction between opposing and mutually dependent forces, there is a tendency for a physical equilibrium to become established. This obtains when both forces do the same amount of work in unit time, i.e. when they work at the same rate or have the same intensity. Accordingly, there is equilibrium at the earth's surface when the exogenetic and the endogenetic processes, when uplift and denudation, subsidence and deposition, take place at the same rate; and not only when—as is generally assumed—both processes have died out, and their intensity is consequently zero.

The visible results of endogenetic and exogenetic influences at the earth's surface are *forms of denudation* on the one hand, and, on the other, the *correlated deposits** which are formed simultaneously. The two stand in a similar relation to the forces producing them as do the cut surface and the sawdust that are formed when a log is pushed against a rotating circular saw. It is clear that very different forms develop according to whether denudation is acting more slowly than uplift, more quickly, or at the same rate; and so whether the ratio of the intensity of exogenetic to that of endogenetic activity works against the former or in its favour, or whether there is a state of equilibrium. Therefore it is possible to see plainly in the forms of denudation not merely the results of endogenetic and exogenetic transference of material; but even more that they owe their origin and their development to a relationship of forces, *to the ratio of intensity between exogenetic and endogenetic processes.*

The physical character of the morphological problem comes out clearly. The task before us is to find out not only the kind of formative processes, but also the development of the ratio of their intensities with respect to one another. None of the usual geological or specifically morphological methods is sufficient for the solution of this problem. As is now self-evident, it requires the application of the methods of physics.

Which physical methods are concerned, and at what stage in the morphological investigation they not only may, but must, be applied, follows from the nature of the three elements which together form the substance of morphology.

**Correlated deposits:* Beds correlated in time with the denudation which has produced the material for them.

2. BASIS, NATURE AND AIM OF MORPHOLOGICAL ANALYSIS

These three elements are:

1. the exogenetic processes,
2. the endogenetic processes,
3. the products due to both, which may here be collectively called the actual morphological features.

It is well known that *the exogenetic forces* are at work over the whole earth, in all climates. They consist of two processes, the onset and the course of which are fundamentally different. The one has already been characterised elsewhere as a process of preparation: *the reduction* of rock material.* By this is to be understood all the processes which lead to a loosening of the solid rock texture, to disintegration of the rocky crust so that it becomes changed into a mobile form, transportable. Climate and the type of rock are what decide in the first place the nature and the course of the reduction, the rendering the material mobile. The structure of the earth's crust, which determines the world distribution of rock types, is therefore just as important for morphology as the world distribution of different climatic conditions.

Rock reduction alters the composition and texture of the material. It does *not* produce denudational forms. These do not appear until the reduced material has been removed: only when matter has been taken away from a body does it change its shape. Only if its composition and texture are altered does there arise what the mineralogist calls a pseudomorph. *Earth sculpture is due to exogenetic transference of material. The sum total of this constitutes denudation.* Rock reduction is an essential preliminary to its occurrence; the rock material must have acquired a sufficiently mobile condition before there can be transport at all, either of its own accord (spontaneous) or by the aid of some medium. The processes of denudation are, one and all, *gravitational streams*, obeying the law of gravity. Climatic conditions and type of rock influence the details of their further course. Their effects, therefore, are of different magnitudes in areas of differing climatic nature and differing geological make-up; nevertheless—and this must here be stressed in view of widely held misconceptions—they are *not of different types.*

All the processes of denudation have, as gravitational streams, a non-uniform character—which, as will be shown, is in contrast to the processes of reduction. That is their fundamental property. Their commencement, their course of development, take place before our eyes. They can be observed in all their phases, and their systematic investigation

**Aufbereitung (reduction):* This is the equivalent of weathering and disintegration combined.

is thus possible everywhere on the earth, not only qualitatively—as has already been more or less fully done—but quantitatively, a matter which so far has been hardly attempted. Here there is a wide field of inductive research, as yet unworked. And it promises results of as great or even greater importance to morphology than those which that recent branch of knowledge, soil science, has already produced with regard to rock reduction.

All the *actual morphological features* can, like the exogenetic processes, be directly observed, and are thus an object of inductive research. However, their limits must be extended far more widely than is customary. It is by no means enough to determine and to characterise the forms of denudation as they actually appear in their various combinations; the *stratigraphical* relations of the *correlated strata*, formed simultaneously, are of just as great importance. Their thickness and the way in which they are deposited on the top of one another, how they are connected with their surroundings in the vertical and in the horizontal direction, their stratification and especially their facies, reflect both the type of development in the associated area of denudation, and its duration, and they supplement in essential points the history recorded there. The position of this record in the geological time sequence depends entirely upon investigation of the correlated strata and their fossil content. As a rule far too little weight is given to working on this stratigraphical material. Because of this, our knowledge about the actual morphological features is correspondingly scanty. True, we must take into account that these are not, like denudation, for instance, subject to one uniform set of laws; but that they are peculiar to each individual part of the crust which will have undergone a special development of its own. *They are individual in their character.*

This individuality is essentially dependent upon the way in which the activity of the *endogenetic processes*, particularly crustal movements, varies from place to place. Since crustal movements cannot be directly observed—with the exception of earthquakes—information about them is deduced from the effects they have produced. These, however, are to be regarded merely as indications that crustal movements are taking place, and bear much the same relation to them as the shaking of a train does to its forward motion. The passenger, now and then looking out from the window of the moving train, recognises its movement by the jolting and by the shifting landscape visible outside. The geologist recognises crustal movements by earthquakes and by disturbance in the stratification of rocks. The latter shows the earliest time at which the crustal movements took place at the given spot, and the total amount of disturbance produced up to the time of observation. But the conditions of bedding give practically no clue as to what, during the period of movement, was the distribution of the separate effects which add up to this total resulting disturbance; neither do they indicate the intensity of the crustal movement—whether it was rapid or slow—the changes of intensity in successive intervals of time—whether increasing or decreasing, nor the course of the movement—whether continuous or by fits and starts, nor whether it is still continuing or came to an end in times past. Yet all these have often been assumed in a perfectly arbitrary manner.

Crustal movements cannot be observed directly, and no adequate tectonic method is known for ascertaining their characteristics. Thus, in studying land forms, it is not permissible to make definite assumptions as to their course and development, and to base morphogenetic hypotheses upon them. Moreover, it is perfectly clear that of the three elements—endogenetic processes, exogenetic processes, and the actual morphological features—dependent upon one another, in accordance with some definite law, like three quantities in an equation—it is the crustal movements which correspond to the unknown, about which statements can be made only as a final result of the investigation, not as one of the premises. On the other hand, it has been shown that each of the other quantities can be established by observation, completely and with certainty for each individual case. Their dependence upon one another, in conformity with some definite law, has already been recognised, and it will subsequently be further developed. The equation—to continue the simile—permits of an unequivocal solution. *Morphological analysis is this procedure of deducing the course and development of crustal movements from the exogenetic processes and the morphological features.*

The function of this analysis of land forms, and its aim, is therefore *geological, or more exactly, physico-geological.* The first thing to be done is to state clearly and solve the problems of the origin and development of denudational forms. The present book treats this complex of problems as a matter of physical geology, to which it organically belongs in virtue of its nature—as was emphasised at the beginning. Having found the outlines of its solution, we have a base from which to move forward to the ultimate aim of morphological analysis, as indicated above. Far-reaching possibilities open up from this point. As has already been more fully set out elsewhere[2], the analysis of land forms gives totally different information about essential features of crustal movement from that which geologico-tectonic research can find out from the structural characteristics of the earth's crust. The two methods, the morphological one here developed, and the geologico-tectonic, supplement one another; at no point is one a substitute for the other. Only the two together supply that sum of basic facts which affords a prospect of solving the main problem of general geology, *the causes of crustal movements*. This is a matter which does not seem to allow of a solution by any purely tectonic method; and so far any attempt to solve it upon this far too narrow basis has been of no avail.

3. CRITICAL SURVEY OF METHODS

(*a*) Cycle of Erosion

A first attempt in the direction of morphological analysis was the theory of the Cycle of Erosion developed by W. M. Davis[4]. Familiarity with it is here assumed. The cycle theory has, it is true, a purely geographical aim, viz. the systematic description of land forms on a genetic basis, which has been aptly called their 'explanatory description'. Thus there was no direct attempt to discover more about endogenetic processes. Results in this direction have been obtained only by the way, and they were not meant for further use except so far as they were considered to serve the explanatory description[5]. This totally different orientation of the setting of the problem does not in any way alter the importance for morphological analysis of the *principle* of the cycle concept. *That principle is the idea of development*, in its most general sense: a block, somehow uplifted, presents one after another, in systematic sequence, forms of denudation which result from one another; and their configuration depends not only upon the denudation which is progressing in a definite direction, but also—as Davis himself suggested, though merely as a conjecture[6]—upon the character (the intensity) of the uplift.

What has found its way into morphological literature as the cycle of erosion is what Davis expressly defined as a special case of the general principle, one which was particularly suitable to demonstrate and to explain the ordered development of denudational forms. It is postulated that a block is rapidly uplifted; that, during this process, no denudation takes place; but that on the contrary, it sets in only after the completion of the uplift, working upon the block which is from that time forward conceived to be at rest. The forms on this block then pass through successive stages which, with increase of the interval of time since they possessed their supposedly original form, i.e. with increase of developmental age, are characterised by decrease in the gradient of their slopes.

They are arranged in a *series of forms*, which is exclusively the work of denudation and ends with the peneplane*, the peneplain. If a fresh uplift now occurs, the steady development, dependent solely upon the working of denudation, is interrupted; it begins afresh, e.g. the pene-

**Peneplane:* This spelling is used to indicate the character of the relief, with no cyclic connotation intended, to differentiate the term from 'peneplain', which is used in the Davisian sense.

plane is dissected. A new cycle has begun; the traces of the first are perceived in the uplifted, older forms of denudation. Thus it has become usual to deduce a number of crustal movements, having a discontinuous jerky course, from the arrangement by which more or less sharp breaks of gradient separate less steep forms above from steeper ones below.

It was possible to draw this conclusion, in such a general form, only because the above-mentioned special case, chosen by Davis mainly for didactic reasons and developed in detail on several occasions, is usually taken as the epitome of the cycle of erosion, and is quite generally *applied* in this sense. Both Davis himself[7] and his followers have made and still make the tacit assumption that uplift and denudation are successive processes, whatever part of the earth is being considered; and investigation of the natural forms and their development has been and is being made with the same assumptions as underlie the special case distinguished above. There is, therefore, a contrast between the original formulation of the conception of a cycle of erosion and its application. Davis, in his definition, had in mind the variable conditions not only of denudation, but of the endogenetic processes; in the application—so far as we can see, without exception—use is made only of the special case, with its fixed and definite, but of course arbitrarily chosen, endogenetic assumption. And criticism, with its justified reproach of schematising, is directed against the fact that the followers of the cycle theory have never looked for nor seen anything in the natural forms except the realisation of the special case which Davis had designated as such. Thus even opponents of the American doctrine have taken their stand not against the general principle of the cycle of erosion, but against its application; and they referred merely to the one special case that alone was used[8]. Thus there seems throughout to have been a misunderstanding with regard to the cycle of erosion: its originator meant by it something different from what is generally understood. The way in which the theory is applied, the trend of the criticism it has received, hardly permit any doubt of this. Thus it is necessary to consider more closely the application of the *cycle of erosion* and the criticism directed against it.

As a method, the theory of the cycle of erosion introduces a completely new phase in morphology. Deduction, so far used only within the framework of inductive investigation, or as an excellent method of presentation, has become a means of research. Starting from an actual knowledge of exogenetic processes, the cycle theory attempts to deduce, by a mental process, the land-form stages which are being successively produced on a block that had been uplifted, is at rest, and is subject to denudation. Not only is the order of the morphological stages ascertained by deduction, but also the forms for each stage; and the ideal forms arrived at in this way are compared with the forms found in nature. There are two points in this method which must be considered critically: (*a*) deduction as a means of investigation; and (*b*) the facts on which the assumptions are based.

To begin with, it is obvious that the ideal forms, which are supposed to develop on *a stationary block*, can be deduced successfully only if there are no gaps in our knowledge of the essential characteristics of the denudational processes. Should this pre-requisite not be fulfilled, the deduction is nothing but an attempt to find out from the land forms alone both the endogenetic and the exogenetic conditions to which they owe their origin. It is like trying to solve an equation having three quantities, two of which are unknown; we can expect only doubtful results. The American school may be justifiably reproached with not considering it their next task to eliminate one of the unknown quantities by systematically investigating the processes of denudation all over the world. On the whole, their part in throwing light upon the exogenetic processes has been a very modest one. Yet this is not a decisive blow to the cycle concept. For amongst the 'exogenetic' assumptions made, there is no principle which has not been verified by experience, and criticism by opponents has been unable to show any mistakes in this field[9].

Till very recent times[10], no one has even seriously examined the second or 'endogenetic' assumption of the applied cycle of erosion, namely, uplift and *then* denudation. On the contrary, morphologists of every school have generally started from the same assumption as soon as they came to discuss the problem of the development of land forms. Even the opponents of the American school have done this, and indeed tacitly still do so, even in those cases when they have completely disregarded any endogenetic influence on the forms of denudation, and have done no more than consider how the individual land forms might have arisen purely from the work of denudation: for instance, the way in which they depend upon rock material. Invariably they have started with a given fixed altitude for the crustal segment considered; that is, they have begun by considering uplift already completed and followed by a period of rest.

Thus, up to now, it cannot be said that there has been any really well-grounded criticism directed against the factual assumptions of the cycle of erosion.

The second point was the use of deduction as a method of morphological research—though of course in addition to induction and essentially based upon it. A. Hettner utterly and roundly repudiated the use of this for morphology[11]. However, no such conclusion would be drawn from

Hettner's remarks and his arguments. In these he deals with the concepts of Davis—which he considers are not precisely enough formulated—especially the concept of morphological age; further, he points out that inadequate attention has been paid to the character* of the rock and to the exogenetic processes; and finally he considers the application of the theory to specific cases. At one point only does he touch upon the problem of method, and that is when he levels the reproach against the cycle of erosion that it rests upon inadequate assumptions as to the exogenetic processes. That reproach has already been considered above. But so far as the erosion cycle is concerned, this is only *one* side of the question; for, as has already been shown, it makes further very definite assumptions about the endogenetic processes. Apparently Hettner considers them to be correct and admissible, since he does not examine them also. But the possibility that the deductive method used for the cycle of erosion may be based upon inadequate assumptions does not permit the passing of judgment upon the applicability of deduction itself to the whole sphere of morphological research[12]. In addition, this statement may be made: *In morphology, as in any other branch of knowledge concerned with physical problems, deduction as a means of research is not only permissible, but also imperative; unless we wish to renounce the greatest possible exactitude and completeness in the results, and to exclude our branch of learning from the rank of an exact science, a rank which it both can and should acquire in virtue of the character of the questions with which it deals.* It is merely a matter of finding out where, in the process of investigation, we should resort to the method of deduction; and above all making sure that correct and complete data are then provided for it. The provision of these is, as before, exclusively the domain of inductive observation; it only can accomplish this, the deductive process never. It is by no means the deductive character of the method itself which makes it impossible to go along the American way of applying the cycle of erosion, but the incompleteness and, as will presently be shown, the incorrectness of the assumptions made. Thus opposition to the deductive method as a tool for use in morphological investigation has been unable to do serious harm to the theory of the erosion cycle, and it is not to be expected that it will ever succeed in doing so.

We now turn to the assumption made about endogenetic processes when applying the concept of the cycle of erosion.

(*b*) Relationship between Endogenetic and Exogenetic Processes

Exogenetic and endogenetic forces begin to act against one another from the moment when uplift exposes a portion of the earth's crust to

**Character of rock:* A comprehensive term, implying stratification, cavity filling, rock partings, pore character, diagenesis, cement, metamorphism, impregnation, reconstitution and recrystallisation.

denudation. So long as uplift is at work, denudation cannot be idle. The resulting surface configuration depends solely upon whether the endogenetic or the exogenetic forces are working the more quickly. Were there no denudation, a block, however slowly it is rising, might in course of time reach any absolute height; and its increase in altitude would be limited solely by the physics of the act of formation, provided that it is inherent in this not to continue indefinitely. It is rather like the way in which an impassable limit has been set to the increase in height of volcanoes by the extinction of volcanic activity, which often comes to an end prematurely, as soon as a certain height has been reached, because lateral effusions replace the summit eruptions. However, it is from the outset that exogenetic breaking-down at the earth's surface works against endogenetic building-up, i.e. denudation works against uplift, in-filling by sediments against subsidence. It is easily to be understood that an actual elevation can come into existence only if uplift does more work in unit time, and so is working more rapidly, than denudation; a hollow appears only when subsidence takes place more quickly than sediment is supplied, than aggradation. *This state of affairs forms the substance of the fundamental law of morphology: the modelling of the earth's surface is determined by the ratio of the intensity of the endogenetic to that of the exogenetic displacement of material*[13].

A brief survey of the earth's surface shows that this ratio very often changes, or has changed, to the prejudice of the exogenetic forces; the accumulation of a volcanic cone is possible only because it takes place more rapidly than the removal by denudation of the accumulated material. Faults can become visible as unlevelled fault scarps, for instance in the zone of the rift valleys of East and Central Africa[14], only when the formation of faults takes place more rapidly than levelling by denudation. Generally speaking, the origin of any outstanding elevation, any mountain mass, is bound up with the assumption that mountain building is more successful, i.e. works more rapidly, than denudation. Thus the varied altitudinal form of the land shows that in many cases the work of denudation is lagging behind the endogenetic displacement of material, here more, there less, or has done so in the past. *The one consistent feature, however, common to every region, is that the activity of exogenetic happenings is subordinate to that of endogenetic processes.* This relationship, most impressively brought to the observer's notice by the different kinds of relief and the different altitudes occurring at the earth's surface, forms the basis of morphological analysis. For if the exogenetic forces are less active than the endogenetic movements, then their effect, the earth's whole set of land forms, must also in its main outlines accommodate itself to whatever law has its visible expression stamped by crustal movement upon the face of our planet. Any change in kind or in intensity which these movements undergo must therefore—as has long been known—leave its traces upon the landscape.

If the intensity of denudation consistently lagged behind that of the endogenetic movement, then in course of time a block rising in such a way could, even in spite of the exogenetic processes, reach any absolute height; though it would of course do so more slowly than if the earth were not subject to denudation. But the relationship is not an unchanging one. For, like all other gravitational streams, the processes of denudation increase in intensity with the gradient, in a definite manner to be discussed later; and the gradient increases with the increase in *vertical distance* between summit and foot of the uplifted portion of the crust. This is true provided the *horizontal distance* between the two points is not at the same time proportionately increased; and, as a rule, it is not. Thus it was possible, even some decades ago, to formulate it as an empirical law, fundamentally correct, that intensity of denudation increases with absolute height, other things being equal. This means a shift in the relationship of the endogenetic to the exogenetic rate of working, in favour of the latter, and an ever closer approximation to physical equilibrium. The actual attainment of the equilibrium could be prevented only where there was no limit to the *increase* in endogenetic movement, so that rising blocks would gain in height indefinitely. The insignificance of the altitudinal modelling of its surface bears witness to how little such conditions obtain on the earth, an insignificance which is not clearly brought out, with distinctness, till comparison is made with the dimensions of the earth as a whole.

The above short survey shows that it is essential, when investigating the origin and development of denudational forms as they appear at the earth's surface, to *ascertain the relationship between the intensity* of the endogenetic and of the exogenetic processes, in short, between uplift and denudation; and it is necessary to follow out how this changes as time goes on. None of the present methods used in morphology brings us nearer to achieving this end; none even attempts to do so. The assumption generally introduced, that uplift and denudation were successive processes, or could at any rate be treated as such, has stood in the way. In this respect the only difference between the cycle theory and its opponents is that Davis made the above assumption in order to provide a specially simple case, of particular use in illustrating the cycle concept; but, at the same time, he kept well in mind[15] the importance of concurrent uplift and denudation. To be sure, this was a notion of which he scarcely ever made use, and his followers never. Those of the other school, no less schematically, start in every case from the same assumption; they, moreover, have occasionally tried to justify the general correctness and permissibility of such a course[16]. It is as if they made use of a device familiar in school physics, which is merely a makeshift for presenting in a physically correct manner the resultant of processes acting concurrently. This is a grave mistake in method.

It is permissible to proceed in this way only in the case of *uniform forces* which, in successive units of time, produce effects that remain of equal magnitude. If, in a diagram such as Fig. 1, the co-ordinates *ab* and *bc* represent the effects of simultaneous, uniform forces, the straight line *ac* represents the resultant effect *during* the whole process. In order to ascertain this, it is sufficient to follow the events first from *a* to *b*, then to *c*.

(*c*) The Differential Method

It is quite different, however, when forces acting simultaneously are *not uniform*, i.e. are changing their intensities in successive units of time and are therefore doing different amounts of work. To find out the resultant during the whole process, it is here necessary *to follow* the course of Nature *continuously*, as was made possible in physics, where such problems are constantly cropping up, only by the invention of the differential calculus[17]. To make this clear, let us remind ourselves of the problem: to find the trajectory of a missile fired horizontally from a point *a* (Fig. 1). As the effect of the firing, it would reach *b*; but at the same time, under the influence of the force of gravity, it would drop down by the amount *bc*. To find point *c*, which the projectile reaches, it would really be sufficient to follow events first from *a* to *b*, then from *b* to *c*, that is to imagine the effects of the firing and of gravity as coming into play successively. The trajectory, however, has not been found in that way. It lies on a curve of some kind between the initial and final points, between *a* and *c*. To determine it, we must find out how the magnitude of the operating forces alters in successive exceedingly small units of time. This can be done by plotting a diagram of forces for each moment, as in our figure, e.g. ab^1c^1, $a^2b^2c^2$ . . . $a^mb^mc^m$, etc., so that the simultaneous effect of the forces is represented by the very minute distances ac^1, a^2c^2 . . . a^mc^m, etc., as if they were uniform during these extremely short intervals of time and took place successively. The error thus made becomes infinitesimally small, if the values chosen for the diagram are made infinitesimally minute; and this method, consisting of an infinite number of infinitely small variables, comes infinitely near the *continuous* course of Nature. This becomes clear if in our figure the triangles ab^1c^1, $a^2b^2c^2$, etc., are, as is necessary, made infinitely small; they then disappear completely in the full line *axc*. Since the forces are now changing from moment to moment, the very small distances ab^1, a^2b^2, etc. and b^1c^1, b^2c^2, etc., are thus of different lengths, and so the infinitely small resultants ac^1, a^2c^2, etc., are also of different lengths and at different inclinations. Strung out after one another, they do not form a straight line, but bend in a curve: the trajectory *axc* which was to be found. *This is the differential method. It is the only way that leads to our goal, which is the exact representation of the resultant of several simultaneous processes that are not acting uniformly during their course.*

The forces which take part in modelling the land do not act uniformly. It has already been shown that this is so for denudation, and it is a matter of course for crustal movements. They begin from the position of rest,

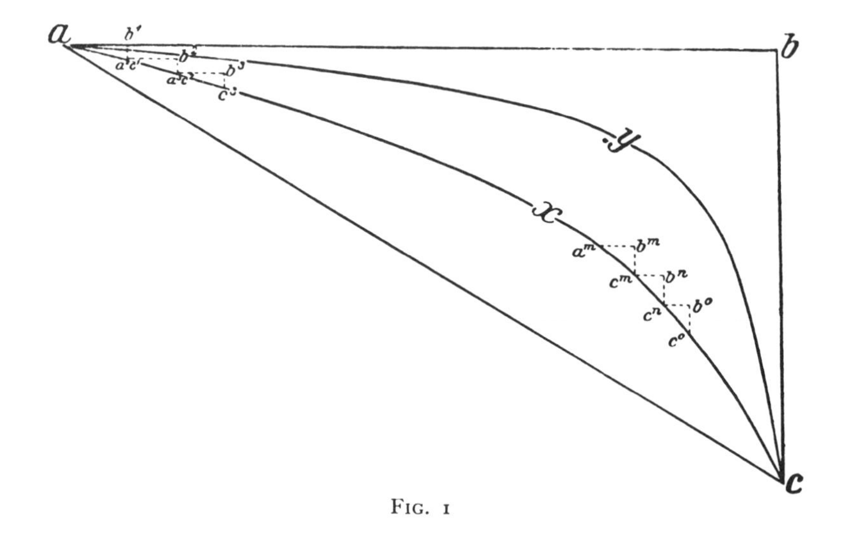

Fig. 1

so they must then be accelerated; and they end with the position of rest, having then suffered deceleration. Whether or not the starting point and the end coincide with a position of absolute rest is of no importance. If indeed we think of the alternation of uplift and subsidence, as has so often, if not as a general rule, taken place on the earth, then the position where subsidence changes to uplift (or vice versa) is the position of rest. *In any case, the movement of the crust is a non-uniform process, which becomes uniform only temporarily during its course, but can never begin uniformly with any definite velocity.* It is not superfluous to stress this obvious factor in view of the often inadequate conceptions which are widespread as to the physics of motion, and even of the fundamental concepts of physics.

To illustrate the position, let us draw a diagram (Fig. 1), in which the co-ordinates represent the effects of crustal movement (*a b*) and of denudational processes (*b c*). The result of their simultaneous action, the forms of denudation, then appears in the shape of curves between *a* and *c*. Naturally they are all situated within the triangle *abc*; since any curve drawn outside it, as for example, a curve deviating in convex fashion downwards from the straight line *a c*, would signify that denudation had begun before subsequent endogenetic action had exposed the crustal fragment to exogenetic destruction or, in other words, that denudation was the prerequisite for crustal movement. Within the triangle an infinite number of curves are possible, all of which begin at point *a* and end at *c*. Each of these curves represents not a single form, but a *developing series of forms* through which a crustal fragment passes when it is being uplifted and denuded. It is obvious that all the series of land forms which can possibly develop on the earth have one common starting point and one common final form. The former is characterised by the beginning of uplift and denudation (point *a*), the latter by the extinction of endogenetic and exogenetic displacements of material (point *c*). In between lies the endless variety of forms that correspond to the varying ratio of intensity of exogenetic processes to that of endogenetic ones; and they are arranged on an infinite number of curves, each of which represents a series of forms peculiar to the surface development of a crustal fragment which has had a particular course of endogenetic development.

Thus we see that land forms are not, as the erosion cycle postulates, a single developing series, but that they form an infinite number, and that they are arranged [on the diagram], *not in a line, but on a surface*. This surface is enclosed by two limiting curves that, at least as regards total dimensions, represent developing series which, on the earth, are *just not possible*: the straight line *a c*, that is the series of forms arising from the uniform development of uplift and denudation, and the axes *a b*—*b c*, which would be the series of forms that would arise if uplift and denudation succeeded one another. Whether *fragments of the limiting curves* can be observed as component parts in the development of forms on the earth, and which fragments might in that way be realised, will have to be decided by the following investigation.

The relationship of the cycle of erosion (and of methods based upon similar assumptions) to the complex of problems concerning development of land forms now becomes evident. Point *b* in the diagram represents denudation which sets in only after the completion of uplift; and the series of forms that arise purely as the work of denudation on a motionless block is represented by one limb of the limiting curve: the line *b c*. Its starting point *b* by no means coincides with the point (*a*) from which all development of forms on the earth begins. The perfectly arbitrary choice of that starting point is clearly shown. But, on the other hand, there also emerges the fact that although the method adopted for the erosion cycle (in its applied form) is incorrect in principle, yet it must nevertheless lead to the discovery of the correct final form common to every one of the series of denudational forms that develop on the earth, provided that the method was logically correct and not based upon faulty exogenetic assumptions. The theory of the cycle of erosion does satisfy both these requisites, as cannot be doubted even by its opponents. Point *c* in the diagram represents the end-peneplane, the peneplain, the origin of which was made clear by W. M. Davis, and a little later, independently, by A. Penck, both basing themselves upon detailed inductive observation[18]. One could not, however, in this way deduce a single one among the infinite series of developing land forms which are not only possible but actually exist on the earth in the different parts which have had various endogenetic histories. Almost everything in this field still remains to be done by morphology. Not merely must the investigation start at the beginning of uplift and denudation (point *a*); it must take into account not only the simultaneous effects of endogenetic and exogenetic action, but above all their variable intensities[19]. *For this one must turn to the methods of physics, and indeed to such as permit a continuous following of the variable quantities, that is, the differential method.* This method not only can, but *must* be used in investigating the interdependence between the *processes of movement* which take part in the modelling of the land. For mass-transport of eroded material depends upon a gradient, the factors producing which are crustal movements; and mass-movement of denuded material depends upon a gradient, the processes producing which arise from the results of erosion†[20].

This leads to the

(*d*) Present Method of Approach

The main stress is laid upon investigating the ways in which denudation works, and the preparatory processes, the aim being to find out if denudation follows laws which are uniformly applicable and what these are. Thus, in the first place, it is a matter of studying those processes of denudation the course of which depends directly upon the *surface gradient* of the crust and therefore indirectly upon its movements. These

[† Translators' note: The author's distinction between erosion along the line of a watercourse and denudation over a whole surface must constantly be borne in mind. This point of difference between German and English-American usage was discussed by W. M. Davis, *Journ. Geol.* XXXVIII (1930), p. 13.]

fall into two categories: (*a*) denudational processes in which the movement is spontaneous*; (*b*) those making use of an agent that is itself in motion (air, water, ice). Now group (*b*) includes also processes which depend only remotely, or not at all, upon the surface gradient, and which do not show any direct connection with crustal movement. Currents in standing water, particularly in the sea, are examples of this, and air currents. In both cases, lines of movement within one and the same medium are caused by *differences of pressure*. They imprint upon the surface of the crust features having no connection with crustal movements, which have but a limited value, or none at all, for discovering these. Hence this book does not treat of ocean currents; and the effects of wind are considered only in so far as they act with and influence other sub-aerial streams that are dependent upon gravity. Their work must be distinguished from endogenetic influences, even though it is restricted to certain parts of the earth, and is of limited importance in these. It is particularly over arid stretches of land, devoid of any continuous cover of vegetation, that wind effects are encountered; and the lowest base level to which they can possibly extend is the surface of any accumulation of water. For arid regions this means, in the first place, the water table. This is so for basins of inland drainage, the visible surface of which forms the base level of erosion for the bordering slopes; and its position is determined—if not exclusively, yet in the main—by climatic conditions and not by those of crustal movement. Wind may here play a considerable part in the modelling of those slopes; not indeed directly, as observation shows, but indirectly through bringing the base level of erosion down to the level of the (dry season) water table.

Minor forms due to wind action appear all over the world, wherever there is an arid portion of surface exposed to the wind, provided weathering has previously done its work. But even apart from these, it is *climate* which primarily conditions the denuding as well as the depositional activity of the wind. Nowhere indeed are the forms which it produces dominant, not even in the regions of its greatest importance; but they are overprinted upon the dominant forms, forms which there too have been created by denudational processes bound up with the surface gradient.

The movement of ice and snow, and the effects produced by them, present similar relationships. True, their movements follow the surface gradients; yet their existence is a response to climatic conditions. It is unsuitable, therefore, to use forms produced by them for the deduction of crustal movements. That their origin has no direct connection with these is at once evident from the character of their level of reference, the

[* See p. 64.]

base levels of erosion below which their sculpturing cannot possibly reach. For glaciers, that level is the melting lower end; for the stretches lying between them, it is the snow-line. The position of the melting tip depends both upon that of the snow-line, determined exclusively by climate, and upon the mass of each individual stream of ice which, other things being equal, increases with the size of the gathering ground. With stable climatic conditions, the influence of crustal movements is to alter the position of the glacier snout and that of the snow-line, not with reference to the base level of running water, e.g. sea-level, but with reference to the glaciated summits or to the original position of the snow-line. Thus, investigation of glacial phenomena can establish to what extent the observed displacements of the snow-line are due to climatic changes or to crustal movements, but no clue is thereby given as to the nature or course of these events. Hence the whole set of phenomena belonging to glacial geology also falls outside the scope of this book.

And finally, an examination of coastal development has no place here. Knowledge is assumed of the processes which are today considered to be modelling the details of sea-coasts, whether by denudation or by deposition, and of how these processes are believed to transform coasts which are neither rising nor sinking. For the problems here treated, the only phenomena of importance are those from which it is possible to deduce vertical movement of the coast-line, with the necessarily associated horizontal displacement, uplift and subsidence of the solid earth, rise and fall of sea-level. In so far as eustatic fluctuations can be successfully excluded, these phenomena supply information as to the occurrence of the crustal movements, the time of their commencement, and their direction (relative to sea-level); but only in a limited degree to their course of development and changes in intensity. In this respect they fall into the same category as the conditions of rock stratification. This does not remove the necessity for a somewhat detailed consideration of the proved oscillations of the coast-line; for this concerns the *oscillations of the base level of erosion for running water*, which—whatever their origin—have the greatest conceivable repercussions on the modelling of the land, the type of denudation and deposition, and the distribution of these.

That the investigation may be directed aright with reference both to matter and place, we are beginning with a short survey of the earth's crust and its structure, this being the stage for all geological and morphological occurrences.

[*Editor's Note:* Material has been omitted at this point.]

NOTES AND BIBLIOGRAPHIC REFERENCES

The following contractions are used below:

A.J.S.	American Journal of Science.
A.S.A.W.(m.-p)	Abhandlungen der Kgl. Sächsischen Gesellschaft (Akademie) der Wissenschaften, Math.-Phys. Klasse. (Leipzig).
B.G.S.A.	Bulletin of the Geological Society of America.
G.J.	Geographical Journal (London).
G.M.	Geological Magazine (London).
G.R.	Geologische Rundschau (Leipzig).
G.Z.	Geographische Zeitschrift (Leipzig).
J.G.	Journal of Geology (Chicago).
P.M.	Petermanns Mitteilungen aus J. Perthes Geographische Anstalt (Gotha).
Q.J.G.S.	Quarterly Journal of the Geological Society of London.
Z.D.G.G.	Zeitschrift der Deutschen Geologischen Gesellschaft (Berlin).
Z.G.E.	Zeitschrift der Gesellschaft für Erdkunde zu Berlin.
Z.M.G.P.	Zentralblatt für Mineralogie, Geologie und Paläontologie (Stuttgart).

W. Penck, Op. 10
W. Penck, Op. 13
W. Penck, Op. 16
W. Penck, Op. 17 } works 10, 13, 16 and 17 listed on pp. 352–353.

(1) A. Penck, *Morphologie der Erdoberfläche*, Bd. I, p. 2. *Bibliothek geogr. Handbücher*. Stuttgart, 1894.

(2) The evidence for this fact is given in Walther Penck, Op. 17.

(3) It may be assumed that they [i.e. fluctuations in sea level] interfere with the crustal movements, but this has not, so far, been definitely proved. R. A. Daly now believes it possible to draw definite conclusions as to the Pleistocene and modern eustatic fluctuations of sea-level from tropical coral islands and atolls, as well as from recent and sub-recent raised beaches. Even if these fluctuations are confirmed, it is clear that, compared with crustal movements, they are small in amount and so of slight morphological importance, at least as regards the main characters found on the face of the planet, though not necessarily for the development of individual features. (R. A. Daly, The glacial-control theory of coral reefs. *Proc. Amer. Acad. Arts Sci.*, 1915, LI, No. 4, p. 157. Also *A.J.S.*, 1919, XLVIII, p. 136; *B.G.S.A.*, 1920, XXXI, p. 303 and elsewhere. A general sinking of sea-level in recent time, *Proc. nation. Acad. Sci.*, 1920, VI, No. 5, p. 246 and *G.M.*, 1920, LVII, No. 672, p. 246).

(4) W. M. DAVIS, The geographical cycle. *G.J.*, 1899, XIV, p. 481; also *C. R.* 7 *Cong. int. Géog.*, Berlin 1899, II, p. 221. D. W. JOHNSON has edited a selection of the writings of this American scholar under the title of *Geographical Essays* (Ginn & Co., 1909). The collection does not contain that article of DAVIS which is of special methodological importance: The systematic description of land forms. *G.J.*, 1909, XXXIV, 2, p. 300. A full German presentation of his theory is contained in *Die erklärende Beschreibung der Landformen* (B. G. Teubner, Leipzig 1912), and a shorter one in G. BRAUN'S edition of DAVIS' *Physical Geography* (1898) (*Grundzüge der Physiogeographie II, Morphologie*, 2. Aufl., Leipzig 1915), etc.

(5) See particularly *G.J.*, 1909, XXXIV, 2, p. 300.

(6) *Proc. Amer. Assoc. Adv. Sci.*, 1884 and *C. R. Cong. int. Géog.*, 1904, pp. 153, 154.

(7) See amongst other writings of W. M. DAVIS: The Triassic formation of Connecticut, *U.S. geol. Surv. ann. Rep.*, 18.

(8) See also W. PENCK, Op. 17.

(9) See the writings of S. PASSARGE (Physiologische Morphologie. *Mitteil. geogr. Gesellsch. Hamburg*, 1912, XXVI, p. 133 and *Grundlagen der Landschaftskunde* III, Hamburg 1920) and A. HETTNER (*Die Oberflächenformen des Festlandes*, B. G. Teubner, Leipzig 1921, the repetition in a comprehensive manner, with additions, of articles which had appeared earlier in the *G.Z.*). PASSARGE makes the most far-reaching use of deductions without always basing himself upon adequate observations (see review by H. WAGNER in *P.M.*, 1913, p. 176). A. HETTNER, on the other hand, is an outspoken opponent of the method of deduction.

(10) The first critical investigation of the 'endogenetic' assumption which is made in the erosion cycle theory, as usually understood, was published by A. PENCK (Die Gipfelflur der Alpen, *Sitzungsber. Preuss. Akad. Wissensch., math.-phys. Kl.*, XVII, p. 256, Berlin 1919). Observations reaching back to an earlier date were made by the author in the Argentinian Andes; and these supplied information which showed that assumption to be untenable (W. PENCK, Op. 16).

(11) *Die Oberflächenformen des Festlandes*, loc. cit. [note 9], p. 215.

(12) See WALTHER PENCK, Morphologische Analyse. *Verhandl. XX. deutscher Geographentag*, Leipzig 1921, p. 122.

(13) See WALTHER PENCK, Op. 16, p. 389.

(14) The very frequently occurring cases of faults which are later on made visible by denudation are not to be classed as unlevelled fault scarps, for the processes of denudation have encountered rocks which differed in resistance on the two sides. This is not being discussed here.

(15) *C. R. Cong. int. Géog. Washington*, 1904, pp. 153–154; *G.J.*, 1899, p. 7 of the off-print; *Erklärende Beschreibung der Landformen*, loc. cit. [note 4], pp. 146–147, 173, etc.

(16) See amongst other writings that of H. SCHMITTHENNER in his dissertation on *Die Oberflächengestaltung des nördlichen Schwarzwaldes*, p. 59 (Karlsruhe 1913).

(17) 'It is well known that scientific physics has existed only since the invention of the differential calculus. It is only since man has learned how to follow the course of natural occurrences continuously that there has been any success in the attempts made to express in abstract terms the connection between [various] phenomena.' Preface to RIEMANN's lectures on *Die partiellen Differentialgleichungen der mathematischen Physik* (5. Aufl., Braunschweig 1910).

(18) Priority belongs to W. M. DAVIS who, in connection with studies on the Great Plains of Montana, east of the Rocky Mountains, expounded the principle of peneplanation in 1886 (*Tenth Census of U.S.*, vol. XV, Statistics of coal mining). A. PENCK, Über Denudation der Erdoberfläche. *Schriften zur Verbreitung naturwissenschaftlicher Kenntnisse*, XXVII, Wien 1886–87.

(19) For the earliest investigations in this direction see A. PENCK, Die Gipfelflur der Alpen, loc. cit. [note 10], and WALTHER PENCK, Op. 16.

(20) A short presentation of the connection between these is to be found in WALTHER PENCK, Op. 17.

[*Editor's Note:* Material has been omitted at this point.]

Part II

MAJOR CONCEPTS IN GEOMORPHOLOGY

Editor's Comments on Papers 12 Through 46

12 WOLMAN and MILLER
Excerpts from *Magnitude and Frequency of Forces in Geomorphic Processes*

13 STODDART
Catastrophic Storm Effects on the British Honduras Reefs and Cays

14 WOOLDRIDGE and LINTON
Excerpts from *Structure, Surface, and Drainage in South-East England*

15 WOOLDRIDGE
Excerpts from *The Upland Plains of Britain: Their Origin and Geographical Significance*

16 KING
Excerpts from *The Study of the World's Plainlands: A New Approach in Geomorphology*

17 BAULIG
Excerpts from *The Changing Sea Level*

18 BOURCART
Excerpt from *The Theory of Continental Flexure*

19 LINTON
Excerpts from *Problems of Scottish Scenery*

20 PENCK
Attempt at a Classification of Climate on a Physiographic Basis

21 PELTIER
Excerpts from *The Geographic Cycle in Periglacial Regions as It Is Related to Climatic Geomorphology*

22 BIROT
Excerpts from *The Cycle of Erosion in Different Climates*

23 **SCHUMM**
Geomorphic Implications of Climatic Changes

24 **SPARKS**
Excerpt from *Relief (Great Britain Geographical Essays)*

25 **CHORLEY**
Geomorphology and General Systems Theory

26 **LEOPOLD and LANGBEIN**
Excerpts from *The Concept of Entropy in Landscape Evolution*

27 **HOWARD**
Excerpts from *Geomorphological Systems—Equilibrium and Dynamics*

28 **SCHUMM and LICHTY**
Excerpts from *Time, Space, and Causality in Geomorphology*

29 **HACK**
Excerpts from *Interpretation of Erosional Topography in Humid Temperate Regions*

30 **HJULSTRØM**
Excerpts from *Studies of the Morphological Activity of Rivers as Illustrated by the River Fyris*

31 **RAPP**
Concluding Discussion on the Total Denudation of Slopes in Kärkevagge

32 **FOLK and WARD**
Excerpt from *Brazos River Bar: A Study of the Significance of Grain Size Parameters*

33 **HOLMES**
Excerpts from *Till Fabric*

34 **SAVIGEAR**
Excerpt from *A Technique of Morphological Mapping*

35 **HORTON**
Excerpts from *Erosional Development of Streams and Their Drainage Basins: Hydrophysical Application of Quantitative Morphology*

36 **SHREVE**
Excerpts from *Stream Lengths and Basin Areas in Topologically Random Stream Networks*

37 **STRAHLER**
Excerpts from *Dynamic Basis of Geomorphology*

38 **PRICE and KENDRICK**
Field and Model Investigation into the Reasons for Siltation in the Mersey Estuary (synopsis)

39 **YOUNG**
Excerpts from *Deductive Models of Slope Evolution*

40 **NYE**
The Mechanics of Glacier Flow

41 **SPRUNT**
Excerpts from *Digital Simulation of Drainage Basin Development*

42 **DOUGLAS**
Natural and Man-Made Erosion in the Humid Tropics of Australia, Malaysia, and Singapore

43 **LANGBEIN and SCHUMM**
Yield of Sediment in Relation to Mean Annual Precipitation

44 **BROWN**
Excerpt from *Permafrost in Canada. Its Influence on Northern Development*

45 **MABBUTT**
Excerpts from *Review and Concepts of Land Classification*

46 **WRIGHT**
Excerpt from *Principles in a Geomorphological Approach to Land Classification*

Part I was devoted to a discussion of the people who made the most impact in the early development of geomorphic ideas; the ideas that they represent were only mentioned briefly. Part II concentrates on the major concepts put forward by the early workers and indicates how these have stood the test of time, and subsequently been modified and developed during the last few decades. The more recent advances in geomorphology and current trends are referred to. Major concepts and recent developments are illustrated by means of a series of papers, reproduced in whole or in part, which aim to exemplify the topics under discussion. They represent a selection out of a very extensive literature; the intention is to indicate the wide range of fields in which geomorphology is now involved. Methods of analysis that are currently being developed are also sampled.

Part II is divided into six main sections. The first five cover the major concepts developed during and since the four phases already discussed. The first section is devoted to the opposing concepts of catastrophism and uniformitarianism, which were the two earliest schools to develop. They have recently been revived in a modified form.

The second section is devoted to the debate concerning peneplanation and pediplanation, which can also be stated as the difference between down-wearing and back-wearing. This argument is based on the conceptions of Davis and Penck, respectively; it has been developed subsequently by their followers and others who have built on the foundations that they laid.

The third section is a discussion of stability and mobility, which in one sense is a further development of the previous section. Davis argued for stability in his ideal cycle, whereas Penck was more concerned with the mobility of the crust. Davis's concepts led to the important phase in the 1920s and 1930s when the study of denudation chronology occupied a considerable number of geomorphologists. With the recognition of the mobility of the crust in both the horizontal and vertical senses in the growth of ideas concerning plate tectonics and ocean-floor spreading, denudation chronology, as previously conceived, has lost its impact, and mobility is now a factor to be investigated. Today the chronology of planation surfaces is becoming a key aspect of plate tectonics.

The fourth section deals with the significance of climatic and structural controls on landscape development. This type of geomorphic analysis had an early origin and has continued to provide a study field of increasing depth. The importance of climate in landform development has traditionally been studied by European geomorphologists, although others have also contributed to it, particularly those working in the semiarid American West, and in Africa, Australia, and South America. Important contributions in the field of glaciation and glacial geomorphology were made mainly in North America, Antarctica, and Europe. The more recent development of periglacial geomorphology has added a new dimension to climatic geomorphology. The part played by rock type and structure has been less intensely studied, apart from some specific rock types. The study of the particular landforms associated with limestone has given rise to a very large and specialized literature on karst phenomena, which is often associated with J. Dvijic, who worked in the classical karst country of Yugoslavia. Less attention has been paid to structural control in other rock types, although considerable attention has been given to the analysis of various types of sediment, particularly fluvial, and sedimentary structures, including studies of till fabric.

The fifth section is concerned with the study of systems in geomorphology. It is a viewpoint rather than a concept, although the two are related. A systems approach focuses attention on interrelationships among different elements of landscape and the many variables upon which the processes and the response to them depend. The landscape is viewed as a developing and changing entity composed of mutually dependent parts. The importance of scale of study, both spatially and temporally, is stressed, and tempo of change plays an important part.

The sixth section is subdivided into the following seven subsections, each devoted to a modern trend of study: (1) process, the study of which is an important aspect of modern dynamic geomorphology; (2) material, which forms the basis of study, both from the viewpoint of description and classification and of information concerning the processes that deposited the material or were affected by it; (3) morphology, studied as the end result of the operation of processes, with important feedback relationships between the two elements and with material; (4) quantification, an important element in many recent geomorphic studies, being based on both statistical analysis and other mathematical methods; (5) models, which provide a systematic framework for investigations in a wide range of fields, inasmuch as there are many types of models suited to different fields of enquiry, ranging from hardware models to theoretical and mathematical ones; (6) environmental considerations, which provide an increasingly important framework for geomorphic investigations by bringing together both the purely geomorphic processes and the mutual interaction between these processes and human action, both inadvertent and intentional; (7) applied geomorphology, as a natural extension of item 6 in that it deals with the deliberate interference by humans in natural geomorphic events and situations. Examples of work in all these fields are presented.

CATASTROPHISM AND UNIFORMITARIANISM

The arguments concerning catastrophism were based essentially on the literal interpretation of Genesis, with the creation of the earth occurring in 4004 B.C. Great weight was also given to the Noachian deluge, which was held responsible for the drift deposits, later attributed to the spread of the glaciers during the Ice Age. In contrast were the uniformitarian ideas advocated first by Hutton and Playfair, and later supported strongly by Greenwood. Early workers in the field tended to be very dogmatic and upheld extreme positions. More recently geomorphologists have come to realize that natural processes are rarely so consistent. The principle of equifinality is recognized as an important one in geomorphology, whereby different processes can produce a similar result, and one cause or process rarely operates alone. It is realized that nearly all landscapes are both polygenetic and polycyclic.

The papers included in this section are concerned with more recent work on the intensity with which geomorphic processes operate. This trend could be called the neocatastrophic approach. The distinction must be made between truly catastrophic events, which cause profound, once and for all changes, and those, such as the features discussed by Bretz, that are only larger than usual examples of processes that act continuously, such as stream action. The intensity with which processes operate over time varies greatly; the more intense the event, the more rare its occurrence.

In 1923, J. H. Bretz published an account of the effects of the sudden emptying of a large ice-dammed lake. The magnitude of the event can be partially assessed from the landforms that this immense, but short-lived flood created. The dry waterfalls, potholes and other forms dwarf all similar features created by even the largest rivers of today and challenge the imagination. Nevertheless, the consensus of opinion now agrees that this must have been a catastrophic event of very great magnitude, requiring special conditions. Similar but smaller events still occur fairly commonly, such as the *jokulhlaup* that occur in Iceland and elsewhere when glacially dammed lakes suddenly drain away, producing very large temporary discharges.

Paper 12, by M. G. Wolman and J. P. Miller, provides a very valuable quantitative assessment of the total effect produced by events of varying magnitude and frequency. The effects can be assessed either in the amount of work done or in terms of the landforms created. It is necessary to balance the very large and often capricious effects of the really extreme events, which occur very rarely, with the smaller effects of moderate events that occur with increasing frequency as their magnitude diminishes. Observations over a considerable period of time suggest that it is the less extreme events of moderate intensity that accomplish the maximum amount of work because they operate more often and for longer sequences. On the other hand, some events can only be achieved by really extreme occurrences. The analogy of the dwarf, the man, and the giant with which Paper 12 ends puts the matter very succintly and vividly. Paper 12 also illustrates the great value of long-continued observations and the desirability of obtaining quantitative date.

In Paper 13 D. R. Stoddart discusses the effect of extreme events on the ecology and total environment of a low coral island in the West Indies. Stoddart, who is an expert on the geomorphology and ecology of coral islands in all three major tropical oceans, is concerned with the effects of a hurricane. He shows that in such an environment it is essential to consider the effects not only on the morphology, but also on the organic life and the total environment. The local nature, both in time and space, of such events is significant.

The conclusion can be reached that neither a strict uniformitarianism nor a strict catastrophism should be assumed. The action of both endogenetic and exogenetic processes is subject to considerable fluctuation in intensity. The longer the time span under consideration, the greater the range within which the processes vary. On a very long term basis, on the order of tens or hundreds of millions of years, there are the contrasts between the periods characterized by glacial events and the much longer periods during which the earth appears to have enjoyed a more equable climate. These events have a cycle of about 200 to 250 million years. Within the "ice-age" periods are shorter-term events, which give rise to the alternation of glacial and interglacial periods and stadial and interstadial phases. The glacial–interglacial cycle is on the order of 90,000 to 100,000 years, and the smaller modulation's are from 500 to 50,000 years. There are similar, but even shorter and smaller, fluctuations, giving rise to the oscillations that took place during the retreat of the last major ice sheet in the late and postglacial periods. In rivers there is the mean annual flood, the 5-year flood, the 10-year flood, and also floods of increasing magnitude and rarer occurrence. The frequency decreases and the magnitude increases logarithmically, as pointed out by Wolman and Miller. Many of the extreme events may be set off by trigger action, or the passing of thresholds, an aspect that will be considered later in relation to the systems approach to geomorphology.

DOWN-WEARING AND BACK-WEARING: PENEPLANATION AND PEDIPLANATION

The concept of peneplanation was named by Davis, although he did not originate the concept. He considered it to represent the old-age stage of his ideal cycle. Peneplanation played an important part in the study of denudation chronology, which attempted to establish the stages of landscape development in an area undergoing intermittent uplift. The process of peneplanation was not, therefore, completed at each substage. The evidence on which denudation chronology was based was related to the formation of partial peneplains. Each partial peneplain ceased to be formed when the base level fell, initiating the next one in the series. This method of landform study was pioneered by Henri Baulig and widely adopted in France and Britain (see page 129), where in the uplifted areas the Tertiary period produced little other evidence of geological events, owing to the restricted deposits of this age.

There are many problems inherent in the study of denudation chronology, not least of which is the problem of peneplanation and the recognition of partial peneplains. These are considered to be the flatter portions of the landscape. There is also the problem of identifying the formative processes, whether they be subaerial or marine. That different geomorphologists in neighboring areas can reach opposite conclusions as to the formative process gives an indication of the problems of interpretation when only a small part of the total landscape is used in the analysis. The problems of peneplanation are discussed much more fully by G. F. Adams in his Benchmark volume, *Planation Surfaces* (Dowden, Hutchinson & Ross, Stroudsburg, Pa., in press).

The success of a study in denudation chronology depends on the evidence available. The work of Wooldridge and Linton, an excerpt of which is given as Paper 14, on the development of the landforms of southeast England is one of the most successful studies of denudation chronology. It is written by two of the most influential geomorphologists of the period, and deals with an area in which datable deposits were available to give a time scale. The drainage pattern could also be interpreted successfully, on the basis that streams adjusted to structure must have developed a subaerial system, whereas those unrelated to structure were probably initiated from a marine-trimmed surface, veneered by more recent sediment. These points are considered in the summary and conclusions reprinted, in which reference is made to peneplanation as recognized by Davis, in the form of the summit surface of possible Miocene age.

In Paper 15, Wooldridge carried the analysis of the upland plains and partial peneplains a stage further to cover much of Britain. He puts forward the arguments for and against the use of such evidence to establish the changes of base level upon which it depends. This paper gives an indication of the state of denudation studies in the middle of the twentieth century. Since that time relatively little further work has been done in this field, owing to the often unsatisfying conclusions reached. This is inevitable when so much of the evidence no longer exists owing to erosion. The interpretation is necessarily ill supported by definitive field evidence, thus leading to ambiguous and inconclusive results. It is generally assumed that the subaerial processes responsible for the upland plains discussed by Wooldridge were formed by the down-wearing associated with peneplanation.

The opposing view of back-wearing, rather than down-wearing, has been advocated strongly by Lester C. King. He based his conclusions on work done mainly in Africa, where large, very flat plains form a conspicuous element of the landscape. Paper 16 is a short excerpt from his study of the world's plainlands. King argued against Davis's term "normal" to describe the humid, temperate climate, as he considered that the semiarid areas had more justification to be termed normal. He argued that back-wearing by scarp retreat is an important process that leads to pediplanation, which results in large flat surfaces, several of which can be flattened simultaneously as a series of scarps retreat across the landscape, stepped one above another. Base level is not a vital control in pediplanation as it is in peneplanation. King's work grew out of that of Walther Penck, who also advocated parallel retreat of slopes under certain conditions.

In King's major work, *Morphology of the Earth,* he extended his ideas of landform development to cover the entire earth's surface, including the ocean basins, and summarized his views on the cycle concept in relation to the process of pediplanation by scarp retreat. This process is discussed with particular reference to Africa and in terms of the cyclic concept, although he is very critical of the "normal" cycle of erosion as enunciated by Davis. King's own interpretation, he claims, allows intercontinental correlation of cyclic land surfaces formed by pediplanation. This possibility now receives important support from systematic analyses of sea-floor spreading and plate tectonics.

STABILITY AND MOBILITY

The concept of stability grew out of the early work of Edward Suess and his epock-making volumes *Das Antlitz der Erde.* It was first translated with revisions in 1906 and subsequently by H. B. C. Sollas and W. J. Sollas. Suess argued for the importance of eustatic changes of sea level, related to phases of transgression and regression, acting on a stable earth. It was thought that the stability of the globe was such that these major eustatic changes in base level controlled landform development widely across the whole earth. The following brief quotations from his book make some of these points: "The formation of the sea basins produces spasmodic eustatic negative movements" (p. 538). This is now an integral deduction from studies of sea-floor spreading and heat flow. "This recapitulation shows that the theory of secular oscillations of the continents is not competent to explain the repeated inundation and emergence of the land. The changes are much too extensive and too uniform to have been caused by movements of the earth's crust" (p. 540). "The formation of sediments causes a continuous, eustatic positive movement of the strandline. We are thus aquainted with two kinds of eustatic movement; one, produced by subsidence of the earth's crust, is spasmodic and negative; the other, caused by the growth of marine deposits, is continuous and positive" (pp. 543–544).

In this argument for "sedimento-eustacy," subsequent quantitative studies suggest that Suess was in error; "tectono-eustacy," caused sea-floor spreading and plate tectonics, would appear to explain both positive and negative trends. Nevertheless, Suess laid the foundation for the school of thought which held that erosion surfaces were the result mainly of eustatic changes of sea level. The corollary to this argument is that it should be possible to correlate erosion surfaces by altitude from place to place, thus providing a worldwide pattern of sea level changes through time. This optimistic outlook concerning the role of eustacy and stability in the formation of erosion surfaces was reflected in the International Commission set up in 1926 by the International Geographical Union to correlate erosion surfaces around the world.

One active worker in this field in the 1920s and 1930s was Henri Baulig, a French geomorphologist who analyzed surfaces of erosion on the Massif Central of France and correlated them with surfaces elsewhere. The publication from which excerpts are reprinted as Paper 17 was influential in Britain, where Baulig delivered a series of lectures. This paper is the published version. Baulig indicates four possible extents to which eustasism is true, hoping that it will be found to play an important part in landform development. This paper did much to stimulate interest in studies of denudation chronology in Britain from the 1930s to the 1950s. There have been attempts to produce cross-Atlantic correlations of erosion surfaces. Most geomorphologists, however, believe that the correlation of erosion surfaces by altitude alone is not possible, in view of the many uncertainties to which the recognition of such surfaces is subject, particularly with reference to crustal mobility, formative processes, and subsequent modification.

The mobility of the earth's crust is becoming increasingly apparent. The work of Jacques Bourcart on continental flexure, although it is clearly not now everywhere applicable, did draw attention to a very important modification of continental margins. Paper 18 summarizes his views on transgression and regression. The theory of continental flexure can account for both submerged and emerged surfaces, and can explain inland rejuvenation when there is evidence of contemporaneous subsidence at the coastline and offshore. The exact position of sea level with respect to the land surface would depend on the relative position of the axis of flexure with respect to the coastline. A compelling example of this sort of flexing is seen along the Atlantic coast of North America. Another area where this theory could apply is around the western half of Britain, where there are deep basins of sedimentation in the Irish Sea, for example, and uplifted and warped erosion surfaces inland. This is the area considered by D. L. Linton in Paper 19.

The theory put forward by Linton concerning drainage development in Britain is based on the assumption of a mobile crust. He considered that the British Isles represent an upwarped arch, the keystone of which subsided to form the Irish Sea between the two uplifted portions. These tectonic movements would initiate short, steep streams flowing westward to the Irish Sea and long, gentle streams flowing east to the North Sea on the gently sloping eastern limb of the major upwarp. If the hypothesis is correct, it would not be possible to correlate erosion surfaces by altitude, because they would be sloping surfaces that had undergone asymmetrical uplift, rather than being due simply to a fall of base level eustatically. In view of universally demonstrated crustal mobility, the eustatic theory must be handled with caution and skill. Nevertheless, the formation of ocean basins, sedimentation, and glacial fluctuations do cause major eustatic movements of sea level. According to Fairbridge (1961), world sea level has dropped by about 100 m during the Quaternary, independent of glacioeustacy, probably mainly as a result of ocean-basin subsidence. Today that subsidence can be demonstrated in mid-ocean ridge areas and in some marginal basins.

CLIMATIC AND STRUCTURAL CONTROL ON LANDSCAPE DEVELOPMENT

It has been thought for a long time that landforms are related to the climatic environment in which they are formed. W. M. Davis devised a series of ideal cycles related to different climatic environments, including those related to semiarid, glacial, and coastal situations. One of the earliest attempts to classify the climatic criteria that can best be related to the processes that form specific morphological features was made by Albrecht Penck; it is reproduced as Paper 20. He argues that landforms reflect the climatic criteria, especially that of precipitation type.

A somewhat similar approach is that of Louis Peltier; a short excerpt from his paper on the periglacial cycle of erosion is reproduced as Paper 21. Apart from being an early description of the periglacial cycle, Paper 21 begins with a useful discussion of the climatic variables that control land-forming processes. Peltier suggests the limits, in terms of annual temperature and precipitation, that control various processes. This approach differs from that of Davis, in that the aim is to set the limits within which different processes operate, rather than to follow one set of climatic controls throughout the operation of a whole cycle. Peltier in fact combines both approaches; following the discussion of the optimum conditions in which different processes operate, he discusses the periglacial cycle and shows how the processes might produce landforms that belong especially to this type of climate.

A useful summary of the problems associated with the relationship between climate and geomorphology is given by Pierre Birot; Paper 22 is the introduction and conclusion to his book. It is included to emphasize the important part that French workers have played in developing this aspect of geomorphology; other important names are Jean Tricart, Andre Cailleux, and Jean Corbel.

One major problem involved in the relationships of climate and morphology is the effects of the very rapid fluctuations in climate that have taken place in the geologically short period of the Pleistocene (1.8 million years), during which some evidence suggests up to 20 major fluctuations of climate. Major changes have even been worked out for the mere 10,000 years of the Holocene. These fluctuations have not only affected the glaciated areas, where climates have ranged through humid temperate, periglacial, and glacial, according to the positions of the areas concerned with respect to relief and climatic elements. Latitudinal shifts of climatic boundaries of up to 3000 km within 10,000 years have been mapped. In Britain there is evidence of glacial and periglacial processes only 15,000 years ago, where now conditions are mild and maritime. In the extraglacial areas of lower latitudes there have also been very significant changes of climate, with oscillations between wetter and drier conditions as well as temperature changes. Perhaps the only areas to have had a more or less consistent climate are certain hot equatorial areas, although the area affected by these conditions must have varied, with only small areas having a consistent climate over a long period. Thus nearly all landscapes must be polygenetic in character. The present interglacial type of climate has only been effective during the last 10,000 years, so that the landscape has hardly yet had time to become adjusted to the present climatic conditions, which are also in the process of change. Rapid reglaciation is at present reported in many areas, such as Baffin Island; thus it seems possible that the interglacial phase is about to be reversed with colder conditions taking over, leading inexorably to renewed glaciation.

The geomorphic implications of climatic change are considered in Paper 23, by S. A. Schumm. The author deals particularly with the relationships among climate and sediment yield and the characteristics of river channels. There are several significant effects in connection with the lack of adjustment of landforms to their present climatic environment. In some instances the change of climate makes landforms more subject to change; in others, slopes have been reduced by former climatic controls below the gradient that would be in equilibrium under present-day conditions. An example of the former is the oversteepening of slopes under some circumstances by glacial erosion. These oversteepened slopes are subject to adjustment by talus formation, particularly under periglacial conditions.

Periglacial processes of mass movement can operate on gradients lower than those needed for processes operating under humid temperate climates, particularly in well-vegetated areas. Thus many of the slopes of southern Britain and western Europe that were modified by periglacial head (colluvium) formation and mass movement are now flatter than the equilibrium gradient. This has the beneficial result of reducing the danger of accelerated soil removal or landslides when the vegetation, which is naturally forest, is cleared under farming practices that require frequent plowing and prolonged periods of bare soil before crops start to grow and thus to help hold the soil in place. It is probably through the effect of vegetation that climate exerts such an important influence on geomorphic processes in many instances. The problems of soil erosion make this relationship clear in many areas where the natural vegetation is disturbed. Thus in Southern California, low-altitude hills are commonly rounded due to periglacial action, but the valleys are V-shaped, so that vegetation destruction leads to catastrophic landslides.

The effect of rock type and structure on landforms is an important field of enquiry. This aspect is discussed by Bruce W. Sparks in Paper 24 in which the author demonstrates the fundamental distinction between highland and lowland Britain, and the major part that rock type and structure play in the landscape of the country. On the whole, this aspect of geomorphology has received less attention than some, apart from the special features associated with karst landforms, which are produced by the solubility of limestone in weak carbonic acid formed by the addition of carbon dioxide to rainwater. The strong control of rock type and structure on morphology is so obvious in so many areas that perhaps too little detailed work has been devoted to this aspect of geomorphology. The scarplands of western America, as of southeastern Britain and western Europe, illustrate the control of rock type and structure on landform very clearly. The weak shale and clay depressions separate the ridges formed of harder rocks, providing the dip slopes and the scarps that are so conspicuous. Where the folded structure is more complex, as in the Appalachians or in Britain in the Isle of Wight and Dorset, the tight folds have in places led to an inversion of relief, with the anticlines now forming valleys between synclinal ridges and hills.

A considerable literature has grown both in Britain and the United States on the relationship between structure and drainage. The different types of drainage pattern, including antecedent, superimposed, and consequent, were early recognized, as has already been mentioned. Drainage pattern in relation to structure is also an important aspect of denudation chronology, as it provides a possible means of differentiating between subaerial and marine surfaces or erosion. The former leads to adjustment of drainage pattern, with strike streams developing as subsequents along the outcrops of weaker strata, whereas marine surfaces are often associated with epigenetic drainage. The superimposed or epigenetic system may develop on a cover of younger marine or continental sediments so that eventually it is superimposed unconformably onto the underlying discordant structure.

SYSTEMS IN GEOMORPHOLOGY

A system can be defined as a structured set of objects or attributes. The objects or attributes are variables that are interrelated in that they work together as a system. Thus an interdependence of variable is characteristic of a system. The system, or whole, is greater than the sum of its parts. Systems can be (1) isolated, (2) closed, or (3) open. The first have boundaries closed to the passage of both mass and energy. The second have boundaries closed to the passage of mass but not energy. The third have boundaries open to the passage of both mass and energy. Most geomorphic systems are strictly open, but the earth as a whole is a closed system, in that it exchanges energy, but not mass, apart from meteorites and spacecraft, with outer space.

Paper 25, by R. J. Chorley, and Paper 26, by Leopold and Langbein, discuss some aspects of the applications of the systems approach to geomorphology. Leopold and Langbein apply the concepts of entropy to the development of landforms; Chorley is concerned with the contrasts between open and closed-system thinking in relation to equilibrium conditions and the Davisian cyclic approach to landforms.

Systems can be classified in terms of their internal complexity or their structure. Chorley and Kennedy (1971) suggest four stages of complexity: (1) morphological systems, which consist of a network of structured relationships between the constituent parts; (2) cascading systems, which are dynamic in that they consist of throughputs of energy and mass, the system being defined by the path followed by the energy or mass; (3) process-response systems, which concern the links between the two previous types, the relationship between form and process; and (4) control systems, which are of greater complexity in that the relationships of the process-response system are controlled by an intelligence. This type is, therefore, beyond pure geomorphology, but enters into applied geomorphology.

Morphological systems consist of the relationships among physical properties of natural features. They occur on a wide range of scales and many different types occur; at present morphological systems are frequently defined in statistical terms. Regression and correlation are particularly useful, as these methods express relationships among two or more variables. In many relationships, threshold problems occur, in that one relationship applies below the threshold and a different one above it. For example, slopes that are being undercut may get gradually steeper until the strength of the material is overcome and sudden failure produces discontinuity in the relationship. One of the clearest morphological systems is the drainage basin, which can vary in size according to the order of the basin. Drainage basins can be described in terms of many variables, such as slope, shape, area, and other criteria. The relationships are expressed numerically, rather than functionally, in a morphological system.

The cascading system is dynamic in character. It is defined as a structure or system within which the output from one subsystem forms the input into the next one. Other characteristics include a regulator within the subsystem. The regulator may direct part of the mass or energy into a store within the system or create a throughput, which becomes the subsystem output. Cascading systems interlock with morphological systems to form process-response systems. The regulators in the system are important geomorphically and may be represented by morphological variables, for example, the soil in transit across a slope or the sediment in a beach or in a river. It is at this point that man can often intervene to make the system into a control system.

The regulators are of three types: (1) threshold regulators, which include surface infiltration capacity; (2) dispositional regulators, which control the distribution of energy or mass, for example, the route taken by water entering a subsystem; and (3) presence or absence regulators, such as whether a stream is present or absent at the foot of a slope. The first group is highly subject to human interference, and thus forms a link with control systems. All regulators provide links with morphological systems, thus producing process-response systems. The regulators can be thought of as valves in a control system, and they can be manipulated. They can divert energy and mass into stores in the cascading system. These stores include water storage in many forms, such as soil moisture, channel flow, glaciers, and groundwater.

Each major process produces its own dynamic cascade, as potential energy is expended on the water and material involved in the system. All these systems are parts of larger ones. In analyzing a system, inputs, outputs, storage of energy, and mass must all be measured; an understanding of the regulators is also necessary. Many different measuring methods are required to obtain the data, partly because the inputs and outputs are continuous, but variable, and observations are often only made at points in both space and time. Regulators can sometimes be expressed in the form of mathematical equations when their operation is fully understood in numerical terms. Detailed fieldwork or laboratory analysis is often necessary to obtain the value of essential constants. Theoretical analysis is also required to establish the mathematical relationships. There is often a lack of empirical data on the processes involved.

Process-response systems consist of linkages between morphological and cascading systems. The process of the cascading system changes the morphological variables. An important element of the interchange is the dynamic feedback loop, which can be positive when it is self-generating or negative when it is self-regulating. Three important principles are involved: (1) the system operation is controlled by the magnitude and frequency of inputs into the cascades; (2) negative feedback operates to create equilibrium between the morphological variables and a steady-state throughput; and (3) time-directed changes can occur if input changes take place, or if there is an internal degradation of the system as it continues to operate.

The importance of equilibrium is discussed in the Paper 27, by A. D. Howard, which is also concerned with structural influences on the landforms and the effects of time on equilibrium. The relaxation time is relevant and is related to feedback. Positive feedback loops must in the end be self-destructive, but may continue for a long time, as shown by processes of glacial erosion. Negative feedback loops are more important and lead to the concepts of grade and equilibrium. The natural feedback system is very complex and several subsystems are often linked.

Time plays a complex, but very important role in systems. This is a matter discussed in Paper 28, by S. A. Schumm and R. W. Lichty. The whole paper is reproduced in the Benchmark volume, *Planation Surfaces*, edited by G. F. Adams (Dowden, Hutchinson & Ross, Stroudsburg, Pa., in press). Time and scale effects are intimately linked, a point discussed in detail in the paper. If there is a progressive integration of organization in the system, it will lead to an irreversible evolution, a point considered in Paper 26. This aspect is more common in organic systems, whereas inorganic systems, including landforms, tend to become less organized with time, thus leading to an increase in entropy. The character of reality alters according to the time scale on which it is viewed. The large extensive landscape viewed over a long period of time becomes modified; slopes, for example, decreased in gradient, but not all aspects need necessarily change, such as drainage density. In time the input and output variables change slowly. There is also the problem of exogenetic and endogenetic processes, which also change their mutual interrelationships through time. As systems of landforms develop through time they tend to disintegrate, and correlations become weaker as entropy increases.

In considering changes over time, it is sometimes possible to substitute space for time, a concept known as the ergodic principle. For example, a sea cliff progressively deserted by the sea as a result of accretion spreading along the cliff foot provides a time sequence of modification with impeded cliff foot removal of debris. A retreating waterfall can also provide a sequence of slopes that fulfill, the ergodic principle. The development of drainage on land vacated by a retreating glacier is another example.

The introduction of the systems approach into geomorphology provides a means whereby morphology can be related to processes in a useful framework of quantitative interrelationships. The scale of space and time has also been shown to be a very important aspect of many geomorphic systems.

The systems approach is a modern trend in geomorphology and could have been included in the following section. It has been discussed separately in view of its role as a framework in which geomorphic study can be set, and which provides links among the three elements, process, material, and morphology.

MODERN TRENDS

Process

As with many other subjects, geomorphology is branching out into a wide range of fields. It is linking with other subjects at the limits of the various paths along which it is advancing. For example, geomorphology links with pedology in the study of soil characteristics. It links with sedimentology in the study of till fabrics, sand size distribution studies, and the measurement of particle shape. It links with photogrammetry in the study of landscape from the air and with remote sensing in general; some geomorphic features, such as periglacial patterned ground, show up much more clearly from the air. It links with engineering geology in the study of liquid limits, soil instability, and similar characteristics that give rise to mass movements of different types. It links with oceanography in the study of the effects of waves on the coastline, and in the recording and analysis of sea-level changes. It links with mathematics in the use of mathematical methods of analysis, including statistics, topology, differential equations, time-series analysis, and many other techniques. It links with physics in the study of phenomena such as autosuspension, kinematic waves in glaciers and floods, and the physics of sediment movement in flowing media. It links with economics in the various aspects of applied geomorphology, such as floodplain zonation, cost–benefit analysis of water resources, or coastal defense schemes and reclamation projects. It links with psychology in the study of hazard awareness and landscape perception and appreciation. It links with history in the mutual use of data concerning time and environment; historical documents and maps can be employed, for example, in assessing changes along a coastline. It links with archeology in the use that can be made of archeological sites in assessing geomorphic change, for example, of sea level and the form of earlier coastlines; in turn, geomorphology can give useful information to archeologists in locating former settlements and interpreting former climates and landscapes. It links with chemistry in the study of chemical changes in water, for example, in the study of karst drainage and denudation. It links with botany in the study of the effect of vegetation on slope processes, salt-marsh and sand-dune development, which is closely associated with the specialized plants of these environments, and in the use of pollen analysis. It links with zoology in the study of the erosional effects of animals on slopes and in the study of coral reefs.

All these different links provide a wide range of different techniques with which to study the many processes involved in the creation and modification of landforms. One of the most important developments in geomorphology has been the growing number and sophistication of studies into the processes that operate to shape the earth's surface. Although there were notable early studies of process, much of the work in this field is of recent origin. Now processes in all fields of geomorphology are being monitored: the flow of glaciers, the movement of sand by desert winds, the recording of slope movement, longshore drift of material along coasts, periglacial processes, fluvial processes, and many others.

The excerpts given as Paper 29, from a paper by J. T. Hack, provide a link between the section on systems and the present section. The whole paper is available in the Benchmark volume, *Planation Surfaces,* edited by G. F. Adams. The author shows that the landscape is a dynamic system and that it is essential that process be adequately considered in arriving at a correct view of the development of landscape. Such process studies are essential to a full understanding of the dynamic view of landscape development advocated by Hack, who argues for the open-system model and dynamic equilibrium. Paper 29 should be considered in the light of Paper 28 by Schumm and Lichty. The scale, both in space and time, is a vital consideration in the study of both process and morphology, but particularly in the study of process. From this point of view it is essential to bear in mind the tempo of change. In some processes their operation is so slow that over the length of time that can be accurately recorded it is not possible to come to meaningful conclusions. It is under these conditions that the ergodic principle can prove valuable, provided it can be used effectively.

Nevertheless, some processes operate fast enough to have a relaxation time which is within a time scale that can reasonably be measured. An example is the effect of wave action on beach profiles, because the profiles respond within a matter of hours or days to the changing waves. The swash slope will adjust to changing waves over one tidal cycle; thus beach changes can be studied under varying conditions so that the effect of different wave types can be assessed.

Coasts, however, are of a much larger extent, and the tempo at which they change is much slower, although it varies greatly from one area to another. In some areas there is still evidence of interglacial or even pre-Quaternary features on the present-day coast, indicating very slow change. Some changes may, on the other hand, be relatively rapid in comparison with those taking place under the action of other processes, such as some slope processes, especially in areas of low relief.

Excerpts from two important works that well exemplify the value of detailed and careful studies of process in the field are included. Paper 30 is from the well-known work of F. Hjulstrøm on the fluvial geomorphic processes operating in the Fyris River in Sweden. This work, published in 1935, provides a model of careful field and laboratory investigation. The study of the relationships among stream velocity, particle size, and the processes of erosion, transport, and deposition has provided some fundamental relationships that have given a much clearer picture of the way that rivers deal with their load. The study includes an appraisal of all types of river load, including bed load, suspension, and solution load. In studying the processes responsible for the river's work, Hjulstrøm has learned a great deal concerning river morphology and the processes by which the river shapes its bed and creates both its longitudinal and cross profiles.

Paper 31 is a result of fieldwork by A. Rapp, from the same school in Upsala. This work was undertaken in the far north of Scandinavia in the periglacial environment of Kärkevagge. It provides a valuable model of field studies in a subarctic environment, in that it is one of the first attempts to make a genuine quantitative study of all the processes that operate to wear down a slope under harsh conditions of the subarctic climate in a remote area, far from the disturbing influence of man. Rapp used a great many different field techniques to establish the relative importance of the different processes that operate under these conditions. His conclusions are summarized in the table in the excerpt. He recorded the operation of soil creep, rainwash, and solifluction, the last being particularly associated with permafrost conditions that render the soil impermeable at depth. He also recorded rock falls and paid special attention to the geomorphic work of avalanches. The interesting conclusion reached was that the most effective agent of removal is running water in the form of transport of load in solution. It shows that volume and distance carried must both be taken into account in assessing the relative importance of different processes. Although creep and rainwash move by far the largest volume of material, they move it such a short distance that their relative effectiveness is much reduced. In contrast, the concentration of matter in solution is small, but it is carried a long way fairly rapidly by the actively flowing slope drainage system.

There is room for many more quantitative studies of this type in different climatic regions and in different types of rock and relief. Quantitative studies of the action of processes in the field are being increasingly carried out and many more examples could be given. They involve all aspects of geomorphology, including fluvial, slope, glacial, periglacial, karst, arid, and coastal processes. A great variety of different field techniques have been developed, some of which are very sophisticated. These include such procedures as the use of tracers on beaches and in rivers, the monitoring of movement of ice in glacier tunnels, the measurement of creep on slopes, as well as many other contemporary processes.

Material

The subject of sedimentology has recently been greatly expanded, with several journals now devoted entirely to it. There has also been a growth in the study of material by geomorphologists, which is providing very useful information for various objectives. Our purpose is to study material from the viewpoint of assessing its response to process. An example is the differentiation of frost-susceptible soils from those that are non-frost-susceptible on the basis of the size distribution in the sediment. The growth of ground ice in periglacial environments has also been shown to be closely linked with the characteristics of the material, with silts being particularly favorable to ground-ice formation. Clays are too fine, because water cannot be drawn into them, whereas sand and gravel are too coarse, because capillary attraction is less effective. Such distinctions can have important practical applications.

Another use made of the study of sediment is the differentiation of bodies on the basis of their size-distribution parameters, such as the mean, median, sorting, skewness, and kurtosis and the relationships among these parameters. The nature of the sedimentary parameters can then be related to the operation of the processes by which they were deposited. The recognition that graded bedding is associated with the action of turbidity currents is another example of the association of sedimentary character with a specific process of deposition.

Paper 32, by R. L. Folk and W. C. Ward, was one of the first to draw attention to the significance of grain-size parameters in studying the various depositional processes and sediment characteristics. In this paper the graphical moment measures were first proposed, following a rather simpler, earlier scheme proposed by D. L. Inman in 1952. The Folk and Ward method has been used a great deal subsequently. More recently, with the advent of computers, it is often easier to calculate the mathematical moment measures, some of which use more points from the cumulative frequency curve. Nevertheless, whichever method of calculating the moment measures is adopted, it is generally agreed that sedimentary size distribution provides valuable evidence both in the study of process and of morphology, through a consideration of material. Both the overall gross morphology and detailed forms can usefully be studied.

The analysis of the bedding characteristics and sedimentary structures of the deposit can also yield useful information. The study of the structural patterns in glacial stratigraphic sections, for example, can be related to the type of movement and flow that gave rise to them. The distinction between a rippled bed, a planar bed, and a bed with dunes enables information concerning the flow regime to be deduced in relation to the size of the sediment. Fossil forms also provide valuable evidence of past processes and depositional environments.

Considerable work has also been done on the shape and surface texture of sedimentary particles. This type of analysis ranges from the measurement of the shape of pebbles to studies under the electron microcope of individual sediment grains at very high magnifications of 10,000 times and more. A number of methods have been devised for shape measurement. Some give ordinal results; others provide a quantitative numerical index, for example, Cailleux's well-known roundness index.

Paper 33, by C. D. Holmes, is an example of another fruitful field of material study, and is one of the most quoted papers in this field. The author's work is still in many respects the most thorough in the field of till-fabric analysis. It has long been recognized that the movement of material by flowing ice induces a preferred orientation and dip of elongated stones within the till. The measurement of the preferred orientation and dip has been used to provide evidence of the direction of former ice flow, but there are sampling problems and the analysis of the results is not simple, as detailed studies have shown.

Another aspect of structural work developed recently is the study of till fabrics in the field under currently moving ice masses. The work of G. Boulton is particularly noteworthy in this field. His studies have given considerable insight into the question of how fabrics are formed and how they can be related to the characteristics of ice flow and the position of the material within the moving ice. Thus till-fabric analysis can give valuable information on the processes of glacial deposition as well as ancient ice flow.

The study of the characteristics of material, including those of individual particles, their combined size distribution and their sedimentary characteristics, all provide valuable data that can be used to study both the processes whereby they were transported and deposited and the landforms to which they gave rise. The study of material thus provides a link between the study of process and the study of morphology.

Morphology

The study of morphology is fundamental in geomorphology, inasmuch as it is the essential raw material and subject matter of the discipline. In the early period of development of our science many of the geomorphologists were acute observers and some were extremely good landscape artists, among whom Powell, Davis, Gilbert, and later Linton must rank high. As with so many other aspects of the subject, the recording of morphology is now achieved on a less subjective and more quantitative basis, providing no doubt a more accurate and detailed view of land morphology, but regrettably less artistically pleasing.

The modern method of recording the form of the land surface is known as morphological mapping. This technique grew out of the necessity for delimiting those areas of the landscape thought to be significant from the point of view of denudation chronology. The flatter elements of the landscape were considered to be relics of previously partially base-leveled surfaces, thus indicating stages of landscape evolution where the landscape was undergoing intermittent changes of base level. These "inherited" elements could be identified by the presence of a concave break of slope at the upper edge, where the flatter surface impinged against a steeper slope, possibly in some instances an old cliff line. The lower margin of the flatter element was delimited by a convex break of slope, leading to the steeper element where rejuvenation allowed dissection to cut back into the older, higher, and flatter surface. Thus the basis of morphological mapping is the delimitation of units of the land surface with uniform slope characteristics by mapping the breaks and changes of slope between the units by specially designed symbols.

The technique of morphological mapping was developed by Savigear (1952) and Waters (1958) in the area around Sheffield, England. A portion of the paper by Savigear is reprinted as Paper 34. It indicates the nature of the method and symbols used and the advantages and uses of the technique. Ideally the system can be applied objectively, although some types of landscape are easier to map by this method than others. Where breaks of slope are sharp, the method is useful, as in the Sheffield area, but where they are gradual and long concavoconvexities occur, the method is less easy to apply. The method has, however, been found to have application far beyond its original purpose. It has become recognized as an important technique in the delimitation of land units for purposes of land classification, a topic that will be referred to in connection with applied geomorphology and the land unit system.

Other methods of studying morphology have been applied to various specific landforms. The analysis of slope profiles has given rise to a considerable literature, and various complex schemes, such as that of A. Young (1971, 1972), have been devised to plot field-recorded data. The morphology of a number of features has been studied by means of fitting geometrical shapes ot the landforms. This technique has been applied, for example, to the fitting of lemniscate curves and rose curves to drumlin outlines, the fitting of circular arcs to cirque longitudinal profiles, and the fitting of logarithmic spirals to the outlines of beaches in the lee of headlands and spits on the coast. Another method that has been used is the fitting of logarithmic profiles to the longitudinal profiles of a river and parabola to the cross profile of a glaciated valley. Some of these methods have been examined by King (1971). There appears to be a tendency for natural forms to adopt an exponential form, which is also the form of the natural growth law curve.

Quantification

One major change that has taken place in geomorphic study methods has been in quantification. The development of quantification is partly the result of the availability of computers, which enable large masses of data to be rapidly analyzed, employing sophisticated statistical and mathematical methods. There is a danger in this approach to the study of landforms, however, in that in some instances the numerical method seems to become the goal of the study rather than merely the means to a fuller understanding of the genesis of the landforms. This is perhaps inevitable in the early stages of a major change of method, until the value of the various numerical techniques can be properly assessed.

One field of geomorphology in which quantitative work started earliest and has gone furthest is that of drainage basin stream pattern analysis. The work was started by engineer hydrologist R. E. Horton, who in 1945 published a very influential paper; Paper 35 is a short extract. In this paper Horton suggested a method of ordering streams and developed a set of laws that connects the various morphometric characteristics of the drainage basin properties. His law of stream numbers shows that there is a constant relationship on a logarithmic scale between the numbers of streams of different orders, so that they fall on a straight line when plotted on semilogarithmic paper. This and similar laws were first developed by Horton, and have since been developed and applied in many different areas and under different conditions. Horton also investigated the physical bases of his laws in terms of the nature of runoff, infiltration, and similar variables.

The topological relationships of the elements of the drainage pattern have been investigated mathematically by R. L. Shreve. Paper 36 is a further development of the quantitative study of drainage basin characteristics and illustrates the value of a mathematical approach to a geomorphic problem. The author uses a random model to study the physical meaning behind the laws of morphometry originally put forward by Horton. Paper 36 demonstrates the high standard of mathematical competence that is necessary to make the best use of numerical data derived from morphometric measurement.

One of the most influential quantitative geomorphologists in the United States was A. N. Strahler, who was also concerned with morphometry and other aspects of quantitative geomorphology. He slightly modified Horton's method of stream ordering, and developed drainage basin analysis to include the three-dimensional aspect, exemplified by his work on the hypsometric integral.

Paper 37 illustrates yet another of Strahler's uses of quantification in geomorphology, and shows the necessity of placing geomorphic studies on a sound basis of mathematical physics whereby the processes operating can be accurately and effectively studied as dynamic operations. He examines types of stresses and strains with the aim of relating energy, mass, and time into a dynamic system. Studies of this type lead to the formulation of mathematical models, an example of which is given under the section on models. Paper 37 provides a useful introduction to this method of approach, at the same time linking with the section on systems in geomorphology.

Another aspect of quantification that has grown out of morphometric analysis is the application to processes other than fluvial ones, such as karst and glacial features. The development of morphometric analysis has given rise to an increase in the use of multivariate techniques, originally developed by statisticians, frequently for analyses in other subjects such as biology and psychology. The use of factor analysis to reduce a large number of variables to a smaller number of significant factors has been used to some effect, although sometimes the results either appear very obvious or are difficult to interpret. The factors derived from factor analysis can be further used in cluster analysis to provide a means of numerical classification of landforms. This is another method whereby a large amount of data may be reduced to the number of classes required. Such methods of analysis are multivariate and provide a means of assessing relationships in a statistically significant way.

Other methods that have been useful include trend-surface analysis, which can deal with data distributed over an area, and, more recently, time-series analysis, which has provided a means of studying phenomena that vary over time. By this means trends can be identified, and cross spectral analysis can provide a means of correlating two time series. Such studies are particularly useful, for example, in the analysis of processes and responses among variables affecting beaches, on which rhythmic phenomena, such as waves and tides, are involved.

Quantification is, however, only as good as the data that go into the analysis. This often raises serious sampling problems. Sampling is a branch of statistics that has relevance in geomorphic analysis, especially as so many geomorphic problems involve a complex multivariate situation. This applies both to the morphological and process variables, the latter varying greatly both over space and time. Material may also be difficult to sample effectively, owing to the limited availability of suitable outcrops in many instances and to great variability over short distances.

Models

Because of the complexity of geomorphic phenomena, the use of models provides another important approach in modern geomorphology. The term "model" covers a wide range of study methods, some of which will be mentioned briefly in this section and illustrated by examples. The main aim of the model is to simplify reality so that meaningful relationships can be established. Another important attribute of a model is the ability to control the variables so that their effect may be studied one at a time. Reference will be made to some examples of models that achieve these two aims.

The term "model" can also be used to denote a general framework for study. It is this type of model that geomorphology lacks at the present time, a unifying theme that can link the wide range of diverse studies which quite properly come within its orbit. Perhaps the nearest approach to a framework type of model that geomorphology has produced is the cyclic concept of W. M. Davis. This is a good example of a theoretical model, which has the properties of simplicity and the control of variables to a certain extent. Many, however, would consider that the simplicity has been carried too far, so that the application to reality has been lost or rendered too remote to be useful. Another major defect of Davis's cyclic scheme is the problem of time in geomorphology, its implicit concentration on stage of development, which gives a time-dependent constraint. Davis's model was based on the trilogy of structure, process, and stage. Perhaps a new model based on the trilogy of morphology, process, and material might prove valuable. It would provide for the main aim of geomorphology, the understanding of the genesis of landforms through the action of process on material, both solid rock and the loose sediment that forms temporary storage as depositional features.

Another possible framework could be based on the method of study or approach, as shown in Table 1. The four methods of approach are the chronological or historical approach, the dynamic approach, the areal or spatial approach, and the descriptive or empirical approach. In the first group are such studies as denudation chronology, the study of drainage development over time, and the study of glacial chronology, all of which can lead to a regional geomorphic evolution study. The study of process includes both exogenetic and endogenetic process. The areal approach deals with the form and distribution of geomorphic features, including morphometric methods and pattern recognition and study, both in landforms and structures. The descriptive approach is concerned with the recording of the landforms by morphological mapping and with empirical multivariate analyses of the type mentioned in the last section. The interrelationships among the approaches are many and complex, giving, for example, the process-response systems with their complex feedback loops. The use of a broad model such as this provides a means of appreciating the links among the varied studies that are undertaken by geomorphologists.

As examples of different types of models, the following may be briefly mentioned: (1) scale or hardware models, which provide a small-scale replica of reality; (2) theoretical models or deterministic mathematical

Table 1 Methodology of Geomorphology

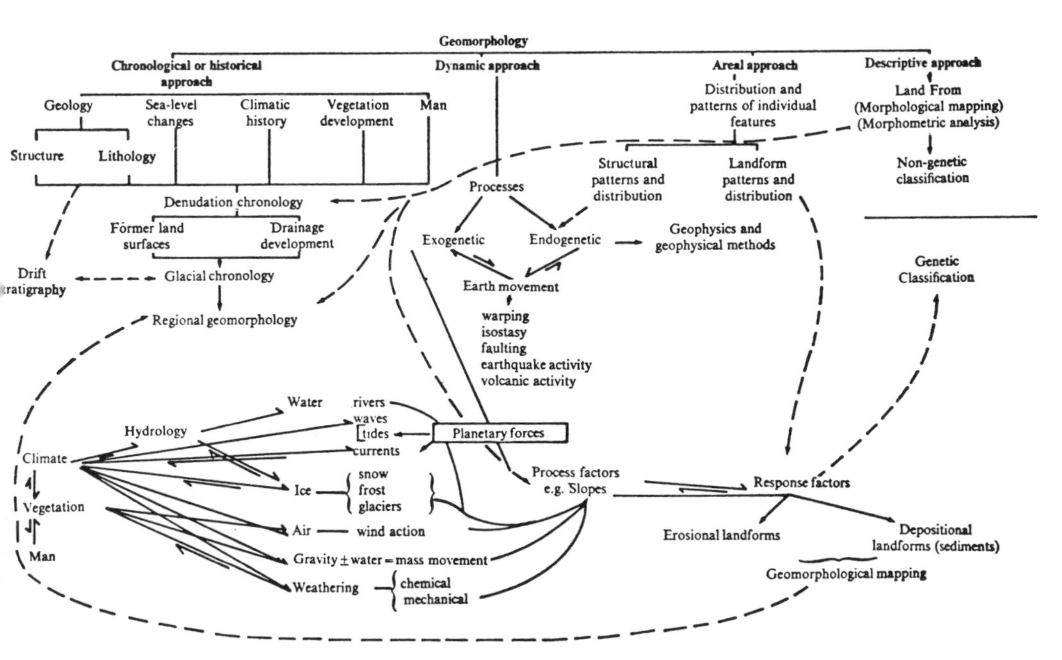

models, which study by iterative methods the stages of development based on certain basic assumptions; (3) dynamic mathematical or physical models, which are usually based on differential equations of dynamic processes; and (4) simulation models, which are often developed for computer analysis in terms of numerical simulation in a simplified setting.

Hardware or Scale Models

Scale models of the hardware type require space and are expensive, but the fact that they are often used by engineers in working out the design of any major project, such as river control or coastal erosion defense structures, indicates the value of such models in a very complex system that cannot be fully analyzed in the prototype situation. Models have also provided much valuable basic research data on the movement of sediment by various processes, such as waves. Models can be either two- or three-dimensional. Those concerned with coastal processes will be cited as examples.

The two-dimensional model studies the processes that influence the development of a beach profile under the action of waves or the movement of sediment normal to the shore. In a model of this type it is possible to control the wave dimensions to clarify the effect of varying wave height, length, and steepness, and the effect of varying material size. Other studies of this type involve the effect of wave run-up on structures and the movement of water within the wave form, all of which can be studied under controlled conditions. Changes in beach profiles can also be related to variations in wave character and to the movement of sediment by the different types of waves. When a third dimension is added, it is possible to study the very important effects of longshore movement of material under wave action, and the growth of the shore forms in plan. The influence of groins on longshore transport can also be studied. The effect of tidal processes can be included, such as the closing of a tidal inlet through the opposing effects of wave and tidal action.

Scale models can also be used to reproduce a particular stretch of coast. Such a model usually has to be designed partially on a trial-and-error basis to establish the optimum vertical exaggeration to give results nearest to those recorded in the prototype. A vertical exaggeration is necessary in a tidal model, but should not be required in a pure wave model, because wave steepness is an important variable.

The model study from which Paper 38 is an excerpt is an analysis of the effect of training works undertaken in Britain on the river Mersey to remedy siltation and improve navigation to the port of Liverpool. Such models can only be successfully built when there is good control of the prototype by field surveys of depths, currents, and changes in morphology. It is necessary to ascertain that the model reproduces the current patterns and morphological changes recorded in the prototype. When this has been achieved, control works and other improvement schemes can be tested in the model and their effects studied. This model produced results that could explain the siltation in the upper estuary, and it showed the importance of correctly simulating the density structure of fresh and salt water. The situation is a complex one, and the conclusions show that unexpected results can occur in places remote from the original works.

Some scale models are very large; the Coastal Engineering Research Center near Washington, D.C., for example, has wave basins that can produce waves up to 2 m in height and of full length, but still with the advantage of control. Models have also provided valuable basic data on the transport of sediment by flowing water, the character of sedimentary structures under different flow conditions, and the pattern of water movement in a river. The transport of material by wind has also been successfully studied in models, including the effect of moisture on the transport of sand by wind. The processes associated with the action of turbidity currents have also been successfully modeled. In the case of flow in glaciers, simulation has proved more difficult, but can be successfully modeled with the use of a centrifuge, as has also been used in some geotectonic studies.

One major problem of model study is that of scale. Scale must be correctly applied to the geometrical as well as the dynamic aspects; inasmuch as different scale factors apply to the two aspects, it is not possible to build a model that is correct from all points of view, according to some model experts. Hibbert's *Theory of Scale Models* shows how all the dynamic quantities of length, mass, and time and their derivations can be scaled in the correct relations. Nevertheless, there are still problems in relation to the scaling of time, because even when it is correctly compressed, the results of the model are not valid. These problems lead to difficulties in correctly applying the results from model studies to the full-scale prototypes.

Theoretical Models

One type of theoretical model is based on repetitive procedures; small steps are taken one at a time to arrive eventually at the longer-term changes that result from the repetition of small steps. The model can be set up on the basis of the known or assumed effects of the operation of specific processes on an assumed initial morphology.

Paper 39 exemplifies this approach to model study in the work of Anthony Young on slope development. This work is similar to that originally carried out by Walther Penck, and subsequently by Frank Ahnert (1970) and others. Young begins with a very simple morphology, a uniform straignt slope of 35° between two horizontal planes. In the simplest situation studied he also assumes that no material is removed from the slope. The steep part of the slope is divided into equal units, and certain assumptions are made concerning the action of creep, rainwash, and removal of material in solution. Each assumption is then carried out step by step on the profile, according to the result of simple but often-repeated calculations, until the modified form of the slope can be clearly seen. The results of this highly simplistic model are of considerable interest and reveal that the form of the slope as it changes through time can be related to the processes operating in terms of the simple assumptions. The forms achieved can also be related to comparable forms in nature.

Young considers various complications to these original assumptions, such as the effects of changing base level, of river incision, and of sediment removal from the base of the slope. Models of this type are necessarily time dependent, in that each step causes a change in the same direction and magnitude, where the conditions are the same, so that there is a continuous development. It is not possible to establish a state of dynamic equilibrium by this type of model technique, but it is a useful way of relating process to form in a simplified, but realistic way, because the results can be related to prototype slopes.

Mathematical Physical Models

The theoretical mathematical models just mentioned were based on changes of form; models that seek to reproduce the actual operation of the processes are mentioned in this section. They are dynamic models that are used to study the flow of water, wind, or ice, and how these media erode, transport, and deposit their load. Much valuable work in this field has been done by physicists, who are familiar with the use of differential equations to describe the operation of the processes. Bagnold (1966), for example, has carried out much fundamental work in this field, first in studying the movement of sand by winds and then the movement of sediment by wave action and unidirectional flow in rivers. Such an approach is essential to a fuller understanding of how processes operate to move material and thus to change form. Such an understanding is a necessary precursor to accurate prediction of the effect of processes of different types and intensities, an ability that would be very valuable if it could be accurately carried out.

In applying models of this type it is necessary to make simplifying assumptions, as with other types of models. The slopes are assumed uniform, the grains in transit are considered to be spheres of uniform size, and similar assumptions are made concerning the effects of friction, bed roughness, and bed form. The equations that are developed by this type of analysis can often be applied to real conditions, when the constants that enter the equations have been determined experimentally or by field observations. Thus, as with all models, there is a problem in applying the results to reality, which is much more complex.

A very good example of this type of model, which has had a great influence on glacial geomorphology, is the analysis of glacier flow made by J. F. Nye, one of whose papers is reproduced as Paper 40. In arriving at his equations he assumes that ice is a perfectly plastic substance, a good example of the necessary simplifying assumptions that must be made. He also assumes that the ice is flowing down a valley of constant slope and that the conditions of temperature, accumulation, and ablation are simple and uniform. A value that is needed to come to numerical results is the constant and exponent of the equation that gives the flow law of ice. This equation relates stress applied to the ice to the strain that results from it. The values for the constant, which depend partly on the temperature of the ice, and the exponent in the equation were worked out by careful laboratory experiments made by Glen (1955), who determined the "flow law" of ice.

By using these constants, Nye was able to calculate the flow of ice in a form that could be checked against observations made in the field. Nye's analysis provides a possible physical process whereby the positive feedback relationship can be explained, by means of which a glacier increases the irregularity of its bed by increasing the erosion in the already concave areas where the flow decelerates. Basin formation is thus made possible under these circumstances, until the bed form fits the shape of the slip planes within the ice.

A great deal of additional information of value can be obtained from this type of analysis; for example, it is possible to calculate which parts of the Antarctic ice sheet are at pressure melting point, provided that the rate of accumulation, surface flow characteristics, and temperature are known. Another interesting study is concerned with the movement of kinematic waves down glaciers, analogous to the flow of highway traffic, subjects that Nye has also treated theoretically in mathematical models. The mechanism of the phenomenon is better known than the physical cause giving rise to it.

Simulation Models

Simulation models have become increasingly common since the advent of the computer, which is able to process data very rapidly, thus producing results much more efficiently than is possible in a hardware model once the program is set up and functioning correctly. The variables can then be adjusted as required one by one, so that the effect of each can be assessed. The model attempts to reproduce a system; it is, therefore, necessary to understand the system, insofar as the variables on which it depends can be isolated and successfully reproduced in the simulation model.

An example of the possibilities that can be explored by means of simulation models concerned especially with drainage patterns is given in Paper 41, by B. Sprunt. The author discusses the nature of simulation models, and then illustrates them by reference to several models of drainage development and slope studies. It is useful to compare this approach with that of Horton and Shreve, who also deal with drainage basin analysis but from rather different points of view. It was shown by Leopold and Langbein, in their analysis of fluvial systems, that a simulated drainage network developed by random numbers in a simulation model had the same characteristics and obeyed the same laws enunciated by Horton. Thus they came to the conclusion that the development of drainage basins has some elements of a random system of high entropy.

Another type of simulation model is that developed to study the formation of a complex recurved coastal spit. On the south coast of England in Hampshire, Hurst Castle spit prolongs the mainland coast where it turns abruptly north. The spit is formed of shingle and is characterized by a set of recurves that become progressively longer and closer together toward the distal end of the spit, which also has a distinctive curvature along its length. Because it is formed of shingle, the main ridge of the spit and the recurves are the work of storm waves, but the material that the storm waves throw up to form the ridges is derived by regular longshore drift from the eroding cliffs to the west, through the agency of westerly waves. The recurves are built by the short, steep waves that come from the northeast. Thus it is fairly easy to isolate the types of waves and their effects on the spit. Four wave types are simulated in the model. The frequency with which they operate can be modified so that the effect of this variable can be tested. A realistic spit cannot, however, be produced unless two other variables are also taken into account: the effect of refraction, which causes the bend in the main spit, and the effect of deepening water off the distal end of the spit. This last variable accounts for the pattern of recurves. By adjusting all the variables in turn, it is possible to achieve a best-fit spit outline. The ones that produce this particular form can then be related to those in reality. It is remarkable how closely the model variables can be fitted to the real ones.

These examples of different types of models have been chosen partly to exemplify their wide range in geomorphic investigation. The models include studies of slope processes and drainage patterns, which together form a very important branch of study. Examples have also been drawn from studies in coastal and estuarine geomorphology and from glacial geomorphology. Models have also been used successfully in problems of periglacial geomorphology, for example in laboratory models of frost heaving under controlled conditions. Hardware models have been used to study the movement of sand by wind and in the study of fluvial processes, such as the recession of knick points and problems associated with meandering and braiding in streams.

Environmental Geomorphology

As world population continues to rise and as natural resources of all types come to be increasingly exploited and some vital ones become scarcer, so man is becoming more and more aware of his natural environment and its finite limitations. In the early period of human development people were very dependent on their natural environment; each small isolated group sought the optimum environment and adapted its way of life to fit that environment, mostly living within the natural ecosystem without doing it serious harm. This relationship has been called *determinism;* the environment determined the nature of man's adaptation to it, and he himself exerted relatively little stress on the environment.

Today we live in a world that we have the power to modify to an increasing extent, partly through a high level of technology and partly through a large increase in population numbers. Despite this technological power, however, the environment cannot be exploited and disturbed beyond a certain point without invoking the law of diminishing returns and other adverse results. It could be said that the present is a time of "neodeterminism," in which it is becoming increasingly vital to live harmoniously with nature within the entire ecosystem, as members of it, rather than to impose human modification and exploitation on the ecosystem as outsiders beyond and above it. It is essential to work with nature and not against nature. To achieve this, however, it is also essential to understand nature and the processes that operate within the natural systems. It is the geomorphologist's goal to achieve this. Geomorphologists can, therefore, play a very important part in helping to make wise use of the bounties of nature in a wide range of fields.

One of the most recent developments in geomorphology has been the great interest shown in environmental geomorphology, which is concerned with the impact of man on the landscape, and, conversely, the impact of the landscape in all its aspects on man. The latter relationship includes the problem of perception. These aspects of environmental geomorphology are covered in three volumes in the Benchmark series under the editorship of D. R. Coates.

Useful contributions to the field of perception have been made, for example, in studies of human reaction to natural hazards, such as floodplain occupancy and reaction to hurricane danger and damage. It is not possible to prevent all natural hazards, although attempts have been made to reduce their impact in some circumstances. Trial experiments have been carried out to reduce the strength of hurricane winds by seeding the developing system to prevent its further growth. Floods can be reduced or eliminated by suitable engineering works, such as dams or retaining dikes, levees, and diversions. One of the best methods of dealing with flood hazards and comparable natural events, however, is to avoid settlement and development in those areas that are most subject to their action, such as floodplains and vulnerable coastal areas, such as barrier islands and other low coasts.

A field in which man and natural processes have interacted for a very long time is in the exploitation of land for farming. In Paper 42, I. Douglas shows the importance of human interference in the natural ecosystem in areas naturally under tropical rain forest in Australia and Malaysia. Detailed field observations have shown that the use of the land for farming and the consequent destruction of the natural forest have greatly increased the amount of sediment in transport in streams. The author concludes that the excess sediment is derived from the slopes under cultivation or other types of human land use. The effects of urban land use on sediment yield have also provided quantitative data on the impact of human interference in natural systems, which affects both the character of the stream discharge curves and the sediment yield, usually in an adverse way.

Paper 43, the work of W. B. Langbein and S. A. Schumm, shows that the relationship between sediment yield and environmental conditions is complex. There is a natural variation in the sediment yield as well as the variation resulting from human interference. The natural variation can be related to irregularities in precipitation. The main control by which the processes respond to the variation in precipitation is through vegetation. In these contributions the importance of the environment in the whole complex ecosystem context is clearly demonstrated.

Another field of environmental geomorphology is the study of the impact on geomorphic processes of man-made works, such as large dams, river diversion, and similar measures. The work of Makkaveyev (1972) on the impact of large-scale engineering works on the biotic and geomorphic processes in the USSR exemplifies this aspect of environmental geomorphology. It also indicates the utilitarian approach of many Soviet and East European geomorphologists to the subject. It is to be expected that such large engineering activities will have a considerable impact on natural processes, particularly in the more vulnerable environments, such as the periglacial environment that covers half the Soviet Union and much of Canada and Alaska. The environmental conditions of such permafrost areas impose strict controls on development from the human point of view. In Paper 44 R. J. E. Brown illustrates some of the problems of this environment, which is increasing in importance as the northern lands are increasingly developed. The anxiety created by the building of oil pipelines in Alaska and northern Canada is another example of the problems associated with interference in a delicate periglacial environment.

Many geomorphological studies now have much more than a purely academic interest, because it is essential to know how the environment operates before it is interferred with in any way. The dangers become increasingly great as the projects become more ambitious and extend over a wider area, particularly when they impinge on the more vulnerable geomorphic zones, such as the permafrost areas.

Applied Geomorphology

Applied geomorphology is closely associated with environmental geomorphology. One main purpose of workers in this field is to provide geomorphic information in a form that is of maximum value to planners and others who need to take geomorphic information into consideration for projects such as road building, coastal defense and reclamation, resource exploitation, and other purposes. To make optimum use of the land and its resources, it is necessary to have detailed information concerning its character and capabilities.

One of the methods of applied geomorphology is, therefore, the assessment of land by a system of land-unit classification. Paper 45, by J. A. Mabbutt, provides a review of the concepts of land-unit classification. This is a method that developed out of the scheme for morphological mapping, which has already been discussed. Land-use classification was originally designed to assess land capabilities for military purposes, although the method has much wider applications. In Paper 46, R. L. Wright illustrates some of the values of this method of approach. It is one that can be carried out by means of careful interpretation of aerial photographs, based on field checks of specific soil qualities and slope characteristics. The land is divided into units of constant character on the criteria selected as being the most important for the purpose of the study. These criteria can be modified according to the requirements of the proposed land use.

The system of geomorphic mapping, which seeks to record as objectively as possible in a geomorphically meaningful way the character of the land, has been developed along one line to produce the land-unit system of land classification. A central theme of applied geomorphology is the production of a geomorphic map. An example of this approach, which has been most thoroughly developed and used by East European and Soviet Geomorphologists, is a paper by A. G. Isachenko (1973). To produce a good geomorphic map, it is necessary to have studied the area in depth and to have analyzed the landforms on a genetic basis. A geomorphic map, therefore, is the end result of research into the geomorphology of an area. The landscape must have been interpreted in terms of the processes that have operated to shape both its erosional and depositional landforms. The material of both types of forms must have considered and indicated on the map or on an accompanying geological map.

The scale of the geomorphic map determines the type of detail that can be shown, and this must be adapted to the problems that require solution. In some instances a broad picture of a large area is needed; in others the details of a small construction site may require a very large scale map, with very detailed information on landforms and materials. On many geomorphic maps the erosional landforms are differentiated from the depositional ones by means of color or appropriate symbols; other maps use color to denote the type of process operating.

Applied geomorphology is relevant to the field of coastal processes, in which the geomorphologist, with his appreciation of the interdependence of a whole system covering a considerable stretch of coast, can usefully pool his knowledge with that of an engineer, who looks at the problem from a different point of view. Both points of view are necessary, but above all a knowledge of the operation of the processes on the material to create the morphology is essential; this is the field of the geomorphologist. His contribution can be a very valuable one.

BIBLIOGRAPHY

Ahnert, F. 1970. Functional relationships between denudation, relief, and uplift in large mid-latitude drainage basins. *Amer. Jour. Sci. 268,* 243-263.

Allen, J. 1947. *Scale models in hydraulic engineering.* London.

Allen, J. R. L. 1970. *Physical processes of sedimentation.* Allen and Unwin, London.

Allen, P., and Krumbein, W. C. 1962. Secondary trend components on the Top Ashdown Pebble bed: a case history. *Jour. Geol. 70,* 507-538.

Bagnold, R. A. 1966. An approach to the sedimentary transport problem from general physics. *U.S. Geol. Survey Prof. Paper 422I,* 37 pp.

Boulton, G. S. 1972. The role of thermal regime in glacial sedimentation. *Inst. Brit. Geog. Spec. Publ. 4,* 1-19.

Bretz, J. H. 1923. Glacial drainage on the Columbia Plateau. *Bull. Geol. Soc. America 34,* 573-608.

Cailleux, A. 1945. Distinction des galets marins et fluviatiles. *Bull. Geol. Soc. France 5*(15), 375-404.

Chorley, R. J. (ed.). 1972. *Spatial analysis in geomorphology.* Methuen, London.

——, and Kennedy, A. B. 1971. *A systems approach to physical geography.* Prentice-Hall, Englewood Cliffs, N. J.

Coates, D. R. (ed.). 1972. *Environmental geomorphology and landscape conservation.* Dowden, Hutchinson & Ross, Stroudsburg, Pa.

Corbel, J. 1964. L'érosion terrestre, études quantitatives (methodes–techniques–resultes). *Ann. Geog. 73,* 385-412.

Cvijić, J. 1895. *Karst.* Geografska Monografija. Drž. štamp. kralj. Srbije. Belgrade, 176 pp.

——. 1924. *Geomorfologija* 1. Drž. štamp. kralj. Srba. Hrvata i Slovenaca. Belgrade, 588 pp.

Fairbridge, R. W. 1961. Eustatic changes in sea level, in *Physics and Chemistry of the Earth,* Vol. 4. Pergamon, Elmsford, N.Y., 99-185.

Friedman, D. G. 1970. Computer simulation of the earthquake hazard, in *Geological Hazards and Public Problems,* R. A. Olson and M. M. Wallace (eds.). U.S. Off. Emerg. Prep. Reg. 7 Proc. Conf. San Francisco, 1969. Washington, D.C., 153-181.

Glen, J. W. 1955. The creep of polycrystalline ice. *Proc. Roy. Soc. A 228,* 519-538.

Isachenko, A. G. 1973. On the method of applied landscape research. *Soviet Geog. Rev. Transl. 14*(4), 229-243.

King, C. A. M. 1971. Geometrical forms in geomorphology. *Intern. Jour. Math. Educ. Sci. Technol. 2,* 153-169.

——, and McCullagh, M. J. 1971. A simulation model of a complex recurved spit. *Jour. Geol. 79*(1), 22-36.

Krumbein, W. C. 1966. Classification of map surfaces on the structure of polynomial and Fourier coefficient matrices. *Kansas Comp. Contrib. 7,* 12-18.

Makkaveyev, N. 1972. The impact of large engineering projects on geomorphic processes in stream valleys. *Soviet Geog. Rev. Transl. 13*(6), 387-393.

Mather, P. M., and Doornkamp, J. C. 1970. Multivariate analysis in geography with particular reference to drainage basin morphometry. *Trans. Inst. Brit. Geog. 51,* 163-187.

Melton, M. A. 1958. Correlation structure of morphometric properties of drainage systems and their controlling agents. *Jour. Geol. 66,* 442-460.

Savigear, R. A. G. 1952. Some observations on slope development in South Wales. *Trans. Inst. Brit. Geog. 18,* 31-51.

Sidorenko, A. V. 1972. Geomorphology and the national economy. *Soviet Geog. Rev. Transl. 13*(6), 344-353.

Solntsev, N. A. 1973. On biotic and geomorphic factors in the formation of the natural environment. *Soviet Geog. Rev. Transl. 14*(6), 347-355.

Sparks, B. W. 1960. *Geomorphology.* Longman, London.

——. 1971. *Rocks and relief.* Longman, London.

Stoddart, D. R. 1969. Climatic geomorphology: review and re-assessment. *Progress in Geography,* Vol. I. Arnold, London, 160-222.

Strahler, A. N. 1952. Hypsometric (area-altitude) analysis of erosional topography. *Bull. Geol. Soc. America 63*(11), 1117-1142.

——. 1954. Statistical analysis in geomorphic research. *Jour. Geol. 62,* 1-25.

Tricart, J. 1962. The discontinuity of erosional phenomena. *A.I.H.S. Comm. d'Erosion Cont. 59,* 233-243.

Verstappen, H. T. 1970. Introduction to the ITC system of geomorphological survey. *Geog. Tijdschrift 4*(1), 85-91.

Waters, R. S. 1958. Morphological mapping. *Geography 43,* 10-17.

Young, A. 1971. Slope profile analysis: the system of best units. *Inst. Brit. Geog. Spec. Publ. 3,* 1-13.

——. 1972. *Slopes.* Oliver & Boyd, Edinburgh.

12

Reprinted from *Jour. Geol.*, **68**(1), 54–56, 70–74 (1960)

MAGNITUDE AND FREQUENCY OF FORCES IN GEOMORPHIC PROCESSES[1]

M. GORDON WOLMAN AND JOHN P. MILLER
Johns Hopkins University and Harvard University

ABSTRACT

The relative importance in geomorphic processes of extreme or catastrophic events and more frequent events of smaller magnitude can be measured in terms of (1) the relative amounts of "work" done on the landscape and (2) in terms of the formation of specific features of the landscape.

For many processes, above the level of competence, the rate of movement of material can be expressed as a power function of some stress, as for example, shear stress. Because the frequency distributions of the magnitudes of many natural events, such as floods, rainfall, and wind speeds, approximate log-normal distributions, the product of frequency and rate, a measure of the work performed by events having different frequencies and magnitudes will attain a maximum. The frequency at which this maximum occurs provides a measure of the level at which the largest portion of the total work is accomplished. Analysis of records of sediment transported by rivers indicates that the largest portion of the total load is carried by flows which occur on the average once or twice each year. As the variability of the flow increases and hence as the size of the drainage basin decreases, a larger percentage of the total load is carried by less frequent flows. In many basins 90 per cent of the sediment is removed by storm discharges which recur at least once every five years.

Transport of sand and dust by wind in general follows the same laws. The extreme velocities associated with infrequent events are compensated for by their rarity, and it is found that the greatest bulk of sediment is transported by more moderate events.

Many rivers are competent to erode both bed and banks during moderate flows. Observations of natural channels suggest that the channel shape as well as the dimensions of meandering rivers appear to be associated with flows at or near the bankfull stage. The fact that the bankfull stage recurs on the average once every year or two years indicates that these features of many alluvial rivers are controlled by these more frequent flows rather than by the rarer events of catastrophic magnitude. Because the equilibrium form of wind-blown dunes and of wave-formed beaches is quite unstable, the frequency of the events responsible for their form is less clearly definable. However, dune form and orientation are determined by both wind velocity and frequency. Similarly, a hypothetical example suggests that beach slope oscillates about a mean value related in part to wave characteristics generated by winds of moderate speed.

Where stresses generated by frequent events are incompetent to transport available materials, less frequent ones of greater magnitude are obviously required. Closer observation of many geomorphic processes is required before the relative importance of different processes and of events of differing magnitude and frequency in the formation of given features of the landscape can be adequately evaluated.

INTRODUCTION

Denudation of the earth's surface and modification of existing land forms involve forces which are ultimately controlled by highly variable atmospheric influences coupled with the unvarying effects of gravity. Almost any specific mechanism requires that a certain threshold value of force be exceeded. However, above this threshold or critical limit there occurs a wide range in magnitude of forces which results from variations in intensity of precipitation, wind speed, etc. The problem to be examined in this paper is the relative importance of extremes or catastrophic events and more ordinary events with regard to their geomorphic effectiveness expressed in terms of material moved and modification of surface form. Thus this is a re-examination of the concept of "effective force" in landscape development.

It is widely believed that the infrequent events of immense magnitude are most effective in the progressive denudation of the earth's surface. Although this belief might seem to be supported by observations of some individual events, such as large floods, tsunamis, and dust storms, the catastrophic event is not necessarily the critical factor responsible for the development of land forms. Available evidence indicates that evaluation of the effectiveness of a specific mechanism and of the relative importance of different geomorphic processes in mold-

[1] Manuscript received May 4, 1959.

ing specific forms involves the frequency of occurrence as well as the magnitude of individual events.

Evidence related to the influence of frequent events of small magnitude is far less spectacular than the exciting descriptions of the Johnstown flood or the Galveston disaster. It may also be true that in many instances the importance of the latter actually is directly proportional to their grandeur. The purpose of this paper is not to play down any valid significance of the awesome catastrophes but to demonstrate by means of several examples that a more accurate picture of the over-all effectiveness of various geomorphic processes should include not only the rare extreme events but also events of moderate intensity which recur much more frequently.

The relative amount of "work" done during different events is not necessarily synonymous with the relative importance of these events in forming a landscape or a particular feature of the landscape. The effectiveness of an event of a given frequency in terms of its performance of work is measurable both by its magnitude and by the frequency with which it recurs. Thus the relative amounts of work performed by events such as floods of different magnitude and frequency are measurable in part, at least, by comparisons of the relative quantities of sediment transported. On the other hand, although related to the form of the landscape, the ranking of events in terms of the relative amounts of work performed is not necessarily directly correlated with their relative importance in the determination of river pattern, drainage density, slope form, or other aspects of the landscape. This paper deals first with the significance of frequency and magnitude in terms of "work done" and second in terms of the formation of specific features of the landscape.

Any discussion of the frequency of events of geomorphic significance clearly raises some concern about the length of the available record. On the geologic time scale any record of water and sediment discharge is infinitesimally short. On the other hand, where something is known about mechanical aspects of the process, a record of twenty-five to fifty years, considerably longer than most river records, may be sufficient to provide an adequate sample of a river's regimen of flow for certain kinds of analyses. The significance of the likely omission of some extremely high as well as extremely low values will vary with the measure of effective force used. Thus for the case of effective force measured in terms of competence, a "rare" event not experienced in historic time may have recurred a significant number of times in the geologic record. However, because of their relative rarity, such events are of less significance in analyses concerned with percentages of material moved by events of varying frequency and magnitude.

EROSION AND SEDIMENT TRANSPORT

GENERAL CASE

The movement of sediment by water or air is essentially dependent upon shear stress and, according to Malina (1941), Bagnold (1941), Brown (1954), etc., can be described by the equation

$$q = k\,(\tau - \tau_c)^n\,, \qquad (1)$$

where q is the rate of transport, k is a constant related to the characteristics of the material transported, τ is the shear stress per unit area, and τ_c is a critical or threshold shear stress required to move the material. In its simplest form, equation (1) is essentially a power function

$$q = x^n\,, \qquad (2)$$

where q is the rate of movement, and x is a variable, some responsible stress such as shear, etc., which exceeds the required threshold value. This relation is shown diagrammatically in figure 1, *a*.

The distribution in time of many hydrologic and meteorologic events, such as wind speeds or flood peaks, has been shown to approximate a log-normal distribution (see Chow, 1954, and Krumbein, 1955, for numerous examples). These events may be visualized as cumulative applied stresses acting upon particular segments of the landscape. If the stress is log-normally distributed and continuous (fig. 1, *b*) and if the quantity or rate of movement is related to some power of this stress, then the relation between stress and the product of frequency times rate of movement must attain a maximum. The recurrence interval or frequency at which this maximum occurs is controlled by the relative rates of change of q with the

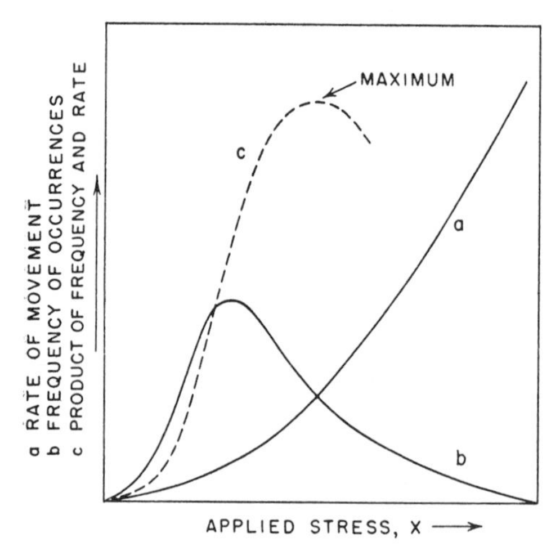

FIG. 1.—Relations between rate of transport, applied stress, and frequency of stress application.

stress x, and of x with time. This maximum, which can be derived mathematically, is shown diagrammatically in figure 1, *c*.

It should be repeated that this generalization holds only if the applied stress exceeds a threshold value. Below this value no work is done in moving material. This means, of course, that both the equation of transport and the frequency distribution must apply to all values of the applied stress above this threshold value. These conditions are met in several of the simple examples that follow. In these, the recurrence interval at which the product of frequency and rate of movement is a maximum (fig. 1, *c*) and the cumulative percentage of the total work performed by successively larger events provide an indirect measure of the relative amounts of work performed by events of different magnitude and frequency.

[*Editor's Note:* Material has been omitted at this point.]

SIGNIFICANCE OF CATASTROPHIC EVENTS

In the preceding discussion it has been argued that forces of moderate magnitude and frequency have greater net effect on land-form development than do intense, short-lived forces associated with catastrophic events. Clearly, such a general conclusion requires qualification to the extent that catastrophic events produce results that are (1) unique in some respect because of magnitude or (2) different in kind from effects of more ordinary occurrences. Several illustrative examples are discussed below.

Landslides and formation of new gullies are common occurrences during exceptional storms. Once formed, a gully continues to grow during more moderate storms, and thus the extreme event may have an enduring effect on drainage pattern and topography.

Changes in dimension and position of stream channels commonly occur during large floods. Many such cases were reported following the record Kansas-Missouri floods of July, 1951. Woolley (1946) mentions several cases where channels were downcut several tens of feet and widened a few hundred feet during cloudburst floods in Utah. These trenches, which are too large for the ordinary flows, persist for long periods after the extreme flood. Channel characteristics of many arroyos are clearly related to flows of different magnitude. Deposition occurs during flows smaller in volume than the losses by percolation into the channel; floods large enough to carry the full length of the arroyo cause scour of the channel bed and banks. Extensive migration of bends and meander cutoffs may also be associated with floods. Jahns (1947) cites a case in the Connecticut Valley where destruction of vegetation by exceptional floods in 1936 and 1938 resulted in accelerated bank erosion during later low-water stages. Channel islands of the Connecticut River were destroyed during the 1936 and 1938 floods, but new islands were deposited in approximately the same places as the floodwaters receded. Aggrading streams like the Rio Grande often undergo spectacular channel changes, called "avulsions." Deposition at stages below bankfull continues until the stream is literally flowing on a ridge. Then, during a flood the stream breaks out of its channel and is re-established in a lower part of its valley floodplain, where the building of a new channel ridge commences all over again.

According to Woolley (1946), channels of larger streams in Utah are sometimes dammed temporarily by coarse debris from tributaries. Rio Puerco and Rio Salado occasionally build fans at their mouths which divert the Rio Grande to the opposite side of its valley. Depending on the flow of the main stem, removal of these features may take years. Breaks in the profile of the main stem below tributaries that contribute flood debris apparently reflect downstream progression of sediment waves. Phenomena of this kind are not restricted to arid regions. Jahns (1947) mentions debris fans in the Connecticut River drainage, and they have been observed in other places.

Scouring of floodplains during exceptional floods is often localized. However, some of the features produced in this way are impressive because of their topographic relief. Jahns (1947), for example, describes scour channels and swirl pits 15–20 feet deep formed during the Connecticut River floods of 1936 and 1938. Like floodplain scour, overbank deposition is sporadic and localized. However, a floodwater bar composed of coarse materials, and several feet thick, may be a major topographic feature of the floodplain. A special kind of case requiring extreme flood conditions is deposition on high terrace surfaces. Such deposition has been reported in several Eastern stream valleys, among them the Connecticut (Jahns, 1947; Wolman and Eiler, 1958) and the Susquehanna.

Extreme floods accomplish transport of material that is impossible by more ordinary flows. For example, Woolley (1946) reports that huge boulders weighing more than 100 tons have been moved long distances on gentle slopes by mudflows in Utah. Countless comparable examples, from southern California, Arizona, and many other places, could be cited. With regard to the sizes of materials transported, the effects of floods appear to be directly proportional to their magnitude. This implies that alluvial fans owe many of their properties to extreme rather than moderate flows, although the frequency relations of stream flows and mudflows have not been adequately defined.

The previous discussion of equilibrium beaches emphasized seasonal variations. In many places during the summer, beaches build seaward, are composed of finer material, and have flatter slopes than during winter when they are eroded by higher waves. Where marine cliffs are fronted by broad beaches, erosion of the cliff may occur only during periods of extreme wave action. Barrier islands, which are more or less stable under ordinary conditions, may be eroded below mean tide level during extreme storms. Similarly, along shores of low coastal plains, building of beach ridges several feet above mean low water occurs only during infrequent events of great magnitude.

Because of their relative familiarity, an exhaustive enumeration of examples of the effect of infrequent catastrophic events on the landscape has not been given here. Placed in the context of the earlier discussion, these examples principally illustrate the known thesis that the rare or infrequent events become increasingly important as the threshold stress (competence) required to move the available masses of material increases.

[*Editor's Note:* Material has been omitted at this point.]

REFERENCES CITED

Bagnold, R. A., 1941, The physics of blown sand and desert dunes: London, Methuen and Co.

Brown, C. B., 1954, Sediment transportation, chap. xii; *in:* Rouse, H. (ed.), Engineering hydraulics, p. 769–857: New York, John Wiley and Sons.

Chow, Ven Te, 1954, The log-probability law and its engineering applications: Proc. Am. Soc. Civ. Eng., v. 80, Sept. No. 536.

Jahns, R. H., 1947, Geologic features of the Connecticut Valley, Mass. as related to recent floods: U.S. Geol. Survey Water Supply Paper 996.

Krumbein, W. C., 1955, Experimental design in the earth sciences: Am. Geophys. Union Trans., v.

Malina, F. J., 1941, Recent developments in the dynamics of wind-erosion: Am. Geophys. Union Trans., pt. 2, p. 262–287.

Wolman, M. G., and Eiler, J. P., 1958, Reconnaissance study of erosion and deposition produced by the flood of August, 1955, in Connecticut: Am. Geophys. Union Trans., v. 39, p. 1–14.

Woolley, R. R., 1946, Cloudburst floods in Utah, 1850–1938: U.S. Geol. Survey Water Supply Paper 994.

13

Reprinted from *Nature*, **196**(4854), 512–515 (1962)

CATASTROPHIC STORM EFFECTS ON THE BRITISH HONDURAS REEFS AND CAYS

By D. R. STODDART

Department of Geography, University of Cambridge

INTEREST in the effects of catastrophic storms on reefs and reef islands has been stimulated by investigations at Jaluit Atoll, Marshall Islands, following typhoon *Ophelia*[1,2], and at Ulithi Atoll, Caroline Islands, following another storm of the same name. Hurricane *Hattie*, which passed across the British Honduras reefs on October 30–31, 1961, presented an unusual opportunity to add to these Pacific investigations, since, during two expeditions in 1959–60[3] and 1961[4] I had completed field mapping of most of the sand cays of this coast, photographed all the cays from the air, and also observed living reefs, both on the three outer atolls (Turneffe, Lighthouse and Glover's Reefs) and the 150-mile long barrier reef. Accordingly, the hurricane-damaged areas were re-surveyed during February–June 1962. This article presents some of the main conclusions from the re-survey; the detailed reports on individual cays will be presented elsewhere.

Hurricane *Hattie*[5] was first identified as a tropical storm near San Andres Island, Nicaraguan coast, on October 27, 1961: it moved steadily northwards and intensified until the early hours of October 30, when it turned westwards in the neighbourhood of Grand Cayman Island and made for British Honduras. Heavy swells began to break on the outer reefs of British Honduras during October 29–30, and the first high winds reached Lighthouse Reef about 1600 h (Belize time) on October 30. The storm centre crossed Sandbore Cay, at the north end of Lighthouse Reef, at about midnight, October 30–31, when the minimum pressure recorded was 27·4 in. At this time the cay was covered to a depth of 7–8 ft. by a surge associated with the storm. The M.V. *Tactician*, 45 miles to the south-west in the barrier reef lagoon, also recorded 27·4 in. as the eye passed over at 0500 h, October 31. A minimum pressure of 28·5 in. was recorded at Belize, 20 miles north-west of the storm track, at 0600 h. The storm surge reached Belize at about this time, and by 0800 h sea-level at Belize had risen 10–15 ft. above normal, subsiding by 1030

h. No wind-speed records are available at Stanley Field, Belize, after 0500 h, when the anemometer ceased recording after gusts up to 115 m.p.h., but sustained speeds of 150 m.p.h. within 15 miles of the storm centre, with gusts up to 200 m.p.h., are considered likely. Hurricane *Hattie* was thus one of the most severe Atlantic hurricanes of 1961, and the most severe in the British Honduras area since Hurricane *Janet* in 1955. Intense wind and wind-driven wave conditions combined with the high storm surge to the immedate north of the storm track to inflict severe damage to reefs and reef islands over a limited segment of the British Honduras coast.

Effect on Reefs. Damage to living coral reefs was variable and in places catastrophic. The windward reefs of Lighthouse Reef, immediately south of the storm track, suffered little damage, and the groove-buttress system remained intact. Damage to the windward reef of Turneffe was greatest near the north end, where fresh reef debris now forms a submerged shingle carpet on the reef crest, but the groove-buttress system is still largely intact. On the barrier reef, however, between English Cay and Rendezvous Cay, over a zone extending for 5 miles north of the storm centre, 80 per cent of the reef corals have disappeared. Patch-reefs which previously were coated with surface-breaking brown and orange *Acropora* now rise, greenish white, devoid of corals, to near sea-level. Along the reef-front all trace of the groove-buttress system has disappeared, and cannot be traced until one reaches Cay Glory, 10 miles south of the storm centre, where reef remnants indicate a pre-existing lineation. Reef species most affected were *Acropora cervicornis*, not seen living along the whole of the central barrier reef north of Curlew Cay; *Acropora palmata; Porites porites;* and numerous more fragile species, including *Porites furcata, Cladocora arbuscula, Manicina aerolata, Siderastrea radians, Eusmilia fastigiata, Favia fragum, Isophyllastrea rigida, Mycetophyllia lamarckana*, and even *Agaricia agaricites*. Consistently, the only species to survive in any quantity was *Montastrea annularis*, in massive hemispherical colonies. Other corals of similar habit, such as *Diploria labyrinthiformis, D. strigosa, D. clivosa, Siderastrea siderea, Solenastrea bournoni* and *Porites astreoides*, also survived in lesser numbers. The survival of the more massive, slower-growing species and destruction of fragile rapidly growing ones has also been observed at Low Isles, Great Barrier Reef of Australia, by Stephenson and others following the minor 1950 cyclone[6]. Many corals have been inverted or rolled

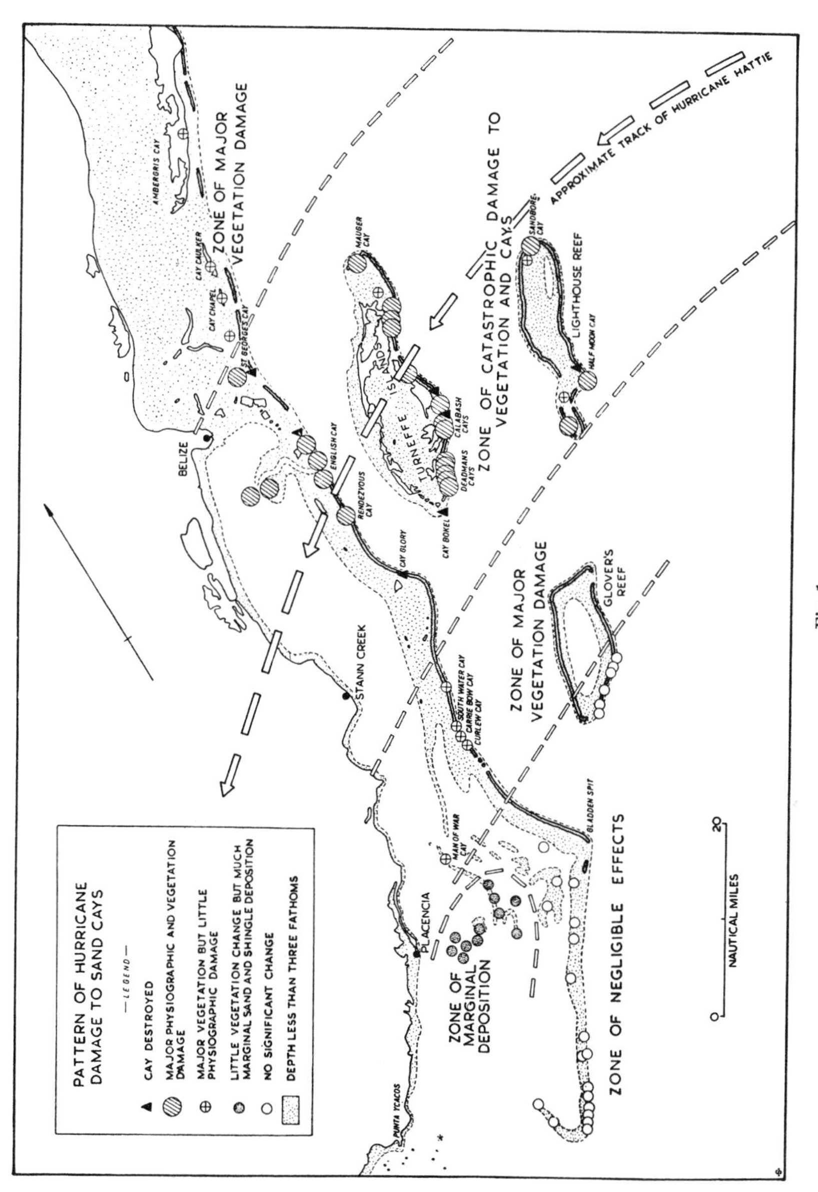

PATTERN OF HURRICANE DAMAGE TO SAND CAYS
— LEGEND —
CAY DESTROYED
MAJOR PHYSIOGRAPHIC AND VEGETATION DAMAGE
MAJOR VEGETATION BUT LITTLE PHYSIOGRAPHIC DAMAGE
LITTLE VEGETATION CHANGE BUT MUCH MARGINAL SAND AND SHINGLE DEPOSITION
NO SIGNIFICANT CHANGE
DEPTH LESS THAN THREE FATHOMS
ZONE OF MAJOR VEGETATION DAMAGE
ZONE OF CATASTROPHIC DAMAGE TO VEGETATION AND CAYS
APPROXIMATE TRACK OF HURRICANE HATTIE
ZONE OF MAJOR VEGETATION DAMAGE
ZONE OF MARGINAL DEPOSITION
ZONE OF NEGLIGIBLE EFFECTS
AMBERGRIS CAY
CAY CAULKER
CAY CHAPEL
ST GEORGES CAY
BELIZE
ENGLISH CAY
RENDEZVOUS CAY
CAY GLORY
CAY BOKEL
TURNEFFE ISLANDS
DEADMANS CAYS
CALABASH CAYS
MAUGER CAY
SANDBORE CAY
LIGHTHOUSE REEF
HALF MOON CAY
GLOVER'S REEF
STANN CREEK
SOUTH WATER CAY
CARRIE BOW CAY
CURLEW CAY
MAN OF WAR CAY
BLADDEN SPIT
PLACENCIA
PUNTA YCACOS
0
20
NAUTICAL MILES

Fig. 1

across the reef-flat, but remain alive. Most of the reef constituents were washed into deeper water round the reef and are no longer visible, though in many places submerged areas of *A. cervicornis* debris indicates disintegration of this species *in situ*. Very little material larger than shingle size has been thrown above high-water-level, except on the east side of Turneffe, where mangrove hinders transportation into deeper water across the reef flat. South of Cay Glory, damage to the reef decreases rapidly; and no change was seen at Glover's Reef or along the southern barrier reef.

Effects on Reef Islands. The degree of damage to any particular cay depends: (*a*) on its location relative to the storm track; (*b*) on the type of cay, whether sand or mangrove; (*c*) on the size of the island; (*d*) on the nature of its vegetation and the extent of human interference. Four zones of cay damage are distinguished in Fig. 1, and will be discussed in turn.

Damage is greatest in a zone extending 15–20 miles north and south of the storm centre, especially on small cays where the natural vegetation has been replaced by coconuts with little undergrowth. In such cases, the cay has either disappeared or suffered severe marginal erosion; surface sand has been stripped to a depth of 1–3 ft., exposing coconut roots; fresh gravel has in some cases been patchily deposited on the cay surface; scour holes and channels have been cut on the cay surface and on the lee shore; and vegetation has been wholly or partly removed. Very large cays in comparable situations suffer only limited marginal erosion; surface erosion and deposition is restricted to a relatively narrow peripheral zone; and while coconut and other trees are thrown down in large numbers, much of the original ground vegetation survives. It is very noticeable that where cays retain much original vegetation, such as thickets of *Cordia sebestena* and *Bursera simaruba*, the effects are different: thus, along the south shores of Half Moon Cay, Lighthouse Reef, the section cleared for coconuts, formerly 7–10 ft. in height, suffered 3–5-ft. vertical erosion, whereas along the section covered with *Cordia* bush, much sediment was heaped against the vegetation barrier, with resultant increases in height of 3–5 ft. Similarly, scour channels were only cut across cay surfaces where the original dense vegetation had been cleared for coconuts or houses, as at St. George's Cay, Mauger Cay and Sandbore Cay.

Cays 20–40 miles from the storm track suffered less severe damage; they were unaffected by the storm surge and were subject to less extreme wave conditions. Physiographic changes were limited to minor shore retreat, exposing fresh beach-rock (South Water Cay, Carrie Bow Cay) or cay sandstone (Cay Chapel); the disappearance or modification of small spits; minor near-shore sand-stripping and root-exposure; and deposition of narrow carpets of sand and shingle inland from this erosion zone. Dominant storm effects were vegetational: on cays planted to coconuts 50–80 per cent of the trees were knocked down, often leaving small hollows filled with ground-water. Damage to other trees was generally restricted to defoliation and snapping off of branches, though in exposed areas large specimens of *Coccoloba uvifera* were seen overturned. Damage was greatest where natural vegetation had suffered most human interference, especially where this allowed unrestricted passage to waves across narrow portions of cays, as at Cay Caulker. Here, scour holes, channels and sand-stripping were well developed. Again, the larger the island the less was the damage sustained inland.

A third zone includes the cays between Placencia and Gladden Spit, barrier reef lagoon, where the main hurricane winds blew from the south across water 15–25 fathoms deep. In most of these cays the vegetational effect were insignificant, but all had sand or shingle deposited on their south and east shores. These deposits take the form of narrow carpets 1–5 ft. thick overlying the old cay surface, broad shingle flats adjoining the old shoreline, and offshore shingle ridges. Some of the offshore deposits are already being destroyed by wave action, the finer materials being flushed out leaving a coarse lag gravel. This zone is not duplicated to the north of the hurricane track, and its existence seems to depend on local conditions.

Finally, the cays of Glover's Reef and of the southern barrier reef and lagoon suffered no vegetational and little or no physiographic change, apart from insignificant shoreline readjustments which may or may not result from the hurricane itself. The Glover's Reef cays lie only 40 miles south of the storm track; while at Punta Gorda sustained gale force winds, perhaps reaching hurricane force, were insufficient to defoliate mangrove or lead to physiographic changes to the cays off Punta Ycacos. This zone is not represented in British Honduras to the north of the storm track, partly because of the absence of cays in that area, partly because of the greater wind velocities and storm surge, and hence greater extent of damage in the northern sector of the storm.

Beach-rock. Old relict beach-rock was remarkably successful in resisting damage, even where cays were completely destroyed, as at Cay Glory. Intertidal beach-rock within 20 miles of the storm centre was swept bare of larger algæ by wave action. Shore re-adjustments on many cays led to the exposure of new beach-rock, including low 'promenades' (Deadman's Cays) with pitted upper surfaces[7], similar to features previously interpreted as due to elevation of reef-rock. Much of this new beach-rock is poorly cemented; but in some cases areas where incipient cementation was previously noted now exhibit one or two lines of hard well-indurated beach-rock. Exposure of fresh areas of lithified sands above present sea-level at Harry Jones, Grand Bogue Point and Big Cay Bokel, Turneffe, and Cay Chapel, barrier reef, indicate that the rock is a cay sandstone rather than an elevated intertidal beach-rock.

Vegetation change. The most conspicuous vegetation changes were: (*a*) defoliation of mangrove within 40 miles of the storm centre; (*b*) the widespread destruction of coconut palms. Mangrove defoliation gradually decreased away from the storm centre, until at 20 miles distance on the south side, mangrove in the centre of large cays escaped defoliation. In early 1962, six months after the storm, some of the *Rhizophora* was again beginning to bear leaves, chiefly on the side of the cays in the lee of hurricane winds and waves, but it is not yet clear how much will ultimately recover. Large-scale fall of coconuts (in excess of 50 per cent) was found within 20–40 miles north and south of the storm track, and gives an indication of direction of extreme hurricane winds. Tree-fall direction at any point is generally constant within 30°–40°. Trees were uprooted, snapped off a few feet above the ground, or lost their crowns. Other trees which suffered heavy damage include *Cordia sebestena, Coccoloba uvifera, Bursera simaruba* and *Terminalia catappa*. All except *Cordia* tended to remain upright, losing leaves or branches; *Cordia* was generally uprooted and the debris piled by waves against obstructions. In the shallow northern barrier reef lagoon, uprooted trees, generally palms, have been carried several hundred yards from the nearest land, deposited on the lagoon floor, and are now visible above sea-level. *Rhizophora* showed great tenacity: not only are the details of mangrove areas almost unchanged by the storm, as shown by air photographs, but also in one case, Big Calabash East Cay II, the island was completely washed away, leaving only a ring of dead *Rhizophora* still in place. Greatest proportionate damage was caused to those species most abundant in exposed situations. *Tournefortia gnaphalodes* and *Suriana maritima*, formerly widespread near-shore plants, are not now to be found in the central zone of cay damage. On devastated cays and areas of freshly deposited sand, the chief colonists during the 4–7 months since the hurricane were, during the re-survey, *Portulaca oleracea*, species of *Euphorbia*, *Sesuvium portulacastrum* and *Ipomoea pes-caprae*.

Human settlement. The cays were previously inhabited by lighthouse keepers, coconut workers and fishermen. The lighthouses at Sandbore Cay (Lighthouse Reef) and Cay Bokel (Turneffe) were destroyed by the storm, and that at Mauger Cay (Turneffe) was put out of action. Only the Half Moon Cay and English Cay lights remained in operation. Of the four cays previously inhabited at Lighthouse Reef, three have been abandoned; of the six settlements at Turneffe, all are abandoned; and several cays on the barrier reef are no longer inhabited. The main source of livelihood, apart from the lighthouses—the coconut industry—suffered very heavy damage, and all the loading installations at Calabash Cays were destroyed. It will be at least seven years before coconuts planted now begin to bear; but it is unlikely that private or public capital will be available for coconut investment following the succession of recent hurricanes. The larger islands near the storm centre, which were inundated by the storm surge, suffered contamination of their fresh-water lenses. Ground-water on these cays is now brackish and almost undrinkable. Since previous inhabitants, however, relied almost exclusively on rain-water vats for water supply, this is of less immediate practical importance than might be expected.

Birds. Two bird colonies in the devastated area were not markedly affected by the storm. The Half Moon Cay red-footed boobies, *Sula sula sula*, were not diminished in numbers, in spite of great damage to their nesting area, but the nesting period was delayed. Normally the young hatch in January–February, but in 1962 hatching began in mid-April and continued into May. *Fregata magnificens* was nesting in considerable numbers on Man of War Cay, barrier reef lagoon, in April.

Prospect. Investigations at Jaluit[8] indicate that, over a period of years, cays tend to revert to the position obtaining before the hurricane struck, shown, for example, by landward migration and erosion of shingle ridges, and by vegetation recovery. This is already beginning on some British Honduras cays. On the other hand, as at Jaluit, certain processes are irreversible: the stripping of fine surface material and exposure of roots, erosion of scour holes and scour channels in cay surfaces, deposition of shingle and sand carpets above high-tide-level, all are likely to be permanent. The large-scale destruction of living reef along the central barrier reef has both provided much new debris for cay accumulation and permitted larger waves access to cay shores. As a result sandbores have appeared in areas where none was previously seen, and many cay beaches are steeper. This, too, is likely to be a short-term effect, for studies in Australia indicate a period of 10–20 years for reef recovery following hurricane damage[6]. Future vegetation changes are difficult to assess With the decline of economic interest in coconut plantations, and hence of the incentive to clear, many areas may tend to revert to bush, consequently decreasing the danger of catastrophic damage during subsequent hurricanes. Recovery will be most delayed in areas where the cay was completely destroyed. and has re-appeared or is likely to re-appear, as a shifting unvegetated sandbore. A second re-survey in 3–4 years, as at Jaluit, could be instructive in assessing the long-term effects of hurricanes on reefs, reef islands and their human utilization. At present the most important practical conclusion, within the general zonation of damage, is the severity of change where human interference with island vegetation has been most intense.

I thank the Royal Society for supporting the re-survey expedition, and Miss E. L. Pruitt, Geography Branch, Office of Naval Research, Washington, for transatlantic transportation. Previous expeditions were supported by the Cambridge Expedition to British Honduras 1959–60, the Office of Naval Research, and the Coastal Studies Institute, Louisiana State University. I also thank Prof. J. A. Steers, Dr. F. R. Fosberg and Dr. Adrian Richards for help with the present investigation.

[1] Blumenstock, D. I., *Nature*, **182**, 1267 (1958).
[2] Blumenstock, D. I. (edit.). *Atoll Res. Bull.*, **75**, 1 (1961).
[3] Stoddart, D. R., *Geog. J.*, **128**, 161 (1962).
[4] Stoddart, D. R., *Atoll Res. Bull.*, **87**, 1 (1962).
[5] Dunn, G. E., and associates. *Monthly Weather Rev.*, **90**, 107 (1962).
[6] Stephenson, W., Endean, R., and Bennett, I., *Austral. J. Marine and Freshwater Res.*, **9**, 261 (1958).
[7] Steers, J. A., *Geog. J.*, **95**, 38 (1940).
[8] Blumenstock, D. I., Fosberg, F. R., and Johnson, C. G., *Nature*, **189**, 618 (1961).

14

Reprinted from *Structure, Surface and Drainage in South-East England,* George Philip and Son, Ltd., London, 1955, pp. 148–159; originally published in *Trans. Inst. Brit. Geog.,* **10,** 148–159 (1938)

STRUCTURE, SURFACE, AND DRAINAGE IN SOUTH-EAST ENGLAND

S. W. Wooldridge and D. L. Linton

[*Editor's Note:* In the original, material precedes this excerpt.]

SUMMARY AND CONCLUSIONS

We have ranged too far in time and space to render it either easy or useful to recall in detail all the conclusions we have reached. Some of them, indeed, are important only in their local context but others throw light on the physical history of the area as a whole and to these we may give brief attention.

We have seen that early growth of the Mid-Tertiary structures both major and minor, took place during a long period before the main tectonic phase of late Oligocene or early Miocene times. In all the stages of growth, the form and structure of the buried Palaeozoic floor were important. The chief tectonic boundary in the area coincides fairly closely with the line of the Lower Thames valley. Southwards of this line where the Palaeozoic floor is at great depth, the cover rocks are strongly folded in a belt which ranges through Wessex and the Weald. Throughout this belt the closely spaced individual folds dominate the structure and relief, though on small-scale maps they appear subordinate to the broad anticlinal warping which separates the London and Hampshire Basins. North of the Thames valley line, where the floor is at small depth, there is no strong folding but only rather indefinite and discontinuous flexuring.

The beginning of the erosional history of the area is indicated by the unconformity between the Chalk and the early Tertiary rocks. A full cycle of erosion ran its course after the emergence of the Chalk sea-floor and, as a result, the upper divisions of the Chalk were removed from the greater part of the area. It is probable, indeed, that the Chalk cover was completely breached locally in the Wealden and South Midland areas, for pebbles derived from the Lower Greensand occur locally in the Eocene pebblebeds.

The sub-Eocene surface, essentially a peneplain trimmed locally by the waves of the Eocene sea, was exhumed in later times from beneath its cover of early Tertiary sediments and forms a distinctive facet of the Chalk country on the margins of the London and Hampshire Basins.

The Eocene deposits and the Chalk floor on which they rest were warped and buckled in the Mid-Tertiary movements. We can make a resonably accurate reconstruction of the form of the folds before they were eroded: the most difficult problem in the physical history of the area is the reconstruction of the sequence of events following the folding. Much of the evidence available for its solution is embraced in that most famous and extensive view of southern England which we obtain from the tower on the summit of Leith Hill.

Here, just above 1,000 feet above sea-level, our prospect embraces a dozen counties. The natural horizon line is some 40 miles distant, but the actual view, given maximum visibility, falls short or ranges further than this with the accidents of relief. What renders the view signal and notable is the fact that the major hill ranges of the south country lie, for the most part, near the horizon circle and thus frame the regional picture. To the north-west the skyline is the level crest of the Oxfordshire Chilterns, continued north-eastwards to the Dunstable Downs and, on a clear day, even beyond the Hitchin gap to the Chalk ridge south of Royston. Westwards and south-westwards the view is restricted by the high bastions of the Lower Greensand at Hindhead and Blackdown, but behind them the western Downs above Selborne present a crest as level and featureless as that of the Chilterns. It is continued in the summit line of the South Downs and the high central Wealden crests of Ashdown Forest. In the northern foreground, seen in in much greater detail, the North Downs present a similar even summit.

In these facts lies the evidence for the theory that the south country has been sculptured from an uplifted plain of which relics are preserved in our highest hill summits. A plane, 800 feet above sea-level, would, indeed, pass close to all these summits. Leith Hill (965 feet), Blackdown (918 feet) and Hindhead (890 feet) would exceed it, as would also the highest points of the North Downs (877 feet), the South Downs (880 feet) and the Chilterns (852 feet). Crowborough Beacon (792 feet) just fails to reach it, as do many parts of the Chalk crest-lines.

This plain was regarded by Davis as the product of the age-long waste and river action which ensued in Mid-Tertiary times, after the strong earth ripples, distant ground swell of the Alpine storm, had thrown the south country from Kent to Somerset into parallel ridges and furrows and thus given birth to the first decipherable drainage system in the area. These first rivers produced the plain; thereafter, as it was uplifted, they dissected it, bringing into being the graceful lines of the present landscape with its broad lowland vales separated by scarp-bounded uplands through which the rivers, older than the hills they traverse, wind in narrow defiles.

Davis thus recognized an ancient land surface. In its later development, physiographic work in the south-east has become a study of this and other surfaces of differing dates and various degrees of preservation. The hill-top plain of southern England may be reckoned, in this sense, a Miocene surface; the product of atmospheric action during that Miocene period which, alone among the major time divisions of the geological past, is unrepresented by stratified rocks in Britain.

If we are to use the south-east as a clue and a measure for other British regions we must take note of the fact that another well-marked surface can be traced in its landscapes. Younger than the main hill-top plain is the broad bench, cut largely in Chalk, on which the remnants of marine Pliocene deposits are widely distributed. This was not treated by Davis for, though Pliocene submergence was in his day known to have occurred in Kent, the full extent of the sea was not then recognized. The character of the deposits themselves, their occasional fossils, and above all the evidently wave-cut flat on which they rest and which is recognizably preserved even when the deposits have been removed, enable us to trace the former extent of the sea. In making the reconstruction shown in the map (Fig. 18) we were guided by a further important consideration – the relation, or rather lack of relation, of much of the drainage to the structure of the underlying rocks. When England, south of the Thames, was first raised into closely spaced ridges, the main lines of drainage must have followed the furrows between them. But that is not at all what we find over much of the main belt of folding to-day. Frequently and systematically, north-south rivers cross the east-west fold axes at their strongest development. This is notably true in Sussex and throughout the Wessex Chalk area. It was a major defect of the Davisian hypothesis that it failed to account for such features. It is true that the first cycle of erosion would have obliterated the anticlinal crests, but if the present drainage were merely successor to that of the first cycle it could not, save by the local accident of river capture, have so developed as to cross the line of strong anticlinal folds. What is required to account for the facts is a new start for the drainage such as must have followed the spreading and subsequent uplift of a sheet of sand and shingle over parts of the old plain. It is worthy of note that it was Davis's student and chief successor, Douglas Johnson, who first hinted to us that an extension of the Pliocene sea into Wessex would satisfactorily account for drainage features otherwise hardly explicable.

Throughout the greater part of the area the Pliocene shelf or platform is backed by a low rise separating it from higher ground – part of the Mid-Tertiary plain. This rise fixes the position of the coastline. It is a feature of the greatest significance that the platform maintains a very uniform general level of about 600 feet. It slopes seaward from between 650 and 700 feet at the base of the rise to about 550 feet or a little lower. This shows beyond question that the late Tertiary uplift of much of South-East England was uniform, i.e. unaccompanied by faulting or bending. On the Essex coast, however, and northwards into East Anglia, Pliocene deposits contemporary with and later than those of the London area lie but little above sea-level; in Holland they are far below it. Thus is exemplified one of the later expressions of a long-established and recurrent tendency for the east coast region to be tilted towards the North Sea depression.

Excluding this eastern zone, the physiographic development of the area has consisted in the uplift and dissection, stage by stage, of the Pliocene sea-floor and the adjacent parts of the Mid-Tertiary plain. The stages of this uplift are marked by the minor platforms and terraces, essentially old valley floors, of which the district presents a complex but coherent record.

The fullest record is afforded by the London Basin, for here the sequence of river terraces can be related to the true glacial drifts. There is clear evidence that, following the retreat of the Pliocene sea, the ancestor of the Thames followed a line far to the north of London and marked to-day by the drift-filled depression of the Vale of St. Albans. Its southern tributaries carried Lower Greensand chert, which is found in the high-level hill-top gravels of the North London area and southern Essex. These gravels lie at about 400 feet above sea-level, north of London, resting on a surface which declines gently eastwards to about 300 feet in central Essex.

From its first northerly course the Thames was expelled in two stages of glacial diversion. The first was the work of a Chiltern ice-sheet, while the second was due to the advance of the main or great eastern ice-sheet deploying in its great terminal lobe from the neighbourhood of the Wash. It is difficult to trace the successive courses of the Thames across Essex, where the main drift sheet deeply buries the sub-glacial landscape, but there is a strong suggestion that the east coast estuaries, from the Stour southward to the Crouch, are lineal descendants of successive Thames exits to the North Sea.

In parts of the Middle Thames region, which lay beyond the furthest extensions of the great eastern ice, the shift in the course of the Thames has been steadily southwards and a full terrace sequence is displayed between about 400 feet O.D. and the present flood-plain. The higher terraces are attributable to the phase of the early or Chiltern glaciation or to the inter-glacial period which followed it.

In particular, the Winter Hill Terrace, the lowest member of the higher group, has yielded flint implements which place it clearly within the First Inter-glacial period. In interpreting the physical history of these higher terraces and of their lower-level analogues, well developed in the Lower Thames valley through London, it is clear that we must reckon with a base-level oscillating in harmony with the waxing and waning of successive ice-sheets and although the terrace sequence as a whole implies 'uplift', or an overall fall of base-level, the record is punctuated in the lower valley by episodes of submergence and aggradation.

The South Coast area, comprising the Sussex coastal plain and the Hampshire Basin, reveal a closely similar terrace sequence to that of the Lower Thames valley, though exact correlation of apparently equivalent terraces can hardly yet be proved by precise evidence. The area supplements the Thames valley record in one important particular since it yields clear evidence of one of the high sea-level phases, represented by the Goodwood raised beach, which may be regarded as marking the date of the great aggradation evidenced in the High or 100-foot terrace of the Thames.

In the South Downs and the Wealden area the full sequence of denudation-chronology is at present less completely known. The South Downs dip-slope displays the dissected remnants of a number of benches between 475 and 180 feet O.D., which may well be of marine origin. In the Weald there are indications of a general platform level at 400–500 feet O.D., which may well be the equivalent of the 400-foot platform prominent north of London and also represented in the South Downs and the Hampshire Basins.

Much more prominent and regionally significant to both the geologist and the geographer is the 200-foot platform – the Ambersham stage of J. F. N. Green – which is well marked throughout much of the Weald and present also in the Hampshire Basin and the central parts of the London Basin. This marks the most widely spread episode of river planation since the emergence of the Pliocene sea-floor. In a broad sense it is probably contemporary with the Winter Hill stage in the Thames valley. It was uplifted and to some extent dissected before the advance of the main or eastern ice-sheet into Essex.

We have seen that the essentially unwarped attitude of the Pliocene platform shows that strong differential movement has not occurred throughout most of the area since the Mid-Tertiary folding. For this conclusion the mutual relations and ready inter-regional correlation of the lower platforms and terraces afford independent support. It appears, indeed, very probable that the later stages of land sculpture have been largely under the control of eustatic changes of sea-level. This conclusion is not applicable, however, to the eastern margin of the area where the evidence suggests the periodic resumption of down-warping towards the North Sea depression. The western hinge of the warped area in Essex appears to lie along a roughly north-south line through Braintree.

The complex history thus summarized constitutes merely the closing stages of the long geological history of our region – the story of the shaping of its land-surface. As such it is inescapably part of the field of study of the geologist and it throws light on the physical history of much of the rest of Britain. But if we now revert briefly to the theme of our introduction, it will be clear that these complex episodes, seen in terms of their imprint on the face of the country, provide an important key to the character of that country as the home of man.

We may see the matter in fitting perspective if we take as our point of departure Mackinder's famous paper of 1887 on the 'Scope and Methods of Geography', read to a rather puzzled and perhaps slightly resentful audience at the Royal Geographical Society.[85] This was essentially the founding document of modern British geography. It was concerned to show that physical geography and what was then called 'political (i.e. human) geography' were two related stages of one investigation not to be divorced from one another, and that the approach to physical geography must be genetic. As an example Mackinder traced the broad structural features of South-East England, the work of erosion upon them and their relation to the pattern of drainage. Let us quote parts of his description:

> 'Let us try to construct a geography of South-Eastern England which shall exhibit a continuous series of causal relations. Imagine thrown over the land like a white tablecloth over a table, a great sheet of Chalk. Let the sheet be creased with a few simple folds, like a tablecloth laid by a careless hand. A line of furrow runs down the Kennet to Reading, and then follows the Thames out to sea. A line of ridge passes eastward through Salisbury Plain and then down the centre of the Weald. A second line of furrow follows the valley of the Frome and its submarine continuations, the Solent and Spithead. Finally, yet a second line of ridge is carried through the Isle of Purbeck and its now detached member the Isle of Wight. Imagine these ridges and furrows untouched by the erosive forces. The curves of the strata would be parallel with the curves of the surface. The ridges would be flat-topped and broad. The furrows would be flat-bottomed and broad. The Kennet-Thames furrow would be characterized by increasing width as it advanced eastward. The slopes joining the furrow-bottom to the ridge-top would vary in steepness. . . . The moulder's work is complete; the chisel must now be applied. The powers of air and sea tear our cloth to tatters. But as though the cloth had been stiffened with starch as it lay creased on the table, the furrows and ridges we have described have not fallen in. Their ruined edges and ends project stiffly as hill ranges and capes. The furrow-bottoms, buried beneath the superincumbent clay, produce lines of valley along the London and Hampshire Basins. Into the soft clay the sea has eaten, producing the great inlet of the Thames mouth, and the narrower but more intricate sea-channels which extend from Poole Harbour through the Solent to Spithead, and which ramify into Southampton Water and Portsmouth, Langstone, and Chichester Harbours. The upturned edge of the Chalk-sheet produces the long range of hills, which, under the various names of Berkshire Downs, Chiltern and Gogmagog Hills, and East Anglian Heights, bounds the Kennet-Thames Basin to the north-west. The North and South Downs stand up facing each other, the springs of an arch from which the key-stone has been removed. The same arch forms Salisbury Plain, and its eastward prolongation in the Chalk uplands of Hampshire; but here the key-stone, though damaged, has not been completely worn through. Beachy Head and the North and South Forelands are but the seaward projections of the Down ranges. The fact that the North Downs end not in a single promontory, like Beachy Head, but in a long line of cliff, the two ends of which are marked by the North and South Forelands, may serve to draw attention to a relation which frequently exists between the slope of the surface and the dip of the strata'.

He then enforced his argument and illustrated his general method as follows:

> 'This being the general anatomy of the land, what has been its influence on man? In the midst of forest and marsh three broad uplands stood out in early days, great openings in which man could establish himself with the least resistance from nature. In the language of the Celts they were known as "Gwents", a name corrupted by the Latin conquerors into "Ventæ". They were the Chalk uplands with which we were familiar, the arch-top of Salisbury Plain and Hampshire, and the terminal expansions of the Chalk ranges in East Anglia and Kent. In East Anglia was Venta Icenorum; in Kent and Canterbury we still have relics of another Gwent. The first syllable of Winchester completes the triplet. In later, but still early times, they were the first nests of the three races which composed the German host. The Angles settled in Norfolk and Suffolk, the Jutes in Kent, the Saxons in Hampshire. In still later England, Winchester, Canterbury and Norwich were among the chief mediæval cities'.

In these passages Mackinder shows the high power of graphic generalization which availed him so often and so greatly to the profit of the subject. Yet his account is notably incomplete and in some respects definitely erroneous. At the time when he wrote, Whitaker's *Geology of the London Basin* had been published for fifteen years, but hardly dealt with the physiographic history of the area. Topley's *Geology of the Weald* had been available for twelve years. From this great work Mackinder certainly derived inspiration; it foreshadowed many of the problems we have here discussed and made almost the first thorough attack upon them. But a further eight years were to elapse before W. M. Davis, completing the work of British pioneers, first spoke clearly in terms of the successive *stages* of development of the southern English landscape – of cycles of erosion. It is this method of interpretation which is necessarily absent in Mackinder's account: the 'chisel' has been applied not in one long uninterrupted period of demolition but in successive decipherable stages, each of which has left its mark on the landscape. Not until 1901 did Herbertson, Mackinder's colleague in the Oxford School of Geography which he founded, refer to 'this whole great new science' of geomorphology, realizing how much it was to mean not for the geologist only but also for the geographer.

Fully to utilize the findings of geomorphology in a geographical description of South-East England would be a task beyond our present compass. It would involve a full and careful division of the area into physiographic regions, so that what the theorists of the subject call its natural 'areal differentiation' might be brought out. This we shall not here essay, contenting ourselves with reference[86] to certain attempts to deal with the problem. But we may conclude by calling attention to a few of the salient physical contrasts in the area, which have guided its human development, but which are explicable only in terms of the full physiographic sequence we have traced.

It is, of course, evident that many of the chief contrasts within the area are directly attributable to the character of the rocks. Wherever the Chalk directly adjoins Eocene terrains a striking local regional boundary is to be seen, and where sand and clay make their frequent contacts within the Weald the fact is writ large in the landscapes. Lateral variation within some of the major formations is not less geographically significant. The sudden on-coming of Blackheath pebble-beds, east of Croydon, determines the fashion of the plateau country of South-

East London, while the thickening of the Thanet Sands from London eastwards through Kent is the basis of the distinctive belt of country which follows the line of Watling Street to the coast. Similarly the westward thickening and change of character of the Lower Greensand in the Weald is answerable for marked scenic contrasts expressed not only in land-forms but in settlement and agriculture.

All this and much more is contributed directly by the geological map to the geographical description and explanation of the country; in such terms, somewhat elaborating Mackinder's broad structural picture, we can discern the nature and cause of many regional contrasts. But our analysis is still notably incomplete unless we take account of the sequence of landscape shaping. This may be illustrated by reference to the varying character of the Chalk country, the consequences of the course of drainage evolution we have traced, and the complex legacies of the Pleistocene ice age.

Few types of country have been more carelessly misunderstood or more widely misrepresented than that of the English Chalk. The salient and memorable characteristics of the South Downs have become the basis of a sort of myth concerning the open and down-like character of our Chalk upland country. For the greater part this shows no 'blunt, bow-headed, whale-back downs' but level-topped, thickly-wooded uplands. Some of the statements made of such country suggest that the writers can never have visited it. Concerning the foundation and rise of Roman London we are told that 'a geological map indicates the necessary direction of communication from the Kent ports, that, namely, along the open heights of the northern Chalk downs'. From such an error the geological map itself will save us, if it shows the Clay-with-Flints and associated plateau deposits, with their heavy cold soils, still thickly wooded. There is no reason, indeed, to suppose that even the bare Chalk of the South Downs was free from woodland at the time of man's first advent in the area, but it is quite certain that the summit plateaux of the North and the Hampshire Downs and the greater part of the Chiltern upland were densely and perhaps almost impenetrably wooded. Our analysis has provided a means of discriminating three different types of Chalk country, according as the dominant surface is sub-Eocene, sub-Pliocene or relict of the great Mid-Tertiary hill-top plain. On the latter, chiefly, longest exposed to the weather, is a full growth of true Clay-with-Flints to be found, but the Pliocene platform both in the Chilterns and the North Downs is sufficiently mantled with stony clay or loamy drifts to give rise to very similar surface conditions. Only on the lower dip-slope, stripped of its covering of sandy Eocene beds, do we find soil conditions favouring arable cultivation and presenting opportunities for the early clearing of woodland. It is a zone, also, not too far removed from underground water and sometimes, as south of London, furnished with springs at the Eocene junction. It was country of this type which provided the 'Gwent' of Kent of which Mackinder speaks; nowhere, indeed, can the contrast between the sub-Eocene and the sub-Pliocene facets of the Chalk be more clearly seen than in East Kent, but it is valid throughout the North Downs to near Guildford, though less clearly marked in the Chiltern Hills. Here the contrast between the Pliocene shelf and those higher parts of the plateau which remained unsubmerged is clearly expressed in land-use; the scarp-crest tracts are in woodland and heath, while the Pliocene shelf forms cultivated ground. It remains to be seen how far legible soil differences exist between the two areas: much plateau brickearth comparable with true loess lies upon both surfaces.

The South Downs, now in a state of much more mature dissection, offer no contrasts such as these, though as we have seen, they exhibit a feature of their own in the secondary escarpment of the Belemnite Chalk, whose development was prevented in the North Downs by Eocene erosion.

The wider Chalk tract of Wessex can in no proper sense be treated as a geographical unit. We have noted (p. 70) that Hampshire and Wiltshire have significantly different physical histories. As a result the rivers of Wiltshire run in deep narrow tortuous valleys whose lush meadows and fine trees but heighten the arid monotony of the upland plain. From the first these streams were condemned to a history of slow incision into a massive outcrop of bare Chalk which their activities have made more and more waterless. The Hampshire rivers, on the other hand, found on the emerged Pliocene plain extensive outcrops of soft deposits and variety of structure which have led to the easy opening of wide smooth synclinal vales and a ready integration of the drainage system which is patently reflected in the flowing lines of a landscape whose features blend in a subdued harmony of upland and lowland impossible to match in the western country. Here essentially lay the 'Gwent' of Mackinder's analysis; its distinction from the Wiltshire upland is well marked in the Roman phase, for Romano-British villages cluster on the Wiltshire heights, but give place to villas in the Hampshire valleys.

The superimposition of the drainage of much of the area from a Pliocene cover was, as we have sought to show, a physical episode of the highest importance, and the valley pattern as a whole is only comprehensible in terms of this hypothesis. Nevertheless, its effects upon the country as a human environment are not everywhere regionally important factors. Yet the drainage unity of both Wessex and Sussex evidently derives from the act of epigenesis. This fact is easily lost sight of in Wessex if the structural complexity of the area is ignored or minimized. In Sussex the presence of strong longitudinal folding is plain enough and it at once appears that the unity of the South Saxon kingdom and the pattern of internal divisions into which it was resolved are a reflex of a direct coastal drainage which ignores this structure. Sussex embraces the greater part of the sandstone core of the Weald. Access to it was made easier by the narrow outcrop of the Weald Clay but it was also directly facilitated by the rivers and the sandstone ridges, often marked by ancient crestways, which separated them. The unity of the area, embracing Downs and Weald, owes much to the drainage pattern established after the Pliocene submergence and the medieval rapes of Arundel, Bramber, Lewes and Pevensey are essentially based upon their axial rivers.

Of the direct contribution of glaciation to the making of the country north of the Thames it is needless here to speak at length. If the drift cover were removed, much of Norfolk and Suffolk and parts of northern Essex would lie below sea-level. The range of generally loamy soils afforded by the drifts encouraged settlement and cultivation from the time of the Saxon entry, and by that of the Domesday Survey the East Anglian province stood out as one of relatively high population density. Nevertheless, the Norwich region, as distinguished by Mackinder, can in no sense be regarded as a Chalk 'Gwent' comparable with those of Kent and Wessex. This is a mistake that can only arise if we ignore the drift cover.

The drift cover made a similar direct contribution in the form of favourable soils and readily available water, to the economy of the pre-Roman Belgic kingdom of the Catuvellauni, the head of the confederacy opposing Caesar's invasion. The municipium of Verulamium, successor to a Belgic city and ancestor of St. Albans, was, for a time, in some sense the capital of southern Britain, with London as its port.

Physical history has still further laid the foundation of human development in the London area. The Vale of St. Albans which marks the line of the early Thames is more than a belt of drift deposits: it is a longitudinal hollow lying athwart the course of the Chiltern drainage and serving to trap and divert a large share of the underground flow of Chalk water towards London. Here is the Chadwell spring from which Hugh Middleton's curious aqueduct of 1613 – the New River – brought water to London, and here to-day the water companies of the northern fringe of London sink their wells.

While the presence of the ice in the north-east of the region poured floods of outwash material into the Thames drainage system and added materially to the extent of its gravel spreads, the diversion of the main stream and the successive southward shifts of the channel caused the gravels to cover wider spreads of country ranging much further south than would otherwise have been the case. The outcrop of the London Clay which might well have given rise to an unbroken belt of wooded and intractable clay country was thus considerably diversified. In short, the early Thames and its diverted successors deposited the gravels which form the permeable hill cappings of the area north of London, providing thus the sites of the predominant and characteristic hill villages of the pre-urban countryside. Further, the diversions of the Thames, by bringing into being the lower courses of the Lea and the Colne went far to define the natural 'territorium' of Roman London and laid down the limits of Middlesex, the sole surviving evidence of the enigmatical Middle Saxons, otherwise lost to history.

Finally, we may note that the Lower Thames valley as we know it to-day is a direct legacy of the events we have traced and is itself responsible for some of the major features in the geography of England. It is permissible to wonder whether the course and outcome of the Roman and Jutish invasions of Kent would have taken the form we know if Kent had extended northwards across a broad zone of forested clay country to a Thames whose estuary lay in Essex, north of the Rayleigh Hills. To argue thus is no doubt to open up a vast and vague field of perhaps rather profitless speculation and to use a method which historians in general might condemn. It serves, nevertheless, to bring into relief the high significance of the physiography of the Lower Thames region as we know it. More certainly may we see that Thames-side with its tide-water Chalk quarries and sea-borne coal owes much of its industrial quality and aspect to the events which placed the Thames estuary where it is. In general, the whole geography of the site and situation of Roman, medieval and modern London reflects the long course of the preparatory processes whose sequence we have traced. 'Nature', it has been said, 'prepares the site, all else is the work of man', yet nowhere more than in the Thames valley does a just appreciation of the physiographic fundament illumine the significances of the human pattern growing from it.

In all these features, not less than in the symphonic composition of the hill and valley outlines themselves, the events of the recent geological past live on, serving to emphasize that in the field of human geography not less than in that of geology itself, the whole of the past is necessary to explain the whole of the present.

BIBLIOGRAPHIC REFERENCES

[1] S. W. Wooldridge and D. L. Linton, 'Some Episodes in the Structural Evolution of South-East England. . . .' *Proc. Geol. Assoc.*, vol. 49, 1938, pp. 264–91.
[2] An identical point of view is developed for neighbouring continental regions by C. Stevens in *La Relief de la Belgique*, Louvain, 1938.
[3] A. Stille, *Grundfragen der Vergleichenden Tektonik*, Berlin, 1924, pp. 231–4.
[4] R. A. C. Godwin-Austen, 'On a possible extension of the Coal Measures beneath the South-Eastern parts of England', *Quart. Journ. Geol. Soc.*, vol. 12, 1856, pp. 38 ff.
[5] Cf. W. J. Arkell, 'Analysis of the Mesozoic and Kainozoic Folding in England', *Rep. XVth Intern. Geol. Cong.*, Washington, 1933.
[6] G. W. Lamplugh, 'The Structure of the Weald. . . .' *Quart. Journ. Geol. Soc.*, vol. 75, 1919, pp. lxxiii–xcv.
[7] J. W. Evans, 'Regions of Compression', *Quart. Journ. Geol. Soc.*, vol. 82, 1926, pp. lx–cii.
[8] S. W. Wooldridge, 'The Structural Evolution of the London Basin', *Proc. Geol. Assoc.*, vol. 37, 1926, pp. 162–96.
[9] For a reinterpretation of the Purbeck thrust, see W. J. Arkell, *Geol. Mag.*, vol. 63, 1936, pp. 56 and 87 and vol. 64, 1937, p. 86.
[10] A. J. Jukes-Browne, *The Building of the British Isles*, 4th edn. 1922, 184–8.
[11] H. J. O. White, 'The Geology of the country near Brighton and Worthing', *Mem. Geol. Survey*, 1924, p. 19.
[12] W. Buckland, 'On the Formation of the Valley of Kingsclere and other valleys by the elevation of the strata that enclose them', *Trans. Geol. Soc.*, second series, vol. 2, 1826, p. 119.
[13] H. J. O. White, 'The Geology of the country around Hungerford and Newbury', *Mem. Geol. Survey*, 1907, pp. 79–80.
[14] H. J. O. White, 'The Geology of the country around Basingstoke', *Mem. Geol. Survey*, 1909, pp. 80–1.
[15] See the scale section through East Woodhay in A. J. Jukes-Browne 'The Geology of the country around Andover, *Mem. Geol. Survey*, 1908, p. 4.
[16] A. Ramsay, 'On the denudation of South Wales and the adjacent counties of England', in vol. 1 'Essays', *Mem. Geol. Survey*, 1846, pp. 326–8.
[17] J. B. Jukes, 'On the mode of formation of some of the river valleys in the south of Ireland', *Quart. Journ. Geol. Soc.*, vol. 18, 1862, pp. 378–400.
[18] W. Topley, 'The Geology of the Weald', *Mem. Geol. Survey*, 1875, pp. 27, 285, 286.
[19] W. M. Davis, 'On the origin of certain English rivers', *Geogr. Journ.*, vol. 5, 1895, pp. 128–46.
[20] H. Bury, 'On the denudation of the western end of the Weald', *Quart. Journ. Geol. Soc.*, vol. 66, 1910, pp. 640–92.
[21] S. W. Wooldridge, 'The Pliocene Period in the London Basin', *Proc. Geol. Assoc.*, vol. 38, 1927, pp. 49–132.
[22] A. J. Jukes-Browne, 'The origin of valleys of the chalk Downs of North Dorset', *Proc. Dorset Nat. Hist. and Antiq. Field Club*, vol. 16, 1895, p. 8.
[23] W. R. Andrews, 'The origin and mode of formation of the Vale of Wardour', *Wilts. Archaeol. and Nat. Hist. Mag.*, vol. 26, 1891, p. 258.
[24] C. Reid, 'The Geology of the country around Ringwood', *Mem. Geol. Survey*, 1902, p. 29.
[25] H. J. O. White, 'The Geology of the country south and west of Shaftesbury', *Mem. Geol. Survey*, 1923.
[26] Other opinions have since been expressed on the age and origin of the high-level erosion surfaces in Wales and South-West England. See especially O. T. Jones, *Quart. Journ. Geol. Soc.* 1952, and W. G. V. Balchin, *Geogr. Journ.*, vol. 118, 1952, pp. 453–76, and a summary relating these areas to South-East England, S. W. Wooldridge, *Geogr. Journ.*, vol. 118, 1952, pp. 297–308.
[27] F. W. Harmer, 'On the Pliocene deposits of Holland. . . .' *Quart. Journ. Geol. Soc.*, vol. 52, 1896, p. 753.
[28] H. Bury, 'Some high-level gravels of North-East Hampshire', *Proc. Geol. Assoc.*, vol. 33, 1922, p. 101.
[29] C. Le Neve Foster and W. Topley, 'On the Superficial Deposits of the Valley of the Medway. . . .'. *Quart. Journ. Geol. Soc.*, vol. 21, 1865, pp. 443–72.
[30] H. J. Mackinder, *Britain and the British Seas*, Oxford, 1902.
[31] H. J. O. White, 'The Geology of the country around Winchester and Stockbridge', *Mem. Geol. Survey*, 1912, p. 64.

[32] H. J. O. White, 'The Geology of the country around Alresford', *Mem. Geol. Survey*, 1910, p. 74.
[33] D. L. Linton, 'The origin of the Wessex Rivers', *Scot. Geogr. Mag.*, vol. 48, 1932, pp. 149–66.
[34] H. J. O. White, 'A short account of the geology of the Isle of Wight', *Mem. Geol. Survey*, 1921, pp. 156–9 and 167–8.
[35] H. Bury, 'Some Anomalous River Features in the Isle of Purbeck', *Proc. Geol. Assoc.*, vol. 47, 1936, pp. 1–10.
[36] B. W. Sparks, 'Stages in the Physical Evolution of the Weymouth Lowland', *Trans. and Papers Inst. Brit. Geog.*, 1952, pp. 17–29, and 'Two Drainage Diversions in Dorset', *Geography*, vol. 36, 1952, pp. 186–93.
[37] B. W. Sparks, 'The Denudation Chronology of the Dip-slope of the South Downs', *Proc. Geol. Assoc.*, vol. 60, 1949, pp. 165–215.
[38] J. F. Kirkaldy and A. J. Bull, 'The Geomorphology of the Rivers of the Southern Weald', *Proc. Geol. Assoc.*, vol. 51, 1940, pp. 115–50.
[39] S. W. Wooldridge and D. L. Linton, 'Influence of Pliocene Transgression on the geomorphology of South-East England', *Journ. Geomorphology*, vol. 1, 1938, pp. 40–54, esp. Fig. 6, p. 52.
[40] F. Gossling, 'Wealden Pebbles in the Valley of the River Darent', *Geol. Mag.*, vol. 74, 1937, p. 527.
[41] G. J. Hinde, 'Gravels of Croydon and its neighbourhood', *Proc. and Trans, Croydon Nat. Hist. and Sci. Soc.*, vol. 4, 1897, p. 219.
[42] H. G. Dines in discussion on 'The River Mole: its physiography and superficial deposits', *Proc. Geol. Assoc.*, vol. 45, 1934, pp. 35–69.
[43] F. Gossling, 'Note on a former high-level erosion surface about Oxted', *Proc. Geol. Assoc.*, vol. 47, 1936, pp. 316–21.
[44] C. C. Fagg, 'The Recession of the Chalk Escarpment and the development of Chalk Valleys. . . .' *Proc. Trans. Croydon Nat. Hist. and Sci. Soc.*, vol. 9, 1923, pp. 93–112.
[45] H. Bury, 'Notes on the River Wey', *Quart. Journ. Geol. Soc.*, vol. 64, 1908, pp. 318–34.
[46] J. Prestwich, 'On the Westleton Beds and their extension inland', *Quart. Journ. Geol. Soc.*, vol. 46, 1890, pp. 34, 123 and 155.
[47] J. D. Solomon, 'On the Westleton Series of East Anglia', *Quart. Journ. Geol. Soc.*, vol. 91, 1935, pp. 216–38.
[48] G. Barrow, 'Some Future Work for the Geologists' Association', *Proc. Geol. Assoc.*, vol. 30, 1919, pp. 36–48.
[49] R. L. Sherlock and A. N. Noble, 'On the Glacial Origin of the Clay-with-Flints of Buckinghamshire and on a Former Course of the Thames', *Quart. Journ. Geol. Soc.*, vol. 68, 1912, pp. 199–212.
[50] R. L. Sherlock, 'On the Superficial Deposits of South Herts and South Bucks. . . .' *Proc. Geol. Assoc.*, vol. 35, 1924, pp. 19–28.
[51] J. W. Gregory, *The Evolution of the Essex Rivers*, Colchester, 1922.
[52] H. L. Hawkins, 'The relation of the River Thames to the London Basin', *Rep. Brit. Assoc.* for 1922 meeting at Hull, pp. 365–6.
[53] H. J. O. White, 'On the Distribution and Relations of the Westleton and Glacial Gravels in parts of Oxfordshire and Berkshire', *Proc. Geol. Assoc.*, vol. 14, 1895, p. 27; and 'On the Origin of the High-level Gravels with Triassic Debris adjoining valley of the Upper Thames', *ibid.* vol. 15, 1897, p. 157.
[54] H. J. O. White, 'The Geology of the country around Henley', *Mem. Geol. Survey*, 1908 pp. 87 and 88.
[55] S. W. Wooldridge, 'The glaciation of the London Basin and the evolution of the Lower Thames drainage system', *Quart. Journ. Geol. Soc.*, vol. 94, 1938, pp. 627–67.
[56] F. K. Hare, 'The Geomorphology of parts of the Middle Thames', *Proc. Geol. Assoc.*, vol. 58, 1947, pp. 294–339.
[57] M. S. Treacher, W. J. Arkell and K. P. Oakley, 'On the Ancient Channel between Caversham and Henley and its contained Flint Implements', *Proc. Prehist. Soc.*, 1948, p. 126.
[58] A. Sutton, 'The river systems of Western Hertfordshire', *Trans. Herts. Nat. Hist. Soc.*, vol. 13, 1907, p. 1.
[59] S. W. Wooldridge and J. F. Kirkaldy, 'The Geology of the Mimms valley', *Proc. Geol. Assoc.*, vol. 48, 1937, pp. 307–15.
[60] S. W. Wooldridge, 'The 200-foot Platform in the London Basin', *Proc. Geol. Assoc.*, vol. 39, 1928, pp. 1–26.
[61] H. Dewey and C. E. N. Bromehead, 'The Geology of the country around Windsor and Chertsey', *Mem. Geol. Survey*, 1915.
[62] J. D. Solomon, 'The Glacial Succession on the North Norfolk Coast', *Proc. Geol. Assoc.*, vol. 43, 1932, pp. 241–71.
[63] W. B. R. King and K. P. Oakley, 'The Pleistocene Succession in the Lower part of the Thames valley', *Proc. Prehist. Soc.*, vol. 2, 1936, pp. 52–76.
[64] F. E. Zeuner, *The Pleistocene Period*, London (Ray Society), 1945, pp. 249–50.
[65] S. W. Wooldridge and J. F. Kirkaldy in R. A. Smith *Proc. Prehist. Soc. E. Anglia*, 1933, p. 165.
[66] A. Coleman, 'Some Aspects of the Development of the Lower Stour, Kent', *Proc. Geol. Assoc.*, vol. 63, 1952, pp. 63–86.
[67] H. B. Woodward, 'On some Disturbances in the Chalk near Royston', *Quart. Journ. Geol. Soc.*, vol. 59, 1903, p. 365.

[68] A. W. Woodland, 'Water supply of the Cambridge-Ipswich district', *War-time Pamphlets Geol. Survey*, No. 20, 1924, Part X, General discussion, pp. 1–84.

[69] H. G. Dines and F. H. Edmunds, 'The Geology of the country around Romford', *Mem. Geol. Survey*, 1925, and accompanying map (Sheet 257, new series).

[70] T. V. Holmes, 'The New Railway from Grays Thurrock to Romford', *Quart. Journ. Geol. Soc.*, vol. 48, 1892, pp. 365–72; and 'Further notes on some sections on the new railway from Romford to Upminster', *Quart. Journ. Geol. Soc.*, vol. 50, 1894, pp. 443–52.

[71] W. Fox, 'When and how was the Isle of Wight separated from the mainland', *Geologist*, vol. 5, 1862, p. 452.

[72] C. Reid, 'The Geology of the country around Ringwood', *Mem. Geol. Survey*, 1902.

[73] H. Bury, 'The Plateau Gravels of the Bournemouth Area', *Proc. Geol. Assoc.*, vol. 44, 1933, pp. 314–35.

[74] J. F. N. Green, 'The Terraces of Bournemouth, Hants', *Proc. Geol. Assoc.*, vol. 57, 1946, pp. 82–101.

[75] J. B. Calkin and J. F. N. Green, 'Palaeoliths and Terraces near Bournemouth', *Proc. Prehist. Soc.*, vol. 15 N.S., 1949, pp. 21–37.

[76] J. Prestwich, 'The Raised Beaches and "Head" or Rubble Drift of the South of England. . . .' *Quart. Journ. Geol. Soc.*, vol. 48, 1892, pp. 263–343.

[77] J. Fowler, 'The 100-foot Raised Beach between Arundel and Chichester, Sussex', *Quart. Journ. Geol. Soc.*, vol. 88, 1932, pp. 84–99.

[78] K. P. Oakley and E. C. Curwen, 'The Relation of the Coombe Rock to the 135-foot Raised Beach at Slindon, Sussex', *Proc. Geol. Assoc.*, vol. 48, 1937, pp. 317–323.

[79] J. F. N. Green, 'The Age of the Raised Beaches of Southern Britain', *Proc. Geol. Assoc.*, vol. 54, 1943, pp. 129–40.

[80] L. S. Palmer and J. H. Cooke, 'The Pleistocene Deposits of the Portsmouth District', *Proc. Geol. Assoc.*, vol. 34, 1923, pp. 253–82.

[81] A. J. Bull, 'Studies in the Geomorphology of the South Downs', *Proc. Geol. Assoc.*, vol. 47, 1936, pp. 99–129.

[82] S. W. Wooldridge, 'Some Features in the Structure and Geomorphology of the country around Fernhurst, Sussex', *Proc. Geol. Assoc.*, vol. 61, 1950, pp. 165–90.

[83] J. F. N. Green, 'The Terraces of Southernmost England', *Quart. Journ. Geol. Soc.*, vol. 92, 1936, pp. lviii–lxxxviii.

[84] A. J. Bull and others, 'The River Mole: its physiography and superficial deposits', *Proc. Geol. Assoc.*, vol. 45, 1934, pp. 54–8.

[85] H. J. Mackinder, 'On the scope and methods of geography', *Proc. Roy. Geog. Soc.*, vol. 9, 1887, pp. 141–74.

[86] See S. W. Wooldridge, 'The physiographic evolution of the London Basin', *Geography*, vol. 17, 1932, pp. 110–16 and Fig. 11; also S. W. Wooldridge in *An Historical Geography of England*, Cambridge, 1936, pp. 89–94 and Fig. 13. On the general principles involved see D. L. Linton, 'The delimitation of morphological regions' in *London Essays in Geography*, London, 1950.

15

Reprinted from *Adv. Sci.*, 7(16), 162-165, 174-175 (1950)

THE UPLAND PLAINS OF BRITAIN: THEIR ORIGIN AND GEOGRAPHICAL SIGNIFICANCE

Address by

PROF. S. W. WOOLDRIDGE

PRESIDENT OF SECTION E

1. My aim in this address is to survey some of the problems of the later stages of landscape evolution in Britain and more particularly the evidence afforded by the various well-marked erosion surfaces developed on rocks of widely varying ages. My subject, therefore, falls within the strict and proper purview of Geomorphology, and during the last 30 years work upon it by both geographers and geologists has been very actively pursued in Britain. The time seems ripe for a review of such work; if it cannot provide answers to all the problems arising, it may at least succeed in asking some of the right questions and thus guide the direction of further research.

If in pursuing this subject we find ourselves cultivating the common borderland of geology and geography I find no need to apologise for that here. It is clearly part of the tradition of the British Association that inter-sectional interests and issues should be discussed. For myself, I do not doubt the close relevance of the physiographic findings to the general tasks of geographical interpretation. That relevance has been evident to a large number of geographers and to it we certainly owe the initiation of the 'Terrace Commission' of the International Geographical Union which has been at work since 1926. The instinct which prompted and still maintains the investigation is sound enough. One can draw no tidy division, armed with 'keep out' notices, between the fields and interests of geologists and physical geographers. I have no desire to enter upon one of those methodological disputations of which we, as geographers, are perhaps inordinately fond, and I will content myself with affirming that, if the study of the shape and shaping of the physical landscape is ever expelled, through prejudice or ignorance, from the ambit of geography, our subject will have lost much of its historical birthright, including even the right to its time-honoured name. Welcoming cordially as I do the marginal growth of our subject into the fields of study of 'social man,' I flatly rebut the suggestion that this alone is 'real geography.' Like geomorphology itself, it is part and only a part of a whole. As I see it, one of the emphases at present needed in British geography is a reminder of the virtue of keeping both feet firmly on the ground—'the solid ground of Nature,' if not quite in the sense in which Wordsworth first used this famous phrase. For 'ground' you may read 'land' or, if you insist, 'landscape.' But let us at least be clear that landscape involves more than the so-called cultural landscape. However important a role we ascribe to the latter, we shall make but little of its study without much closer scrutiny of its physical foundations than is now fashionable in some quarters.

2. By a combination of physical chance and human perversity the Tertiary era has almost become a neglected 'Dark Age' in the geological history of Britain. Despite, or perhaps because of, the general wealth of our geological record in these Islands there is as yet no adequate interpretation or even description of their geomorphology. The geologist deals convincingly and at length with what has been called palaeogeography, but the

geographer turning to these findings for the antecedents of the features and phenomena of his study is too often handed a stone in place of the relevant bread. For palaeogeography is necessarily and inevitably an affair, primarily, of long-past sea-bottoms and of coastlines tentatively located. The evolution of the Tertiary land surfaces, not less a proper enquiry in palaeogeography, presents quite different problems, amenable to attack only by the methods of geomorphology.

These problems are rendered difficult of solution by the limited and fragmentary character of our Tertiary sedimentary record. This default of evidence, taken with the fact that many of the British uplands show manifest structural lineaments of ancient date, and present seemingly simple relations to the lowlands, puts a premium on over-simple explanations of the physical features of Britain. They are coloured, too, by the widely received conclusion that the Upper Cretaceous transgression left little of the British area unsubmerged. The widely-spreading sheet of Chalk, or equivalent deposits, uplifted in early Tertiary times, becomes as it were the *terminus a quo* of British landscape evolution, and, to a first approximation, all that seems necessary is to conceive of the stripping of this Cretaceous mantle from upland and lowland alike, thus allowing the older structural patterns to show through. More than a century of detailed work on the succession and structure of Britain enables us to give graphic and circumstantial accounts of the antecedents, climax, and sequel of the great episodes of Caledonian and Hercynian mountain-making. The faulting which accompanied or succeeded the latter is still plainly legible in the marquetry of the British landscape and we can trace, in gratifying detail, the later story of the Triassic deserts and the Jurassic and Lower Cretaceous seas, until the great Upper Cretaceous transgression closes and, as it were, seals off the account. It is, naturally enough, such episodes as these which have always taken pride of place in the tale of British palaeography as written by pioneers like Edward Hull (1) and A. J. Jukes-Browne (2) or in broader outline by geographers seeking an evolutionary background to their canvas. The later parts of the record, excepting only the Alpine folding in the South Country, receive much scantier treatment in the standard accounts, though they are obviously much more important for the student of present surface forms. Jukes-Browne, indeed, made a brave and ingenious attempt to trace the progress of Tertiary denudation; but he was necessarily denied much of the evidence now available and many of his conclusions are no longer tenable. Later Professor L. J. Wills (3) very clearly and fairly summarised the more recent work done in South-East England, but he did not find occasion to deal with the landscapes of Devon and Cornwall which probably hold not a few of the keys to the interpretation of our Tertiary land surfaces.

In the light of the evidence gained during the last 30 years it is now time that geologists and physical geographers faced the fact that one cannot, in any real sense, explain, say, the Welsh mountain mass in terms of Caledonian orogeny, which was a phase not so much in its history but its pre-history. Still less are the land forms of Devon and Cornwall at all closely related to the Hercynian folding ; they bear the plainest impress of Tertiary faulting and late Tertiary planation. To elide, as we constantly tend to do, some 65 million years of geological process, accountable to Tertiary time, is to import a sort of 'antiquarian' squint into our physiographic interpretations. Since Tertiary deposits obtrude so little, and over much of Britain even Cretaceous rocks are missing between the solid basement and the Pleistocene drifts, physiographic accounts perform a sort of leap-frog antic and pass in one breath from tracing the topographic expression of ancient structures to the story of glaciation. Since no deposits intervene, it has been widely assumed that nothing can safely be said of the vast 'lost interval,' or that it is a field, at best, for aimless and profitless speculation. It is this great lost interval in the geological story, and a corresponding area of ignorance in the interpretation of the geographer's landscapes, that the methods of modern geomorphology are fitted to fill. To ignore them and to leave the lost interval intact is as if we sought the roots of some modern politico-social problem in minute and careful studies of the manorial system, ignoring the immediately precedent stages of history. Or it may afford a closer parallel to our case if we imagine an archaeologist reconstructing the original form and stages of decay of some major building of the past. Let it be assumed that it was fashioned of local building stone. Even so the grain or other internal structures brought out by weathering in the building blocks would not form valid parts of the archaeologist's reconstruction. The plan and fashion of the original building and the successive phases of its demise would reflect quite other processes than those which formed or weathered the stone. We must not, of course, press this analogy too far, for, in our case, the great principle of geological continuity is at stake. The later are of one piece with the earlier episodes as parts of a continuous physical evolution. But if as Professor O. T. Jones recently noted (4) the greater part of Britain has been structurally dead since the close of the Hercynian movements it has none the less been morphologically very much alive. If the structural geologist has little or nothing to say of the vast periods which appear to him quiescent, the geographer must needs turn to geomorphology for guidance in his own analyses.

3. In broad terms there are two methods of tackling the problem of the lost interval. The first consists in the analysis of the drainage system and its evolution. It was first applied in this country by W. M. Davis, followed by Cowper Reed, Buckman, Strahan and Lake and, in more recent years, notably by Linton and Bremner. I shall not attempt to deal here with the method and its findings. Arguments of considerable elegance can be based upon it, but they tend, of themselves, to be indecisive unless definite stages in chronological sequence can be established. This the second method supplies, for it aims at the recognition of past base-levels of erosion whether represented by submarine or sub-aerial surfaces. Associated with it is the analysis of composite or cyclic river profiles first applied in this country by O. T. Jones in 1924 and since developed by the late J. F. N. Green. I shall not discuss it at length since it is definitely ancillary to the study of erosion surfaces. Miller has shown the wide margins of error which characterise the method of base-levels determined by simple extrapolation of river curves. Without the check imposed by erosion surfaces and river terraces, reconstruction of past river profiles could hardly yield dependable or definitive results. Far greater importance must be assigned to plateau-like erosion surfaces, for these are actual geographical features which can be mapped with a reasonable approach to accuracy and traced for considerable distances. Only the preliminary question of nomenclature need detain us before we review on a regional basis the considerable body of evidence now to hand.

4. It is usual to refer to the features in question as ' erosion surfaces ' or ' erosion platforms ' and in ordinary working parlance these terms are unobjectionable and too well entrenched to be easily replaced. Yet neither term is free from ambiguity, and since geographers are showing timely signs of a certain tenderness of conscience in the matter of terminology it will not be amiss to employ a little circumspection. It is clear that there are many erosion surfaces, e.g. valley sides and cliff faces, which are in no sense platforms. The term ' platform ' is graphic and self-explanatory, and in some cases apt enough, yet the purist may insist that a platform should be flat or nearly so. In fact, most of the surfaces in question are appreciably inclined, whether by virtue of original slope, subsequent dissection or tectonic warping. Moreover some authors use the term ' structural platform ' where the top of a resistant master-stratum is locally stripped of a conformable overburden. Such a platform is evidently, in some sense, an erosion surface, though not generally in the sense in which ' the platform school ' use the term. It is, of course, obvious enough that the significant features for our purpose are surfaces of low relief inferred from accordant summit levels or preserved as flats truncating or bevelling the structure. It is to be noted that we cannot use a genetic nomenclature without begging the very question we are required to solve. Some of the surfaces are no doubt peneplains in the Davisian sense, but unless that term is given a grossly widened and weakened connotation

it cannot be applied to them all. I do not favour the use of the term marine peneplain. Nor is this all. Extensive surfaces of low relief bevelling structure may be neither peneplains nor strand-flats, but the product of lateral river corrasion, arid planation or advanced corrie sculpture. There is the further difficulty that a surface may be 'fossil'—i.e. an exhumed plane of unconformity, which commonly leaves us in a further doubt whether its original fashioning was submarine or subaerial, or in part both. To press these difficulties to the limit would no doubt be pedantic, yet they are real enough if we seek a definitive and unambiguous glossary definition of erosion surfaces and platforms. I shall not forgo the use of these simple and essentially commonsense terms in reviewing the British evidence, but my title revives a practice of the Geological Survey in Devon and Cornwall and notably of Mr. Henry Dewey. Generically, as geographical features, the chief members of the series of surfaces are 'upland-plains.' In using this term one begs no question at all and merely states an observed fact. Even those who are sceptical of the reality of erosion surfaces must surely grant that upland plains are a conspicuous feature of the physiography of Britain. There is the further advantage that the term is applicable to composite surfaces, peneplains retaining an appreciable relief, or a series of closely-spaced platforms difficult or impossible to separate. I will not pursue further what might easily become an irritating discussion. I submit, merely, that to speak of the 'upland plains' of Britain conveys at once more and less than the 'platforms' of Britain. If anyone thinks that I might as well have said the plateaux of Britain, I will ask leave to express my personal disagreement without further argument.

5. I must attempt now the far from easy task of summarising the evidence at present available under regional headings.

If I begin in South-East England it is not because this is the region I personally know best, but because here Tertiary deposits, and folding of known age, put our problem in its correct time setting. The cardinal feature is the Pliocene platform developed on the Chalk outcrops flanking the London Basin and either bevelling the escarpment or forming a distinct bench on the dip-slopes (5). The deposits which rest on it, originally ascribed to the Older Pliocene have since been shown to yield a Newer Pliocene (Red Crag) fauna (6). The typical elevation of the platform and one faithfully maintained is about 600 ft. but it ranges downwards to 550 ft. or perhaps slightly lower, while what has been taken as a degraded cliff-line feature at its back comes on somewhat below 700 ft. Where the platform adjoins higher ground, as in the North Downs of Surrey or Kent, or in the Chiltern Hills, this ground shows distinctive features both in its drainage patterns and soil cover which are consistent with the conclusion that it was not submerged at this time. On the platform the drainage lines could not arise till the Pliocene sea floor was uplifted, and they must have been ultimately superimposed from the cover of sand and shingle. With the help of this clue the coastline was traced by Professor Linton and myself southwestwards from near Hitchin to the neighbourhood of Dorchester (7). It occurs also in a dissected condition on the South Downs, but we concluded that much of the central Wealden area remained unsubmerged.

Within the Chalk areas the recognition of the Pliocene platform defines also not only the higher tracts but the exhumed sub-Eocene surface. The former areas have been regarded as surviving relics of the Miocene or Mid-Tertiary peneplain. The latter forms a distinctive facet on the lower part of the Chalk dip-slope. Only rarely as in the Western Chilterns does it form a true upland plain. Its existence reminds us that a lengthy period of erosion ran its course before the earliest Eocene deposits were laid down : there is evidence that the Chalk cover was locally breached by early Eocene times. Though the sub-Eocene surface was locally trimmed by the waves of the Eocene sea it was essentially a peneplain, and the possibility that it may prove recognisable in northern or western Britain must be borne in mind.

[*Editor's Note:* Material has been omitted at this point.]

REFERENCES

(1) E. HULL, *Contributions to the Physical History of the British Isles*, 1882.
(2) A. J. JUKES-BROWNE, *The Building of the British Isles*, 4th Ed. 1922.
(3) L. J. WILLIS, *The Physiographic Evolution of Britain*, 1929.
(4) O. T. JONES, 'The Structural History of England and Wales.' *Rep. Int. Geol. Cong.*, Part I. (1950), p. 216.
(5) S. W. WOOLDRIDGE, *Proc. Geol. Assoc.* **38** (1927), p. 49.
(6) C. P. CHATWIN, *Summ. Prog.* (*Mem. Geol. Surv.*) 1927, p. 154. Also *Summ. Prog.* 1930, p. 2.
(7) S. W. WOOLDRIDGE and D. L. LINTON, *Journ. Geomorph.*, Vol. 1 (1938), p. 41, and in 'Structure, Surface and Drainage in S.E. England.' *Inst. Brit. Geog.* Publication, No. 10 (1939).
(8) See 5 above and G. BARROW, *Proc. Geol. Assoc.* **30** (1919), p. 36.
(9) S. W. WOOLDRIDGE, *Proc. Geol. Assoc.* **39** (1928), p. 1.
(10) B. W. SPARKS, *Proc. Geol. Assoc.* **60** (1949), p. 163.
(11) J. F. N. GREEN, *Proc. Geol. Assoc.*, **52** (1941), p. 36.
(12) A. E. TRUEMAN, *Proc. Bristol Nat. Soc.*, **8** (1938), p. 402.
(13) C. REID, 'The Pliocene Rocks of Britain' (*Mem. Geol. Surv.*), 1896.
(14) G. BARROW, *Quart. Journ. Geol. Soc.*, **64** (1908), p. 384.
(15) W. G. V. BALCHIN, *Geog. Journ.*, **90** (1937), p. 52.
(16) C. F. GULLICK, *Trans. Roy. Geol. Soc. Cornwall*, **16** (1936), p. 380.
(17) A. A. MILLER, *Proc. Yorks. Geol. Soc.*, **24** (1938), p. 31.
(18) S. E. HOLLINGWORTH, *Quart. Journ. Geol. Soc.*, **114** (1938), p. 55.
(19) S. E. HOLLINGWORTH, *Proc. Yorks. Geol. Soc.*, **23** (1936), p. 159.
(20) E. H. BROWN, *Inst. Brit. Geog.*, Publication No. **16** (1950).
(21) H. H. READ, 'The Geology of the Country around Banff, Huntly and Turriff' (*Mem. Geol. Surv.*) 1923, Chap. XI.
(22) A. G. OGILVIE, 'Great Britain' (1937), p. 425.
(23) H. C. VERSEY, *Proc. Yorks. Geol. Soc.*, **23** (1937), p. 302.
(24) J. E. RICHEY, 'The Tertiary Volcanic Districts' (*British Regional Geology*, 1935), p. 110.
(25) H. FLEET, *Rep. Int. Geog. Cong.*, Amsterdam, (1938), p. 91.
(26) R. B. MCCONNELL, *Proc. Yorks. Geol. Soc.*, **24** (1939), p. 152.
(27) K. GOSKAR, *Proc. Swansea Sci.* and *Field Nat. Soc.*, Vol. 1 (1935), p. 305.
(28) J. F. N. GREEN, *Proc. Geol. Assoc.*, **60** (1949), p. 105.
(29) F. M. TROTTER, *Proc. Yorks. Geol. Soc.*, **21** (1929), p. 161.
(30) H. BAULIG, *Inst. Brit. Geog.*, Publication No. 3 (1935), and in 'Problèmes des Terrasses' (*Union Geographique International*), (1948).
(31) D. L. LINTON, *Scot. Geog. Mag.*, **48** (1932), p. 149.
(32) H. DEWEY, *Quart. Journ. Geol. Soc.*, **72** (1916), p. 63.
(33) C. REID, 'The Geology of the Country around Newton Abbot,' *Mem. Geol. Surv.*, (1913).
(34) S. W. WOOLDRIDGE, *Journ. Soil Sci.*, **1** (1949) p. 31.
(35) S. J. SAINT, *Barbadoes Agric. Journ.*, No. 3 (1934).

16

Reprinted from *Quart. Jour. Geol. Soc. London,* **106,** 101-103, 126-127 (1950)

THE STUDY OF THE WORLD'S PLAINLANDS: A NEW APPROACH IN GEOMORPHOLOGY

BY PROFESSOR LESTER C. KING, D.SC. F.G.S.

Read 21 June, 1950

CONTENTS

SUMMARY

A review of the older plainlands and plateaus of the earth reveals a similarity in age and elevation sufficient to suggest that their independent origin and development in their present situation is unlikely. It is claimed that continent-wide bevelled surfaces have been produced by "pediplanation". A world-wide correlation of pediplaned surfaces now standing at varying levels above O.D. is suggested by the following grouping: (a) pre-Cretaceous super-continental cycles of erosion ("Gondwana" and "Laurasia" surfaces); (b) pre-Miocene intermediate continental cycles ("African", "Prairie" etc. surfaces); (c) modern continental cycles. The difference in elevation between the fundamental surface (a) and the plains of the continental early-Tertiary cycle (b) is attributed largely to the altitude of the twin Gondwana-Laurasia super-continents at the time of their break-up. It is suggested that global forces, and probably continental drift, must be invoked in explanation of these world-wide pediplaned surfaces.

I. INTRODUCTION

THE monotonous aspect of the great erosional plains, their seeming lack of useful data, and the difficulty of ascertaining their modes of origin and ages have for a long time discouraged their study. Yet the great plains and plateaus (using the term without structural connotation) record in a relatively simple manner the geomorphological history of the continents. Thus, in Africa, the continent almost devoid of folded mountain ranges but renowned for the extent and perfection of its bevelled erosion surfaces, recent investigations have revealed a simple landscape history covering the greater part of the continent. These advances are recorded in the works of Wayland, Jessen, Veatch, Dixey and King, and have been to some extent summarized in a coloured map (King 1950A).

Briefly, the advances which afford the grounds of the further argument here set forth are as follows:—

(1) Recognition of the fact that the plainlands are the result of "cycles of erosion", and that certain of the cycles (and corresponding plains) may be regarded as "continental" in extent: that is, they may be traced from place to place, or correlated in similar situations, continuously or at intervals throughout at least the area of a sub-continent, e.g. South and central Africa (Dixey 1938, 1942).

(2) That portions of more than a single cyclic surface exist, and while there has been some disagreement as to the number and nature of such

cyclic surfaces that may co-exist, nevertheless it is amply demonstrated that surfaces of different ages may survive side by side. Further, that the extent of such surfaces is susceptible of mapping, and several maps have been drawn, e.g. by Jessen (for Angola), Dixey (Nyasaland), King (Natal, Southern Rhodesia, central and southern Africa).

(3) That the mode of development of African land-surfaces has not been by the Davisian " normal cycle " with broad-floored river valleys and convex divides, culminating in the multi-convex " peneplain "; but by Penckian scarp retreat, with narrow river channels, ever-broadening concave pediments, and culminating in the multi-concave " pediplain " (King 1947, 1950A). Penck's *endrumpf–primärrumpf* concept is valid in south-central Africa, not the Davisian peneplain. Land-surfaces in other continents seem, on data available to the writer, to be comparable, and we may even note that Mt. Monadnock, classic example of a residual upon a late-stage erosion surface, is concave in profile, not convex, and differs in no material way from the many thousands of such erosional relicts in Africa.

The co-existence of landscapes in an advanced stage of development, attributable to two or more cycles of erosion, is incompatible with the Davisian conception of " peneplanation " with its emphasis upon vertical lowering of land-surfaces under weathering; but is wholly in accord with the Penckian doctrine of landscape development by scarp retreat and pedimentation. It is, moreover, demonstrated abundantly in the field, not only through the African region where so much work has recently been done, but seemingly also in every continent, and almost every country, in the world. The insistence upon lowering of land-surfaces even at advanced peneplanation leads to serious error in the dating and correlation of such surfaces, as is particularly noteworthy in the Appalachian region.

The text-books have long, and rightly, pointed out that once a landscape is reduced under erosion to very low relief, it virtually ceases to be modified unless some change in the regime is introduced by climatic or tectonic agency. This is a point which we shall investigate further; for the moment, we must be clear, one of the fundamentals of the new approach is that *after land-surfaces have attained a stage of advanced planation, there is no appreciable process at work causing them to be continuously and uniformly lowered.* On this point the field evidence is precise, for surfaces of remote age still survive with little alteration in many parts of the world.

(4) It should be noted, however, that the hypothesis of pedimentation does not exclude some lowering of surfaces locally and intermittently. With the conflict of pediments operating from different tributaries and different drainage basins, and even from different streamlets crossing the larger pediments, the land surface at any given spot will probably, even in extreme old age, have an occasional thin shaving removed from it as one pediment grows laterally at the expense of another. This conflict of pediments is intense and real, even though the lowering of a land-surface of only a few feet (or even inches) over a restricted area may be the entire topographic result. But the factor is nevertheless of fundamental importance, and will be referred to again in the following section.

(5) Much has been written of the ages of existing (as distinct from geologically buried) land-surfaces, and the amount of data available is not inconsiderable. But the value of much of it is vitiated by the non-recognition of certain critical limitations. Dating has been generally by the palaeontological approach, where surfaces have been found with fossiliferous superficial deposits. But the point made in the preceding

paragraph shows how false may be the testimony of such deposits and fossils to the age of a land-surface. A surface may well be bevelled in the Oligocene and bear thin deposits of that age. Yet, when in the Pliocene a new pediment works laterally across it, removing the deposits and slicing off a few inches of bedrock to give a barely perceptible alteration in the planed landscape, new deposits then will conclusively point a Pliocene age. As has already been recorded (King 1949), surfaces belonging fundamentally to a single continental cycle of erosion (the " African ") bear superficial deposits ranging in age from Cretaceous to Pleistocene, and a parallel may be drawn with the palaeontological difficulties encountered in study of the superficial " duricrust " of Australia.

To summarize : from the discussion under (3) it is clear that land-surfaces may, under suitably stable tectonic and climatic conditions, remain virtually without alteration over wide areas for considerably longer periods than has hitherto been generally accepted. Moreover, though some authors, imbued with a preconception of peneplainic vertical down-wearing (even at extreme peneplanation) have asserted the contrary, geological literature is full of references by competent authorities to land-surfaces of early Tertiary, and even late Mesozoic age. These ages were assigned to these features to accord with the geological histories of the regions. References to " Cretaceous " land-surfaces are particularly abundant. It is not likely that all these observers were wrong in their conclusion that the surfaces were planed at a remote date.

Without repeating much matter already published elsewhere (King 1947, 1949), we may say that practically all difficulties have been resolved by the adoption of the " pediplanation " as contrasted with the " peneplanation " concept, and by the distinction drawn between continental or " comparative " ages and local or " actual " ages for planed land-surfaces.

When these concepts are accepted it is seen that the ages of planed land-surfaces are in general much greater than have usually been postulated. The oft-quoted " Miocene peneplain " of Africa, for instance, turns out to be of Mesozoic " Gondwana " age, older than Africa itself; while the " end-Tertiary peneplain " is discovered to be the surface developed in the cycle which was generated at the roughing-out of Africa on the dismemberment of Gondwanaland in the late Jurassic to early Cretaceous period (King 1949). To this day it is the most widespread planation in Africa, though the cycle terminated at the coast, following uplift, in the Miocene.

[*Editor's Note:* Material has been omitted at this point.]

List of References

DIXEY, F. 1938. Some observations on the physiographical development of central and southern Africa. *Trans. Geol. Soc. S. Afr.* 41, p. 113.

——. 1942. Erosion cycles in central and southern Africa. *Trans. Geol. Soc. S. Afr.* 45, p. 151.

KING, L. C. 1947. Landscape study in southern Africa. *Proc. Geol. Soc. S. Afr.* 1, p. xxiii.

——. 1949. On the ages of African land-surfaces. *Q.J.G.S.* civ (for 1948), p. 439.

——. 1950. Geomorphology of the Eastern and Southern Districts, Southern Rhodesia. *Mem. No.* 40, *Geol. Surv. S. Rhod.*

——. 1950A. South African scenery : a text-book of geomorphology. Second edition. Edinburgh.

17

Reprinted from *Trans. Inst. Brit. Geog.*, 3, 12, 44-46 (1935)

THE CHANGING SEA LEVEL

H. Baulig

[*Editor's Note:* In the original, material precedes this excerpt.]

To sum up. The object of the research being to reconstruct the river profile at definite stages of its evolution, only those features are of interest which are likely, first to correspond to such definite stages along a goodly part of the river course, and second to have been preserved at least approximately in their original condition. Terminal surfaces of constructional terraces generally satisfy the first condition, for it takes the river a relatively short time to build up its bed, and the maximum of aggradation occurs almost simultaneously in the different parts of the course ; on the other hand, they are much exposed to destruction and replacement by erosional forms of no particular significance. Wide rock benches, on the contrary, imply prolonged stability of the river profile, which can hardly be of merely local occurrence ; on the other hand, they are all the more likely to escape destruction as they are wider. We thus arrive again at the conclusion that erosional forms, especially when developed in hard rocks, are the best indicators of changes in the relative position of land and sea. This leads us to a new and, I think, more promising line of research, I mean the study of erosional forms without an alluvial cover.

[*Editor's Note:* Material has been omitted at this point.]

Thus geomorphology, while of course availing itself of every assistance that general geology can afford, must rely mainly on its own methods. These, it must be confessed, seldom lead to absolute certainty. They rather tend to develop on each question independent lines of argument, the force of which greatly increases as their convergence is more clearly perceived, even though final agreement may long remain unattainable. Truth in geomorphology, indeed, is seldom more than increasing probability.

Correlation of ancient levels may be attempted on the basis of continuity, a very good basis indeed, so far as it goes. But inevitably gaps will be encountered which are generally explicable on structural reasons, either excessive resistance to erosion, which did not permit of the development of levels, or excessive weakness of the rocks which did not ensure their preservation. When seas have to be bridged over in extending the correlation from country to country, we shall sooner or later have to rely entirely, or mainly, on altitudinal correspondence. Such correspondence between an isolated surface here, and another isolated surface there, obviously means little, for it may be accidental. If repeated, however, in many and distant places, it means much more. If the correspondence is observed again and again, not only between isolated levels, but also *between whole series of successive levels*, it may carry conviction. But what shall we understand by correspondence between whole series of levels ? If the record were complete and completely legible, any series might include any number of main and subordinate terms. But, in fact, no series is likely to include all the terms present in all the others, for both the development and the preservation of each level depend essentially on structural conditions which vary from place to place. On the other hand, it can reasonably be expected that, barring very special conditions of structure, each of the *main* terms shall be represented in each of the supposedly parallel series.[1] As the main levels, because of the great length of time involved in their formation, cannot have been many more than are preserved in the apparently complete record of Lower Languedoc (see above, p. 27), correspondence between the main levels of different series is strong evidence of complete parallelism.

Now, supposing the search to be carried farther, what will the result be? We cannot tell, but we can imagine extreme possibilities and intermediate cases.

First: *Total failure.* No supposed eustatic levels can be traced for any great distance before they pass into deformed surfaces. *A fortiori* no correlation between distant lands is possible. The sea level may have changed repeatedly, but the lands have moved at the same time in such a complicated manner that movements of the sea and movements of the land cannot be clearly distinguished. The morphological history of recent times is then, and may remain for ever, a succession of purely local and independent events, although general trends, when long periods are considered, may be discernible. This conclusion, which many will consider the most probable, cannot, to my mind, be reconciled with such facts as are observable in Languedoc and elsewhere, for these clearly demonstrate *practical stability of both land and sea for long periods, certainly much longer than the phases of relative or absolute instability.* No matter how complicated the detail of events may have been, the main facts are simple and clear, and these should be kept in the foreground.

Second: *Eustatic appearances are mere appearances.* Indications of ancient levels of erosion, it is true, are found apparently undeformed, hundreds of metres above the present sea level, and can be traced horizontally for hundreds, eventually thousands of kilometres; but, as no exact correspondence is found between such levels in different parts of the world, their present altitude must be explained by uniform and independent uplifts of the lands. Such a conclusion, conservative as it may seem, would nevertheless constitute a great novelty, for movements of the lands have almost always been conceived as differential. Moreover, it would raise interesting questions concerning the true nature of such uniform uplifts and their effects upon the sea level.

Third: *Eustatism is real but not universally verifiable.* Ample, rapid, and intermittent shifts of the sea level, although necessarily world-wide, have been registered as such only in stable regions. In districts affected by movements of the land, the record is more complicated, eventually so much so as to become illegible. Not only are such shifts of the sea level compatible with certain deformations of the earth's crust, but they seem to imply them. If so, *eustatic evolution in stable regions depends on deformations, more or less synchronous, of unstable regions, both continental and sub-oceanic.* Geomorphological history

[1] Obviously, any series may be incomplete at the top, the upper levels never having existed for want of sufficient altitude, or having been subsequently destroyed. It may also be incomplete at the base because late erosion has not had time, or power, to plane the district under consideration. Another evident principle is that non-deformation of higher levels implies non-deformation of the lower; but the converse is not true.

may then lay claim to something like universality : not only can correspondences be established between distant stable lands, but the evolution of unstable regions can, in favourable cases, be correlated with that of stable ones.

Fourth : *Eustatism, except for very limited and distinctly unstable regions, is universally verifiable,* so that shifts of the sea level would have to be explained without important deformations of the lands. This, of course, would much simplify the task of geology and geomorphology, while taking away much of its interest. On the other hand, it would set before geophysics a formidable problem, namely, that of depressing the sea bottom by hundreds of metres without at the same time raising at least part of the lands by a commensurate amount.

Whatever the final outcome, if only not entirely negative, I feel assured that research conducted along the proper lines, in a spirit of impartiality and independence, will lead to important conclusions, the bearing of which on geology, geomorphology, and geophysics we can only surmise.

18

THE THEORY OF CONTINENTAL FLEXURE

J. Bourcart

This excerpt was translated expressly for this Benchmark volume by Cuchlaine A. M. King from "La Théorie de la fléxure continentale," Compte Rendu XVIe Congr. Intl. Géog., Lisbonne, 1949, *Lisbon, 1950, pp. 167–168*

In 1926, in a general work, I proposed a unique explanation to account for, in the case of fjords, rias, and certain estuaries, the contrast between the deep erosion of their upper courses with the frequent presence of terraces, which are sometimes rock-cut features, and the often very long, sea-drowned lower courses: that of upwarping of the continent, associated with a downwarp of the ocean bed.

I called the junction between these two reverse deformations the continental flexture. The axial plane of the flexure corresponds with a lack of deformation.

I later extended the hypothesis to explain the complex facts shown along the Quaternary coast of Morocco and Portugal, to interpret the morphology and deposits of the continental shelf, and finally I have looked for the cause of the transgressions and regressions.

It would be an exaggeration to believe that this hypothesis was, at first, simply accepted by the majority of geologists and geographers. These, who were strongly attached to the doctrine of eustatism, would not readily allow one to touch this dogma. Nevertheless, in the recent past, it has been closely examined by Umbgrove in his important work. He has compared the theory with the ideas of Jessen and du Toit. G. Lucas has used it to interpret certain facts which emerged from his study of Algeria and Morocco. G. Zbyszewski has accepted it with minor modifications. Finally, Umbgrove has reminded us that this heretical explanation was already envisaged by T. C. Chamberlin and R. D. Salisbury in their classic work of 1906. *Nihil novi sub sole* . . . (there is nothing new under the sun . . .).

Today it is possible for me to enumerate many geological, oceanographical, and morphological facts that can be explained by this hypothesis, which, consequently, in the sense in which Claude Bernard used the word, should be, despite me, promoted to the rank of theory. I will attempt to summarize them briefly.

1. EXPLANATION OF TRANSGRESSIONS AND REGRESSIONS

Transgressions and regressions are fundamental in geology. E. Haug has distinguished geocratic and thalassocratic periods, characterized, respectively, by cooling and regressions and warming and transgressions, as in glacial oscillations. The regressional phases are associated with coarse deposits resulting from rapid erosion on steep slopes. These are periods of mountain uplift. In thalassocratic periods relief is reduced. Flexure can account for these effects either on an extensive or on a local scale.

2. CAUSES OF TRANSGRESSIONS AND REGRESSIONS

The theory explains various facts concerning the continental shelf, including the structures off the Appalachians and southwest Britain, where the basement beneath the sediments is downwarped seaward. Everywhere, it seems, offshore from the Armorican geanticlines there are major synclinal troughs filled with Cretaceous and Tertiary strata. The evidence of submarine canyons is important. They occur in the Georges Bank area, off Cap Breton, the Gulf of Lyons, and elsewhere. The submarine topography off California also suggested down-warping. The canyons are steep and rocky in places and must have been cut by fluvial action, as marine currents are too weak; turbidity currents cannot erode in limestone or granite. Some canyons can be followed to depths of 2000 m, and such depths can only be accounted for by crustal warping.

19

Reprinted from *Scottish Geog. Mag.*, 67(2), 68-73, 74, 83-85 (1951)

PROBLEMS OF SCOTTISH SCENERY

David L. Linton

[*Editor's Note:* In the original, material precedes this excerpt.]

But the evolution of a landscape is not the same thing as the evolution of the rocks from which it has been fashioned, and the history of Scottish scenery begins, not in those early periods whose names are inscribed on our coloured geological maps but with the moment

> " When Britain first at Heaven's command
> Arose from out the azure main."

For Scotland at that time was truly just North Britain, as yet undifferentiated from the southern half of the island. The " azure main " was without doubt the Upper Cretaceous sea, with its flat floor newly carpeted with chalk of the Senonian stage. Our eastward-flowing rivers first took their courses on this sea-bed when it emerged and became dry land, what time the Danian and Maastrichtian chalks were being laid down in Belgium and Holland. From that day to this, that land surface has remained a land surface, and its rivers, or their dismembered remnants, have continued to run. Successive uplifts have raised it and possibly tilted it further during the sixty million years of Tertiary time that have since elapsed. At the extreme south, the part that now borders the English Channel was considerably distorted about half-way through that period. During those same sixty million years heat and frost, rock rotting by ground water, the erosive and scavenging action of rivers, and gnawing by the waves of the sea, have operated unceasingly, and for a time the powerful bulldozers and excavators of the glacial period made a conspicuous contribution to the task of wasting away what uplift had raised up. Scotland as we know it has been sculptured from the solid during Tertiary time. First the barely consolidated cover of chalk was cut through by the largest rivers, and the older rocks of the undermass were exposed in their valley floors, just as the Belgian rivers Dendre, Senne and Dyle in the country south of Brussels have cut through the Tertiary cover rocks and exposed the lower Palaeozoic schists and the granite of Ath. Later, the chalk would be wasted from the divides, but by that time the main rivers had bitten too deeply into the hard rocks below to be dislodged save by river capture and glacial accidents. In relatively short time, we may believe, the chalk was entirely removed except in the west where it had been armour-plated by the lavas of the Inner Hebrides, and near the eastern coastline where it had not been raised significantly above base-level. The erosive agents then began to discover the differences in resistance of the various portions of the undermass. Then, and not till then, those fundamental subdivisions of Scotland—the Highlands, Central Lowlands, and Southern Uplands—began to take shape as regions of contrasted scenery. The original unity of the uplifted and tilted block began to give way to the diversity we know to-day.

At this point a question inevitably obtrudes. If all Britain has been sculptured in Tertiary time from a single massive block, are there no remaining traces of the form of that block ? The original chalk cover has gone from northern and western Britain, stripped like wax from a sheet of glass. But what of the surface of the hard undermass from

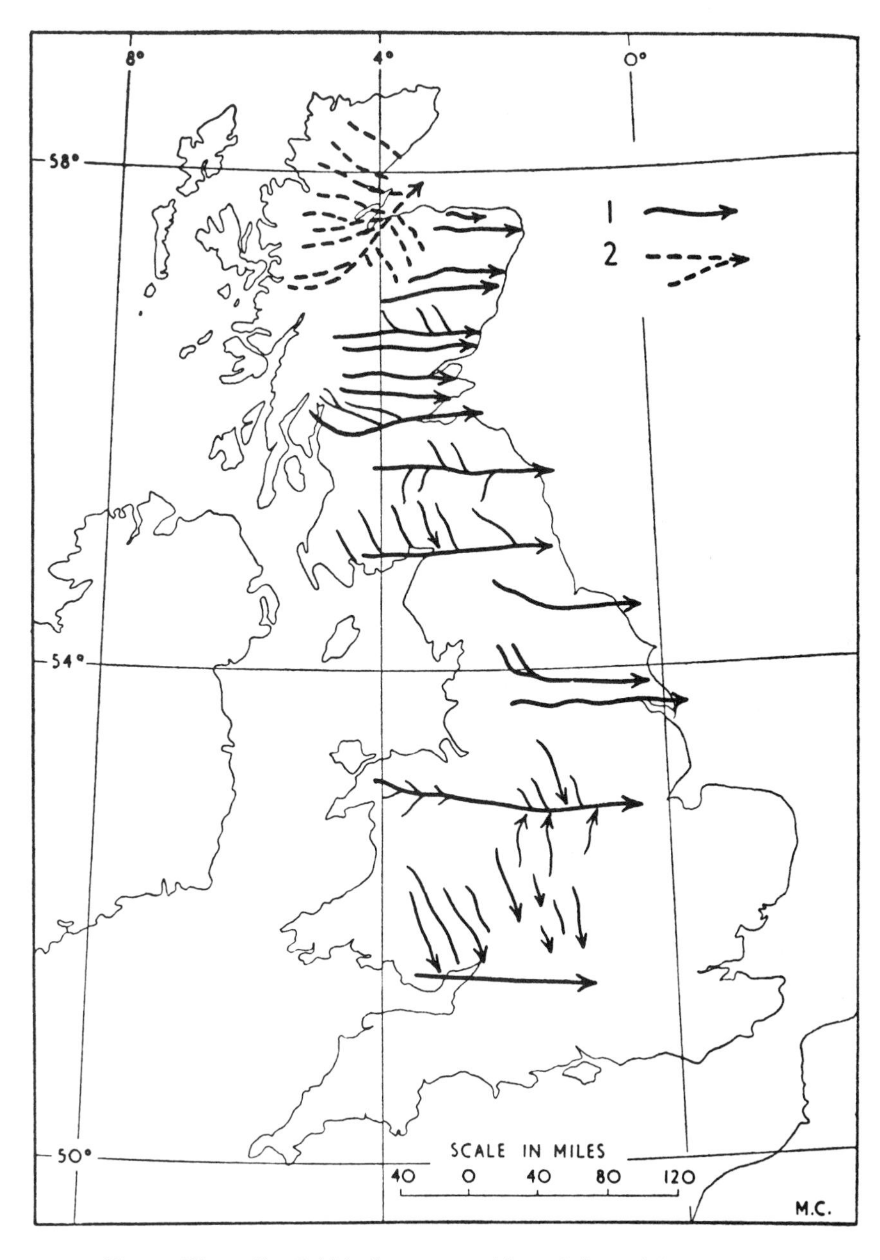

Fig. 1. The earliest British rivers, restored from their surviving remnants.
(1) Rivers presumed consequent on the initial post-Cretaceous uplift and eastward tilt.
(2) The possibly independent or later drainage of the Moray Firth depression.

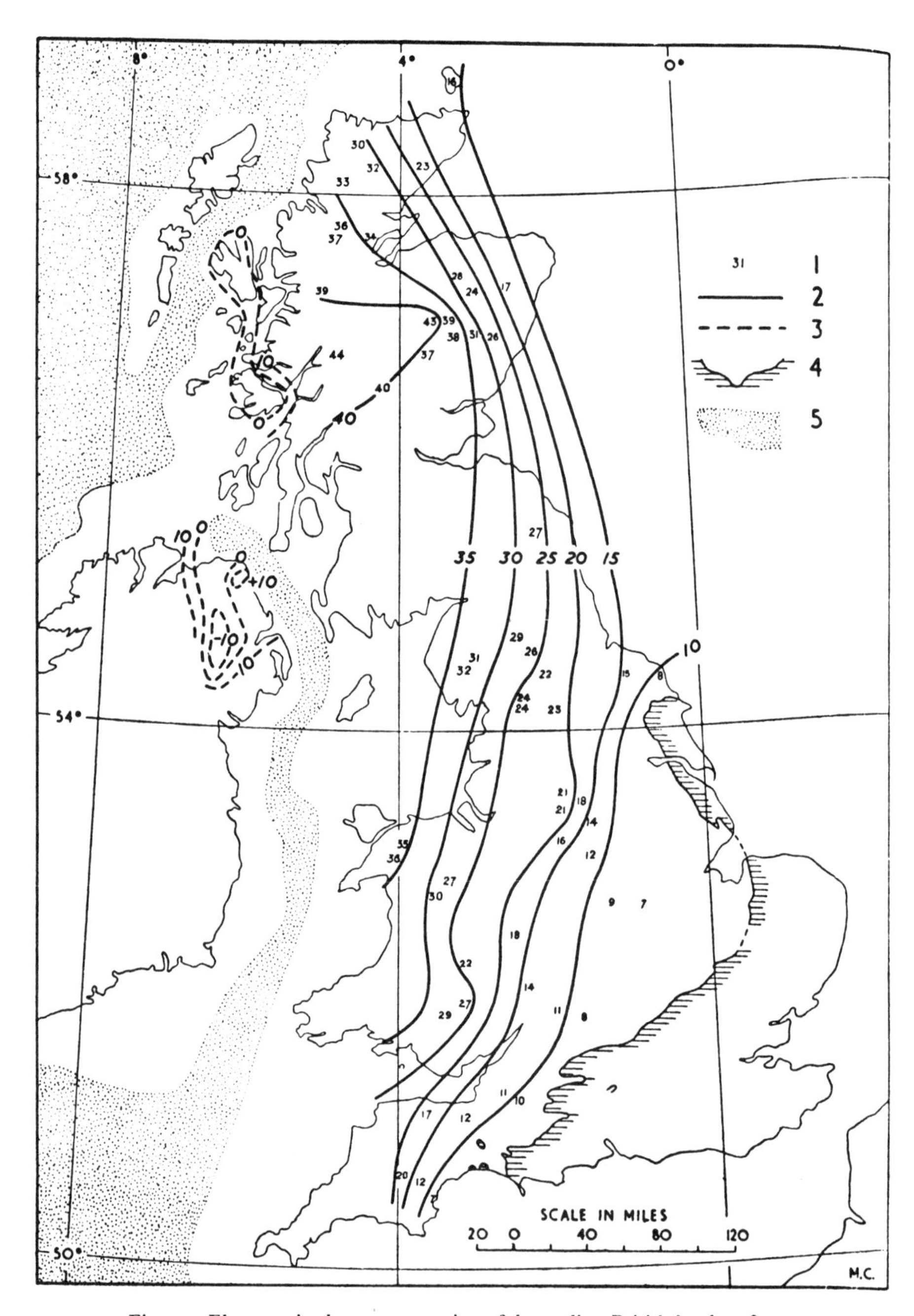

Fig. 2. Elements in the reconstruction of the earliest British land surface.

(1) Heights (in hundreds of feet) of the surviving most easterly high summits.

(2) Contours of a smooth surface that would touch these summits.

(3) Approximate stratum contours of the base of the Antrim and Hebridean lavas, below which thin Cenomanian strata are present in places.

(4) Outcrop of the Cenomanian strata of south and east England.

(5) Deeper parts of the Irish Sea, Minch, etc.

which the cover was removed? There are good reasons for thinking that this surface was very nearly plane. Bailey has argued from the general absence of terrigenous matter in the chalk of western Europe generally, and from the wind-blown character of the sand grains of the Cretaceous sandstones of the Hebrides, that the northern shores of the Chalk sea were fringed by a desert peneplain.[17] The great lateral extension of the Cenomanian sea from mainland Britain to Antrim and the Hebrides, coupled with the mere score or so of feet of the Cenomanian deposits in these areas, testifies eloquently to the negligible relief of the surface—the *sub-Cenomanian surface*—that was laid bare again when the Scottish Chalk was stripped and wasted away, and must effectively be the surface of the resistant masses from which the uplands of Britain have been fashioned. Is it possible to recognise any hint or vestige of this surface in the landscape of to-day? I think it is.

For any portion of this surface to survive, it must have suffered virtually no denudation since it lost its chalk capping. This can only be the case if it has occupied a watershed position ever since. Only a relatively small number of our higher peaks are likely to qualify for inclusion here. But there is a further condition to be satisfied—our rivers tell us that the surface of the block sloped to the east. With these considerations in mind, Fig. 2 has been constructed. The figures represent the altitudes, in hundreds of feet, of our most easterly high peaks, for only these are likely to have reached up to the sub-Cenomanian plane. They have not been selected by intuition or mere inspection but represent the result of successive processes of graphical elimination, which it would be out of place to discuss here. It may be taken that each summit is the most easterly of its altitude in its own latitude and is not flanked by higher summits to north and south along its own meridian. Between these plotted figures 'contours' have been drawn to represent the altitudes 20, 25, 30, 35 and 40 hundred feet (i.e. 2000, 2500, etc. feet). It will be seen that these contours are astonishingly regular. Between the meridians of 2° and 4° west, and a little further west in Highland Scotland, they portray a surface sloping eastward: almost due east in southern Scotland, a little south of east in the English Midlands and definitely north of east north of the Grampians. The confirmation of the testimony of the rivers is independent and striking. We may surely accept as a working hypothesis, to be put to the test whenever methods of doing so can be devised, that the summits mapped just rise to the level of the tilted and uplifted sub-Cenomanian surface. In particular we may note that the quite extensive surface of the Cairngorm summits at about 4000 feet which has frequently been noted as being far above the general level of Peach and Horne's "High Plateau,"[18, 19, 20] may in fact be an actual remnant of the sub-Cenomanian surface.

The eastward gradient shown by the contour lines of Fig. 2 is, of course, the cumulative result of several episodes of elevation and tilting throughout Tertiary time. The initial gradients were doubtless much less, but even so, they rose westwards. How far was this rise continued? Mackinder once speculated that, "north-westward, at any rate, even the hundred fathom line does not mark the limit of the vast tabular block from which Scotland was carved."[21] No evidence lends any support to such a suggestion, but the facts shown in Fig. 2 are entirely consistent with the attractive hypothesis advanced by Cloos in 1939.[22]

Cloos had made a close study of such areas as the Rhineland, the Red Sea region, and East Africa, where large-scale rifting is prominent, and had concluded that this rifting was the consequence of an upward bulging of part of the earth's crust. After the upward bulging has reached a certain development, collapse of the most elevated area is likely to occur and to be associated with vulcanicity. Among the dome-shaped areas which Cloos recognised is the area of Great Britain and Ireland. Britain forms the eastern half of his bulge—and here we may note the significant eastward convexity of the contours of Fig. 2 ; Ireland and the Outer Hebrides form the western half. Between the two he shows a meridional rifted zone whose southern portion is suggested by the roughly parallel-sided trough whose sides descend sharply from the 30- to the 50-fathom line beneath the western Irish Sea. In the northern portion the plateau lavas of Antrim and the Inner Hebrides have been preserved from erosion by being faulted or warped down below Tertiary base-levels. If the sub-Cenomanian surface is indeed represented by the contours of Fig. 2, rising from 2000 feet about Stonehaven to some 4500 feet at Ben Nevis, then this same surface is depressed to about 1500 feet beneath the summit of Beinn Iadain in Morvern and descends to sea level on the shores of the Sound of Mull. At Strollamus in Skye, at Gribun on the west coast of Mull, and extensively in Antrim, the Chalk and Greensand deposits outcrop on the coast. It would appear that the crown of Cloos's arch must have been dropped down by an amount approaching 5000 feet.

It is worth noting that this viewpoint has an important corollary. If our eastward slope broke off toward the west along the margins of the "Hebridean Rift," there must from the outset have been two sets of rivers, the one set flowing eastward to the North Sea depression, and the other westward into the Rift. The succession of high peaks that extends northwards from Ben Cruachan almost to Cape Wrath and roughly along the meridian of 5° W, must surely represent the line of the original watershed ; the peaks are, indeed, very difficult to explain on any other hypothesis. But here a difficulty arises. This zone of peaks is broken through again and again by low passes, mostly less than a thousand feet above sea level and two thousand feet below the peaks. It has long been customary, following the views of Mackinder and of Peach and Horne regarding the south-eastward tilt of the High Plateau, to explain these passes as having been cut by early consequent rivers heading far out over the Minch and the Sea of the Hebrides. Bremner continued to give currency to these ideas in his map of the "Tertiary Rivers of North Scotland" as late as 1942.[23] Sölch, however, has argued against this view. If these valley passes were made by eastward-flowing rivers, where, he asks, are the western mountains from which those rivers came ? Moreover, many of the islands and peninsulas west of the passes show erosion surfaces at levels above 1000 feet which were presumably fashioned by these western

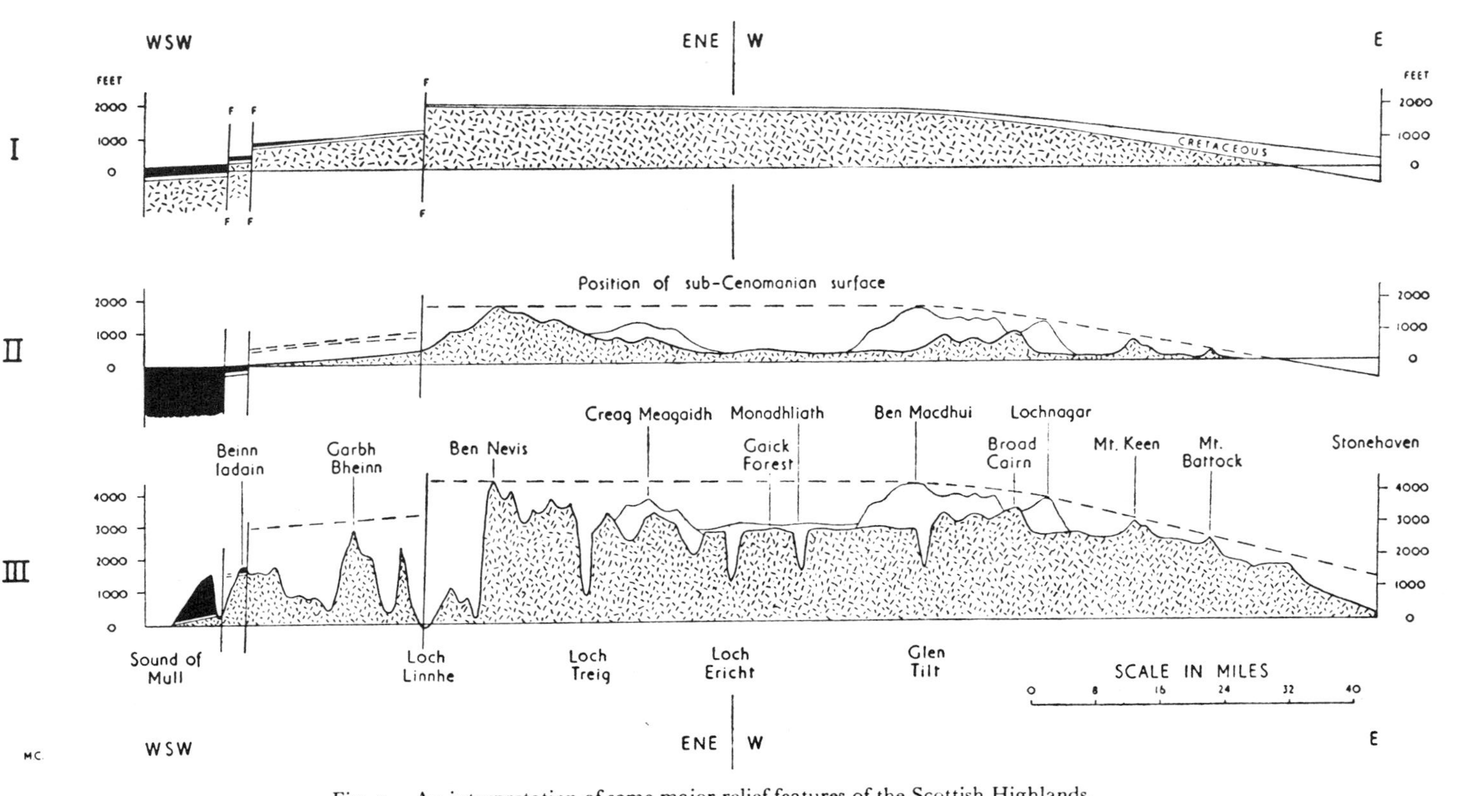

Fig. 3. An interpretation of some major relief features of the Scottish Highlands.

III. An approximate scale section through Morvern, Lochaber and the main Grampian watershed, to show present relief: Cairngorms, etc., projected on to line of section. Black=Tertiary lavas. White=Cretaceous deposits. Stippled=Granite and metamorphic rocks. The fault in Beinn Iadain is visible, the others are presumed.

II. A reconstruction of the same section at the end of the early Tertiary erosion cycle which produced the peneplain termed the Grampian Main Surface, when only part of the warping and faulting had been accomplished.

I. Reconstruction of the same section at a still earlier stage, when the main drainage was initiated on the presumed continuous cover of Upper Cretaceous strata. Warping as in II, but lava effusion and faulting in the west are shown at an early stage.

rivers ; but Sölch finds it difficult to understand how the rivers that could open up these base-levelled lowlands should simultaneously or later have traversed the mountains in relatively narrow valleys. He therefore accepts the zone of peaks from Argyllshire to Sutherland as the zone of maximum uplift and leaves the problem of the through valleys to be explained in other, and possibly various, ways. To this problem we shall later return.

[*Editor's Note:* Material has been omitted at this point.]

REFERENCES

[1] BAILEY, E. B. The Interpretation of Scottish Scenery. *S.G.M.*, 1934, 50 (*5*) : 308-330.

[2] BREMNER, A. The Capture of the Geldie by the Feshie. *S.G.M.*, 1915, 31 (*11*) : 589-596.

—— A Geographical Study of the High Plateau of the South-Eastern Highlands. *S.G.M.*, 1919, 35 (*9*) : 331-351.

—— The River Findhorn. *S.G.M.*, 1939, 55 (*2*) : 65-85.

—— The Origin of the Scottish River System. *S.G.M.*, 1942, 58 (*1*) : 15-20 ; (*2*) 54-59 ; (*3*) 99-103.

—— The Later History of the Tilt and Geldie Drainage. *S.G.M.*, 1943, 59 (*3*) : 92-97.

[3] —— The Glaciation of Moray and Ice Movements in the North of Scotland. *Transactions of the Edinburgh Geological Society*, 1934, 13 : 17-56.

[4] MACKINDER, H. J. Britain and the British Seas. London : William Heinemann, 1902. P. 133.

[5] PEACH, B. N., and HORNE, J. The Scottish Lakes in Relation to the Geological Features of the Country. *Bathymetrical Survey of the Scottish Freshwater Lochs*, vol. 1, p. 457. Edinburgh, 1910.

[17] BAILEY, E. B. The Desert Shores of the Chalk Seas. *Geological Magazine*, 1924, 61 : 102-116.

[18] LOUIS, H. Glazialmorphologische Studien in den Gebirgen der Britischen Inseln. *Berliner Geographische Arbeiten*, 1934, Heft 6, esp. p. 19.

[19] SÖLCH, J. Geomorphologische Probleme des schottischen Hochlands. *Mitteilungen der Geographischen Gesellschaft in Wien*, 1936, Bd. 79, p. 32.

[20] FLEET, H. Erosion Surfaces in the Grampian Highlands of Scotland. *Rapp. de la Comm. pour la Cartographie des Surfaces d'Appl. Tertiares*, Union Géogr. Internat., Paris, 1938, pp. 91-94.

[21] MACKINDER, H. J., *op. cit.*, 1902, p. 133.

[22] CLOOS, H. Hebung, Spaltung und Vulkanismus. *Geologische Rundschau*, Bd. 30, Zwischenheft 4A, Stuttgart, 1939.

[23] BREMNER, A., *op. cit.*, 1942, p. 101.

20

Reprinted from *Climatic Geomorphology,* E. Derbyshire, trans., Macmillan & Co. Ltd., London and Basingstoke, 1973, pp. 51-60

Attempt at a Classification of Climate on a Physiographic Basis

ALBRECHT PENCK

IN place of the ancient's division of the earth's surface into parallel climatic zones based solely on latitude, there have been recent attempts at climatic classification which are based on temperature and precipitation, although the actual delineation of the individual climatic regions has been undertaken from many different standpoints.

GEOGRAPHICAL DEFINITIONS OF CLIMATE

A. Supan (1884) pushed the geographical viewpoint to the fore, by asking which areas possessed a more or less similar type of climate, and presented us first with 34 climatic provinces, and later with 35, which were primarily to be regarded as geographical unities. In fact, they differ only very slightly from the natural regions of Herbertson (1905), who divided the surface with regard to climate and surface relief. R. Hult (1892–3) has further stressed the climatological viewpoint and in a little-noticed work (noted by Ward, 1906) distinguishes 33 climatic zones which are delineated primarily by temperature conditions and secondly by precipitation and wind conditions. Thus he arrived at nine major zones, which he then proceeded to subdivide further into 33 zones which he subdivided again into provinces. The climatological viewpoint was put even more strongly by W. Köppen (1901). His very remarkable attempt at a classification of climate proposes a sharp division of climatic provinces on the basis of temperature and precipitation conditions, and both serve in delineating the boundaries. Biogeographic facts determine his seemingly arbitrary selection.

On the ground it seems possible to use climate (that is, the interaction of all atmospheric conditions) as a basis, for it imprints itself so clearly on the landscape that it becomes possible to distinguish whole climatic regions without having to start from long columns of meteorological observations. The influence of the climate on the character of the land surface is above all dependent on what form the precipitation takes. Whether it ultimately takes the form of rivers or glaciers is entirely dependent on the climate. A. Woeikof (1885–7) has emphatically stated that rivers are products of climate. From the climatological point of view it is important to know whether or not precipitation is completely evaporated thus leaving the land without water.

DIVISION OF CLIMATIC PROVINCES

Three different principal climatic provinces or regions may be distinguished on the earth's surface.

(1) The humid climates, in which more precipitation (N) falls than is removed through evaporation (V), so that a surplus in the form of rivers (F) runs off.

(2) The nival climates, in which snowfall (S) exceeds ablation (A), so that a removal in the form of glaciers (G) must ensue.

(3) The arid climates, in which evaporation absorbs all precipitation and could absorb even more, thus preventing a flow of river water.

We can characterise these three climates with the following formulae

$$(1)\ N - V = F > 0 \qquad (2)\ S - A = G > 0 \qquad (3)\ N - V < 0$$

Our three main provinces are separated by two important boundaries; the first is characterised by a balance between evaporation and precipitation, the other by one between snowfall and ablation. This latter boundary is the well-known snowline (SG), which can be expressed as $S = A$. Its other boundary has been termed the *dry boundary* (TG): $N = V$ is valid for this.

The snowline

The snowline has long aroused interest. It separates the areas under constant snow cover from areas occasionally under snow; this concept, therefore, often occurs in the context of landform evolution. Similarly, there have recently been probing discussions on the exact position of its location and, indeed, doubt has been expressed as to its very existence. In fact, its position is not constant; it changes from year to year, according to variations in ablation and snowfall, but in the course of the year it oscillates around a definite mean position. This is in no way related to a particular isohyet, for the snowline can be found in one and the same area at greatly differing heights, depending on exposure and surface relief, which in one place may be favourable to snow accumulation and in another hinder it. It has therefore become necessary to introduce an ideal height for the climatic snowline of a certain area, instead of local, observed heights. It is the height at which the snow that has accumulated in one year on a horizontal surface exceeds the total ablation. This value is important in comparing the snowlines of various regions; but in the delineation of the nival and humid areas the local snowline plays a considerable role. Under local conditions the same is true for the local snowline as for the climatic snowline on a horizontal surface, namely that above it more snow falls than can be melted in any series of years. Its position is therefore determined by the total snowfall and by the sum of the temperatures above 0°. A diurnal maximum of over 0°, which will melt the surface of the snow, will not decrease the snow cover as long as the meltwater freezes again during the night. Only a continuous period of warmth will *reduce* the snow. Therefore Finsterwalder and Schunck (1887) regarded ablation as more or less proportional to the snow-free period and the average temperature above freezing point during this time, and Kurowski made them proportional to the duration and average temperature of the period above 0°. But the only serious attempt to determine mathematically the relationship between snowfall, average temperature and length of frost-free period gave widely differing results from neighbouring glaciers (Machatschek, 1899) and today we are still a long way from understanding the individual elements of climate which determine the position of the snowline. It is the product of various factors, which are not yet fully understood.

The dry boundary

Less striking than the snowline is the dry boundary of the earth. Towards this boundary the humid regions become progressively poorer in rivers, so that these tend to disappear as the arid region is reached. It is evident that the position of this zone is noticeably influenced by the nature of the ground, just as the snowline is influenced by exposure; rivers will disappear on permeable sooner than on impermeable ground. It follows that arid areas are never completely without rivers; every heavy downpour is accompanied by surface runoff, although this is not a regular occurrence. Here we are concerned with torrents, rather than genuine rivers. Furthermore, numerous rivers flow out of the humid areas into arid regions. Whilst, however, they gradually increase in size in these humid areas, they progressively decline in the arid regions. Although the channels in arid areas are fundamentally different from those in the humid areas it is not always easy to separate the two. However, this difficulty does not prevent us from realising that the dry boundary is one of the most important boundaries on the earth's surface. The determination of its position has not yet been achieved, for we are not sufficiently acquainted with the controlling climatic factors. But a few bases for such a determination are available from an investigation of the relative conditions of precipitation (N) and runoff (F) in humid regions. They show that this relationship is not, as was formerly supposed, characterised in a river by a certain runoff factor, but can approximately be expressed by the following formula

$$F = (N - NO)x$$

where NO represents a critical precipitation value which may vary only slightly for neighbouring rivers and x is a proper fraction. To extrapolate, lack of runoff in the catchment area in question appears when

$$N = NO$$

The value NO therefore indicates the precipitation below which runoff ceases in humid areas. Axel Wallén in his work on central Sweden puts it at 100 mm—the values for Central Europe are around 420–30 mm. Merz (1906) has deduced the value 1100 mm for Central American rivers. As the average yearly temperatures for these areas are 1°, 7° and 24° respectively, it can be seen that the level of precipitation at which the runoff equals 0 increases with the temperature. But we are still a long way away from determining more exactly all individual elements which determine the position of the dry boundary.

PHREATIC ZONES

There are two main zones which can be distinguished in humid climates. In the first zone the precipitation can percolate into the ground and, depending on the ground's permeability, more or less fill it to form ground-water. This is not possible in the other zone because the ground is frozen. Here in the polar climatic province we have ground-ice instead of ground-water in what may be called the *phreatic* zone. The boundary of this ground-ice has aroused repeated interest. Fritz (1878) has represented it on a map which has been often reproduced; it coincides approximately with Wild's (1881) average annual temperature of $-2°$. In the polar climate sources of ground-water as well as ground-water itself are lacking. There is only some surface water which in summer succeeds in percolating through the thawing surface layer; it can move readily along such a surface and numerous movements of a purely superficial nature occur. In this sliding and partly flowing earth layer weathering is mechanical—with regular refreezing the water occasionally present in the surface layers shatters the uppermost rock layers and loosens them. The river is fed mainly by snow-melt and this generally produces considerable quantities of water within a relatively short period of time; a short high water stage in summer and a protracted period of winter low water characterise the polar rivers. The snow cover extends over the land for months, but not long enough to prevent tree growth. It is now known that tree growth is not related to the limit of frozen ground, as was originally supposed.

In climatic provinces characterised by phreatic zones, a greater or lesser part of the precipitation (depending on permeability) soaks into the ground and only joins the river after passage underground; these rivers then are only partly fed by the falling rain. The percolating water loosens the rocks along its path and attacks susceptible rocks with carbonic acid (H_2CO_3): it extends to the upper layer of the regolith and forms the characteristic leached or eluvial layer.

Provinces in phreatic regions

Within such a phreatic region we can distinguish individual provinces based on precipitation distribution. If rainfall is regular throughout the year, then feeding the rivers through precipitation or indirectly by ground-water continues regularly the whole year round and the rivers maintain a fairly constant flow. If, however, the precipitation shows an irregular distribution, exhibiting a clear distinction between rainy and dry seasons, then the rivers show a marked period of high water separated by periods of low water, and even of periodically dry stream beds. Such *occasional* channels are termed *wadis*. These occur at the boundaries of the arid and humid regions. In dry periods, percolation and ground-water supply ceases, and arid conditions develop which interrupt the humid conditions. Areas in which arid and humid conditions alternate annually are termed *semi-arid provinces.*

Towards the polar or nival province the phreatic region exhibits its special feature of the development of a regular snow cover, which can prevent the flow of water seasonally; then, after melting it can augment not only the ground-water but also the rivers. Accordingly, these rivers show a characteristic high level, which occurs according to the lateness of the melting; it occurs later in mountain rivers than in those of the plains and later in the season with greater proximity to the pole. This high meltwater stage often directly follows a period during which the streams are covered with ice. This subnival climatic province is separated from the nival province on its poleward margin by the snowline, and from the polar province by the occurrence of frozen ground. Its equatorward boundary is drawn where occasional snow cover ceases to contribute to the régime of the river. This occurs approximately where there is snow cover for about one month each year; where it lasts for a shorter period it causes no noticeable increase in precipitation storage. The subnival province, therefore, extends neither as far as the regions of permanent snow cover nor as far as the equatorial limit of snow fall, a point which has been closely examined by Hans Fischer (1887). Its boundaries still remain to be determined locally; and in attempting this, similar uncertainties will have to be met as with determining the dry boundary. The limits of the subnival province correspond approximately to the 1°C to 2°C isotherm for the coldest month. In the subnival province, as in the polar province, we can distinguish two subprovinces according to duration of snow cover; one with a predominantly snow-free period and the other snow-covered for most of the time. The boundaries of these two subprovinces coincide approximately with the treeline, and we therefore distinguish both in the polar province and also in the subnival province between forested and unforested regions.

The phreatic regions with a regular distribution of rainfall are typical of the humid climatic province. The latter breaks down spatially into subprovinces generally separated from each other by semi-humid or arid regions, namely the equatorial region with its high temperatures and abundant rainfall all the year round and the temperate region with its considerable annual range of temperatures. Neither prolonged ice formation on the rivers nor the regular appearance of a snow cover occurs in the latter region, although neither frost nor snowfalls are completely lacking. In the temperate humid subprovince, as in the subtropical province, the rivers usually exhibit high water in the cooler part of the year which, however, does not necessarily coincide with the period of most rainfall. During this time, evaporation is at its lowest and consequently the flow during this period is not only relatively, but also often absolutely, at its greatest.

PROVINCES IN ARID REGIONS

Just as the humid region can be divided into subregions in which precipitation percolates into the ground all the year round and those in which this process is temporarily or completely interrupted, so the arid region can be divided into two provinces in which aridity is important for all or part of the year. As we have seen, precipitation is never completely lacking in arid regions; it is always present, but perhaps not in sufficient quantity to be able to supply regularly flowing rivers. However, it can be of considerable importance in the development of torrents and quite a considerable vegetation growth, adapted to the dry climate. In this semi-arid climatic province the rainfall of isolated downpours often partially percolates into the ground, but cannot collect as extensive ground-water, because during the dry season it evaporates. Then it is brought to the surface again by capillary action. The 'percolating water' therefore, has no regular downward passage, as in the phreatic regions, and while it returns to the surface and is evaporated, it leaves behind those substances which it has dissolved at depth. Correspondingly, the leaching of the soil which occurs in phreatic areas is absent and is replaced by deposition of soluble salts, particularly calcium carbonate, in the upper layers of the soil. It is this calcium carbonate which so often makes up the hard surface crusts which are very characteristic of semi-arid regions.

In the completely arid province this rising and percolation of ground-water ceases and, consequently, no hard crusts are formed. The rocky surface is subjected to mechanical weathering only, as it has neither the vegetation cover of the humid regions nor the hard crusts of the semi-arid regions as protection against the wind. Hence the wind plays an extremely important role—eroding here, depositing there. According to temperature conditions the completely arid climatic province may be divided into two subprovinces: a temperate arid region with marked *seasonal* variation of temperature and a subtropical subprovince with a large *diurnal* range. A similar classification is possible for the semi-arid province.

NIVAL REGIONS

The nival region is characterised by accumulating snow deposits, both in the completely nival province where precipitation takes the form solely of snow and the semi-nival province where this is interrupted by rainfall. This rainfall, however, does not contribute to a decrease in the amount of snow; it causes moist surface conditions, which in turn favours a compaction of the snow and with the return of frost this again turns into ice. These hard crusts on the snow cover of our highlands play a particularly important role, but they may also develop in the completely nival regions as a consequence of intense insolation whereby the surface snow melts and the meltwater freezes again at very shallow depths.

In the nival region the land surface is protected from atmospheric weathering. But it is not improbable that a unique weathering process is initiated under the load of the snow and ice cover—and our attention has been drawn to this by Blümcke and Finsterwalder (1890). If a local increase in pressure causes a local fluidising effect on the base of the ice, then a moisturising of the basement rock may result. However, as soon as freezing sets in again, then this freezes too, and this can cause quite considerable frost-shattering forming fine dust particles. But this subglacial weathering of the ground remains far less important than the direct mechanical action of the glacier ice.

The ice erodes the surface and then deposits the eroded material when continual ground melting occurs—whether in parts of the basal ice where movement is minimal (Penck and Brückner, 1909) so that geothermal heat results in the liberation of englacial deposits, or at the periphery where the ice melts marginally.

The glaciers which originate in the nival region usually move out of this region and extend far into the subnival and polar climatic provinces, where melting occurs. The effect of glaciers extends far beyond the limits of the nival regions and the limits of former glaciation do not coincide with the extent of the earlier nival region. Just as the lakes of the glaciers extend out of the nival region, so the rivers of the humid region extend into the arid; therefore, the presence of typical fluvial activity at any one point does not indicate either that the area was formerly or is now a part of the humid region. Rivers entering the arid zone react in the same manner as glaciers entering the humid areas; they are consumed, they lose their water content—partly by direct surface evaporation and partly to the ground from which they gain no supply from ground-water and to which they lose a lot of water through filtering. In every respect they appear as strangers in the climatic province in which they find themselves.

KARST AREAS

If the presence of regularly flowing rivers is not indicative of a humid régime, then conversely the lack of rivers does not necessarily characterise an arid area. There are areas in the humid regions where the permeable nature of the ground not only favours the percolation of rainwater, but also the complete disappearance of whole streams. The karst areas are a good example of this. Numerous further examples can be found on extensive gravel sheets (*schotter*) and sandy landscapes, which soak up both rainwater and rivers. These pseudo-arid regions are distinguishable from the genuinely arid regions by the fact that the lack of surface water is combined with the occurrence of a good supply of water at depth, which succeeds in supplying springs. This spring-feeding water is lacking in truly arid regions: these have only filtered water which often extends far beyond the limit of surface water in allochthonous river beds.

Thus it is not one single feature which characterises a climatic region: rather, the character of a region is the sum of all its parts and it is possible to separate the individual regions by direct observation of these characteristics. This observation is valid for all the provinces discussed here.

REFERENCES

BLÜMCKE, A., and FINSTERWALDER, R. (1890). Zur Frage der Gletschererosion. *Sitzungsber. d. math.-phys. Klasse d. Kgl. Bayer, Adak. d. Wiss.*, **20**, 435

DE MARTONNE, E. (1909). *Traité de géographie physique*. Paris, 205

FINSTERWALDER, R., and SCHUNCK, W. (1887). Der Suldenferner. *Zeitschr. des Deutschen und Osterreichischen Alpenvereins*, 70

FISCHER, H. (1887). Die Aquatorialgrenze des Schneefalls. *Mitteilungen des Vereins für Erdkunde*, Leipzig, 97

FRITZ (1878). *Petermanns Geographische Mitteilungen*, table 18

HERBERTSON, A. J. (1905). The major natural regions. *Geographical Journal*, **1**, 300

HULT, R. (1892–3). Jordens Klimatomraden. Forsok till en indelning af jordytan efter klimatiska grunder. *Vetenskapliga Meddelanden af Geografiska Foreningen i Finland*, **1**, 140

KÖPPEN, W. (1901). Versuch einer Klassifikation der Klimate vorzugsweise nach ihren Beziehungen zur Pflanzenwelt. *Geographische Zeitschrift*, **6**, 593

KUROWSKI, P. Die Höhe der Schneegrenze. *Geogr. Abh.* **5**, 1, 115

MACHATSCHEK, F. (1899). Zur Klimatologie der Gletscherregion der Sonnblick-gruppe *VIII. Jahreshericht des Sonnblickvereins für 1899*, 24

MERZ, A. (1906). Beiträge zur Klimatologie und Hydrographie Mittelamerikas. *Mitteilungen des Vereins für Erdkunde Leipzig für 1906*

PENCK, A., and BRÜCKNER, E. (1909). *Die Alpen im Eiszeitalter*, 951

SUPAN, A. (1884, 1908). *Grundzüge der physischen Erdkunde*. Leipzig, 1st ed., 129; 4th ed., 227

WARD, R. DE C. (1906). The classification of climate II. *Bulletin of the American Geographical Society*, **38**

WILD (1881). *Die Temperaturverhältnisse des Russischen Reiches*. St Petersburg, 348

WOEIKOF, A. (1885). Flüsse und Landseen als Produkte des Klimas. *Zeitschrift der Gesellschaft für Erdkunde*. Berlin, 92

—— (1887) *Die Klimate der Erde*. Jena, 39

21

Reprinted by permission from *Ann. Assoc. Amer. Geog.,* **40**, 219, 221–223, 233–236 (1950)

THE GEOGRAPHIC CYCLE IN PERIGLACIAL REGIONS AS IT IS RELATED TO CLIMATIC GEOGRAPHY

L. C. Peltier

[*Editor's Note:* In the original, material precedes this excerpt.]

Nine different climatic and possible morphogenetic regimes, illustrated graphically in figure 7, may be postulated from the foregoing analysis. Each should be distinguished by a characteristic assemblage of geomorphic processes. Thus, if cycles are to be defined in terms of the agents and regimes which produce them, they should correspond to nine different geographic cycles. Davis has admitted the existence of the moderate cycle (called by him "normal") (1899), the arid cycle (1905), and the glacial cycle (1909). Each of these "cycles" can easily be identified with one of these climatic, morphogenetic regimes. Cotton has recognized a savanna cycle and a semi-arid cycle (1942) which, in a climatic sense, are intermediate between the arid and moderate ("normal" of Davis) cycles. He also mentions briefly a "hot-humid" cycle based on observations by Sapper (1935) and Freise (1938). In order to maintain brevity of name, and in keeping with Cotton's use of the ecological term

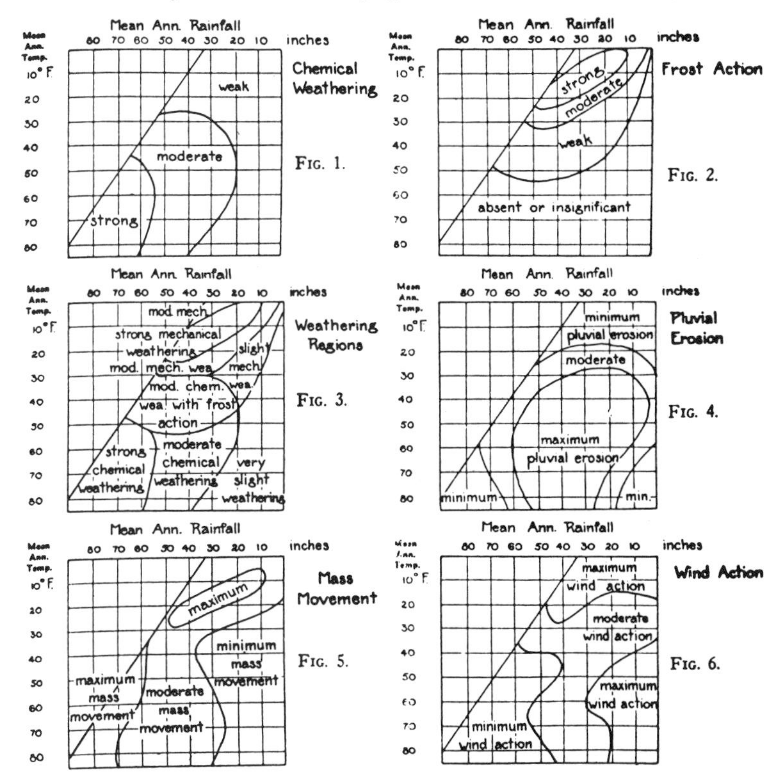

Fig. 1. Fig. 2. Fig. 3. Fig. 4. Fig. 5. Fig. 6.

"savanna" to describe characteristic processes of the "inselberg landscape" of Bornhardt (1900), this "hot-humid" cycle is referred to as the "selva cycle." This term is appropriate because the protective effects of the high tropical forest, or selva, prevent or inhibit erosion on hill slopes by running water either as slope wash or in gullies. A thick soil, vulnerable to mass movement, is therefore able to develop (see Wentworth, 1943) and the characteristic geomorphic features of this cycle may thereby be formed. The most recently described separate regime is called the periglacial cycle by Troll (1948) and includes the cycle of cryoplanation of Bryan (1946). It will be discussed in more detail below. Thus, of the nine climatic re-

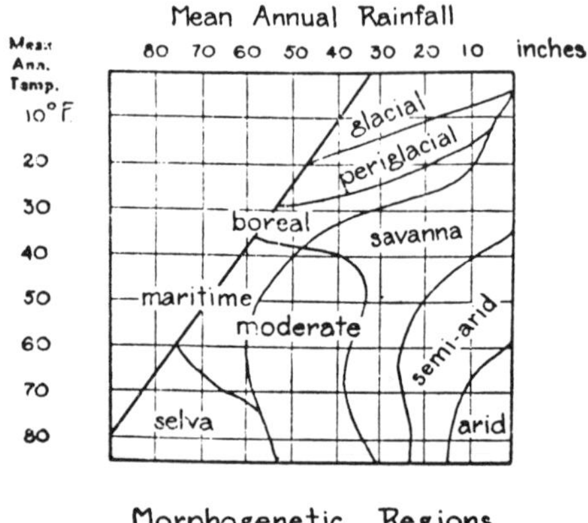

FIG. 7.

gimes, and corresponding cycles, here postulated, seven have already been described as producing unique geomorphic results. For the remaining two regimes, here called the maritime and boreal regimes, no distinct geomorphic characteristics have so far been reported. The probable erosional features of these as well as the other regimes are presented in table 1. Possibly the unique properties of the boreal regime have been obscured by the relatively recent prevalence of the periglacial cycle. Its morphogenetic characteristics may not be independently discernible. The maritime regime, though in many places recently subject to the effects of the glacial and periglacial cycles, may be recognizable by the severity of mass movement. The limiting climatic boundaries of these regimes are shown graphically in figure 7. The climatic boundaries of this graph are in part the same as those given by Penck (1910), Davis (1912) and Troll (1947) whose influence is acknowledged by the writer. Particularly are the glacial, selva and arid regimes of figure 7 parallel to the glacial, humid and arid climates of both Davis and Troll. I should, however, say that many of the ideas here expressed are explicitly stated in the lectures of Professor Kirk Bryan, whose emphasis on climatic morphogeny has led to the formulation here set forth.

If the geographic cycle is considered to be the sequence of events leading to the complete destruction of the geomorphic landscape, irrespective of the details of topographic form which may be produced or the details of the geomorphic processes which are effective, there can be but one geographic cycle. From such a broad viewpoint the other "cycles" here defined are unnecessary. Because the run-off of water from the land surface is, over a long period of time, likely to be the dominant agent in molding the surface, this single cycle is best described as the "pluvio-fluvial cycle." However, if the development in youth and maturity of minor topographic forms, particularly the development of slopes, is to be stressed, then the pluvio-fluvial cycle is an inadequate framework for the description of surface features peculiar to the several morphogenetic regions. Each geographic cycle is here considered as the unique expression of a climatic regime. Each acts through successive stages upon rocks of differing lithologies which are themselves arranged in various structures. Landforms are therefore to be described by an expanded Davisian system: 1) structure including lithology, 2) process as modified in nine morphogenetic, climatic regimes, and 3) stage.

[*Editor's Note:* Material has been omitted at this point.]

BIBLIOGRAPHY

Bornhardt, W. (1900) Zur Oberflächengestaltung und Geologie Deutsch-Ost-Afrikas, D. Reimer, Berlin, 595 p.

Bryan, K. (1946) Cryopedology—the study of frozen ground and intensive frost action with suggestions on nomenclature, Am. Jour. Sci., vol. 244, pp. 622–642.

Cotton, C. A. (1942) Climatic accidents in landscape making, 2nd. ed., Wiley, New York, 354 p.

Davis, W. M. (1899) The geographical cycle, Geographical Jour., vol. 14, pp. 481–504.

——— (1909) Complications of the geographical cycle, Proc. 8th Internat. Geog. Cong., pp. 150–163 (see also Geog. Essays, ed. by D. W. Johnson, Ginn & Co., 1909).

Freise, F. W. (1938) Inselberge und Inselberg-Landschaften im Granit- und Gneisgebiete Brasiliens, Z. f. Geomorph., vol. 10, pp. 137–168.

Penck, A. (1910) Versuch einer Klimaklassifikation auf physiogeographischer Grundlage, Sitz.-Ber. Preuss. Akad. Wiss., Phys.-Math. Kl., vol. 12, pp. 236–246.

Sapper, K. (1935) Geomorphologie der feuchten Tropen, Geogr. Schriften, vol. 7, 154 p.

Troll, C. (1947) Die Formen der Solifluktion und die periglaziale Bodenabtragung, Erdkunde, vol. 1, pp. 162–175.

——— (1948) Der subnivale oder periglaziale Zyklus der Denudation, Erdkunde, vol. 2, pp. 1–21.

Wentworth, C. K. (1943) Soil avalanches on Oahu, Hawaii, Bull. Geol. Soc. Am., vol. 54, pp. 53–64.

22

Reprinted from *The Cycle of Erosion in Different Climates,* B. T. Batsford, London, 1968, pp. 9-12, 133

THE CYCLES OF EROSION IN DIFFERENT CLIMATES

P. Birot

INTRODUCTION

The concept of a cycle of erosion expresses the evolution of slopes towards a level surface. Our purpose is to study this sequence of events in different climates.

At the outset it must be recognised that the term 'cycle' has itself been criticised. The word suggests an evolution that returns to its point of origin. But the Davisian 'cycle' begins with youth, passes through maturity and arrives at old age. It is hence an evolution that takes place in one direction only. Such criticism is certainly justified when one is dealing with the first cycle of erosion to affect a region, for example in the case of mountains which are geologically very young. However, the idea of a cycle is rigorously exact in all those cases where the area concerned has previously passed through senility, or even through the stage of 'maturity'. This is the case for the greater part of the land area of the world. All the ancient shields, all the folded mountain ranges of Mesozoic age (which are much more extensive than are Tertiary ranges), and even Tertiary ranges, have been planed at least once in their existence. We can suggest as an approximate, but fairly reliable, rule that any chain which was folded before the Miocene has already been base-levelled. One can therefore say that, over the greater part of the earth's surface, the landscape has at some time passed through old age; and that following this, renewed uplift has occurred, initiating new cycles.

The other fundamental objection currently made against the Davisian concept is the idea that in many cases orogenic movements and erosion have been contemporaneous and of roughly equal order of magnitude, so that the structures have been eroded *in statu nascendi*. But this is only important in areas of pronounced structural instability. Usually the earth movements come to an end, and then the cycle of erosion proceeds without having passed through the stage of youth in the sense in which we shall later define it.

In a first attempt to study the subject we shall begin in the middle of the evolutionary process (at the stage we may provisionally call maturity), by examining the simplest landscape which is possible. We are concerned, then, with a dendritic drainage pattern with interfluves which decline towards the streams.

The relief is composed of slopes of varying inclination. These slopes are generally convex in their upper part and concave in their lower part. They are covered by a layer of detritus, the thickness of which is approximately constant from top to bottom, and which is protected by a continuous plant cover. This may be herbaceous; more usually it is a tree cover with an undergrowth of herbaceous species. This is what one may call a 'normal' bioclimatic condition, and it implies a fairly humid climate. Obviously reservations may be made about this terminology; nevertheless it is the usual one and remains useful. We shall use the word 'soil' to describe this detrital cover; this is a useful convention and shorter than the alternative 'cover of detritus'. It is true that pedologists apply the term soil to the layer inhabited by organisms; thus in humid tropical regions it is relatively thin. This definition causes pedologists many problems because in fact they do not know at what level in the soil the action of life stops: microorganisms may appear well below the humus level. It seems preferable to treat the decomposed layer as a whole from the completely fresh rock up to the surface. This is in the interests of pedology as much as of geomorphology. The word 'soil', then, will be used for the detrital cover in the broad sense of the word.

The soil thickness also varies with the stage reached by the cycle of erosion. In the stage of maturity, which we are considering at present, its thickness is extremely variable, from a few decimetres in temperate regions, to several decametres in tropical regions. The further the cycle has progressed, the thicker the soil will be. The soil thickness expresses a balance sheet involving the rate of decomposition of the rock (the *Aufbereitung* of Walther Penck) and the speed of removal of debris on the slopes; these are the two processes which model the slope and which will be the subject of the following chapters. The presence of a soil immediately shows that this balance is positive and that its absolute value increases as the cycle of erosion progresses. Slopes will decline, and as the effectiveness of the agents of transport depends on the angle of slope, the thickness of the soil will increase as the rate of decomposition exceeds the speed of transport.

It must be added that as the slope declines, there will be a corresponding decrease in the speed of movement of water across it. The presence of water is an essential and indispensable factor in the decomposition of rocks; temperature changes alone are insufficient. Despite opinions to the contrary, water is the limiting factor in decomposition, even in humid tropical climates. An equilibrium is thus established in which climate affects both its elements, the rate of weathering and the speed of transport.

These agents of transportation move the debris as far as the river bed where, in the conditions of maturity, 'linear' erosion is concerned essentially with its removal. There is also erosion from the bed itself of previously weathered material, the thickness of which does not exceed that of the soil on the slopes. The profile of the bed is thus a provisional profile of equilibrium as described by H. Baulig. Its slope depends only upon the 'necessity' to remove the load provided by the slopes.

This concept of equilibrium allows for small oscillations which may be annual or more frequent, and which lead to temporary displacement on either side of the mean condition. If a catastrophic event occurs, for example the destruction of the forest cover by fire (which could well be a natural occurrence started by lightning), or a sudden landslip, there may be a very rapid increase in the alluvial load carried by the river. At first the discharge and slope of the river are unable to remove this increased load and temporary, localised deposition will occur. In subsequent years equilibrium will be re-established. By contrast, if we suppose an increase in discharge as the result of a particularly wet year without affecting the forest and so avoiding any appreciable increase in the amount of material supplied by the slopes, the ability of the stream to transport material will be increased, and the stream may incise itself a few decimetres, or even as much as a metre. Again this is a temporary oscillation, and the experience of European engineers in the control of rivers is that the bed of the stream does not alter. When the reasons for this stability of the river bed are investigated they are found to be twofold. In the first place it is the result of climatic oscillation on either side of a mean value. There is in addition a more fundamental cause for this stability, a type of automatic regulation of the régime. As in some other physical phenomena, any disturbance sets in train a series of events that tend to cancel its effects. Thus the compression of a gas causes heating which tends automatically to lead to expansion. We may suggest as an example a landslide, causing a section of forested slope to slip down and block the course of a river, so forming a lake on the upstream side. This initiates a series of changes that work to re-establish the profile of equilibrium: (1) alluvial material brought into the lake by the stream fills it, restoring surface flow; (2) headward erosion at the barrier takes place automatically; the slope here is steeper than before the slide and so the river will incise itself.

But a profile of equilibrium is only provisional. As the angle of slope decreases (an automatic result of any cessation of downcutting by the river), the load carried will decrease both in amount and in size. The stream will find itself to be underloaded; it will then cut down into previously decomposed rock, until the reduced slope (since the lower point at the river mouth is fixed) permits the stream to transport exactly the new load—a load increased to some extent as a result of the incision itself. Thus equilibrium is established again. Naturally this alluvial material must not be so thick that it exceeds the height difference between the river banks and the floor of the bed; otherwise deposition will occur. Outcropping of fresh rock, whether on the slopes or on the river bed, is a characteristic of youthfulness under these particular bioclimatic conditions.

To understand how the stage of maturity is reached, and how continuing evolution leads to senility, it is necessary to examine systematically the three fundamental processes, which are: (1) the decomposition of rock into detritus; (2) the transport of this detritus on slopes; and (3) the transport of this detritus in the river bed and the erosion of the river bed itself. So as to select homogeneous initial conditions, we shall mainly be concerned with a comparison of relief features developed on crystalline rocks.

[*Editor's Note:* Material has been omitted at this point.]

Conclusion

The whole problem of the evolution of the cycle of erosion in different climates is dominated by the conflict between peneplanation and pediplanation which gives rise to passionate controversy. Whatever may be the solution reached in any particular case, certain essential logical principles must not be forgotten. The most important is that a steep slope can only retreat parallel to itself, preserving at the same time its angle and its approximate size, if the detritus which it provides can be moved across a gentle slope; this can depend both on the agent of transport and the fineness of the colluvial material. The first of these two factors predominates in a hot and dry climate, and the second in a hot and wet climate. But although both types of evolution lead to the juxtaposition of steep residual relief and areas which have been prematurely levelled, the morphology of these level surfaces is very different in the two types of tropical climate which we have distinguished. True pediments, examples of particularly precocious levelling, imply that the small talwegs have no erosional advantage in relation to the open slopes, a relationship only found where plant cover is discontinuous. Under a continuous vegetation cover, by contrast, the talweg incises itself into the decomposed rock, and a system of slopes is formed which gradually declines towards a level surface.

In temperate climates, where the weathering of large blocks is slow, but where their comminution and their incorporation in a fine matrix allows their movement on slopes of varying degree, the cycle of erosion evolves towards peneplanation. The concave slope always remains gentle, each section just being able to evacuate the material received from the next section upslope; this material always includes a certain proportion of large detritus. It is probable that at least in rocks which have not been severely frost-shattered, the periglacial cycle works in the same way.

23

Reprinted from *Water, Earth, and Man*, R. J. Chorley, ed., Methuen & Co. Ltd., London, 1969, pp. 525-534

Geomorphic Implications of Climatic Changes

S. A. SCHUMM
Department of Geology, Colorado State University

As we view our familiar environment the question arises, did it always appear so? Major climatic changes are known to have occurred during the past million years of earth history, and over vast areas of the earth evidence of ice action dominates the landscape. In this chapter we concern ourselves only with the long-term effects of climate change on the hydrologic cycle and on the landscape, while omitting treatment of the more obvious effects of glacial and periglacial climates.

Any generalization concerning changes of climate can be very much in error for a given locality, but much evidence has been compiled to suggest that during the last million years average temperatures could have ranged from 10° below to 5° F above present average temperatures. In most, although not all, regions higher average precipitation was associated with the lower temperatures of continental glaciation, and average precipitation was, at least for some presently arid and semi-arid regions of the United States, about 10 in. greater. During inter-glacial time and during a brief post-glacial episode higher temperatures prevailed, and average precipitation was about 5 in. less than that of today (Schwarzback, 1963).

The effects of climate changes of these magnitudes are not direct. Rather, with changing climate the relations between climate, runoff, and erosion are altered by significant changes of vegetation. Geomorphic evidence of a climate change is, in fact, evidence only of a change in the hydrologic variables of runoff and sediment yield. Therefore, the relations between climatic, phytologic, and hydrologic variables must be considered before the effect of a climatic change on the landscape can be evaluated.

1. Sediment movement

Modern hydrologic data from the United States have been used to demonstrate climatic influences on the quantity of runoff and sediment delivered from drainage basins. The family of curves of fig. 11.II.1 illustrate the general relation between climate and runoff (Langbein *et al.*, 1949). The curves show that, as might be expected, annual runoff increases as annual precipitation increases. However, runoff decreases as temperature increases with constant precipitation because of increased evaporation and water use by plants.

The relations between annual sediment yield and annual precipitation and temperature for drainage basins averaging about 1,500 square miles in the United States are presented in fig. 11.II.2. The 50° F curve of fig. 11.II.2 shows the relationship between sediment yield and precipitation adjusted to a mean annual temperature of 50° F (Langbein and Schumm, 1958, p. 1076). Sediment

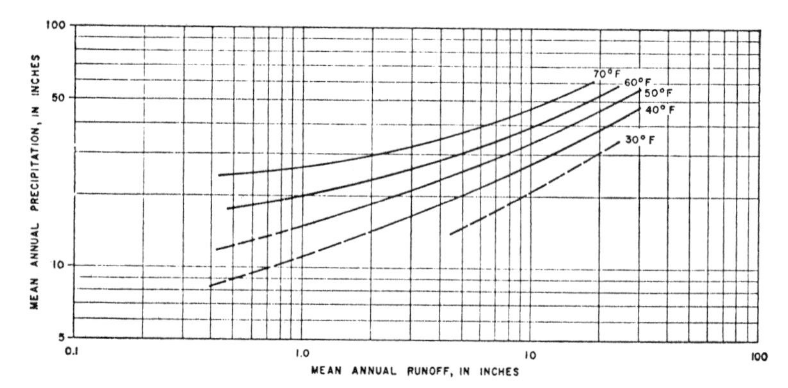

Fig. 11.II.1 Curves illustrating the effect of temperature on the relation between mean annual runoff and mean annual precipitation (After Langbein *et al.*, 1949 and Schumm, 1965).

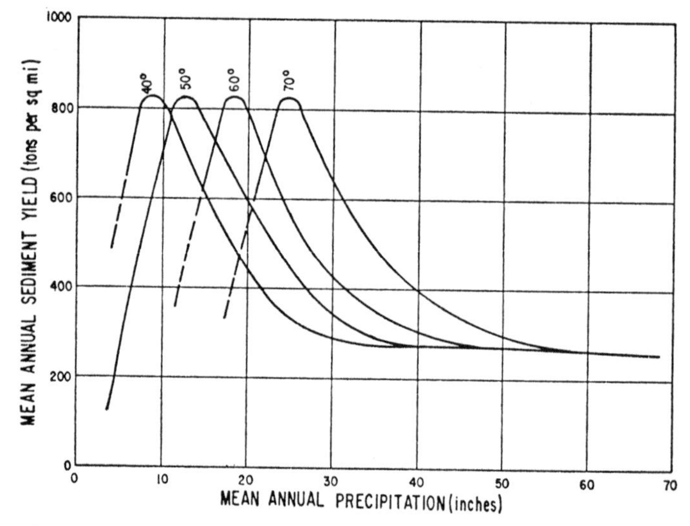

Fig. 11.II.2 Curves illustrating the effect of temperature on the relation between mean annual sediment yield and mean annual precipitation (From Schumm, 1965).

yield is a maximum at about 12 in. of precipitation, but it decreases to lower values with both lesser and greater amounts of precipitation. The variation in sediment yield with precipitation can be explained by the interaction of precipitation and vegetation on runoff and erosion. For example, as precipitation increases above zero, sediment yields increase at a rapid rate, because more runoff becomes available to move sediment. Opposing this tendency is the influence of vegetation, which increases in density as precipitation increases. At about 12 in. of precipitation on the 50° F curve the transition between desert shrubs and grass occurs. Above about 12 in. of precipitation on this curve sediment-yield rates decrease under the influence of the more effective grass and forest cover. Elsewhere in the world, where monsoonal climates prevail, sediment-yield rates may increase again above 40 in. of precipitation under the influence of highly seasonal rainfall (Fournier, 1960).

The sediment-yield curves for temperatures of 40°, 60°, and 70° F are dis-

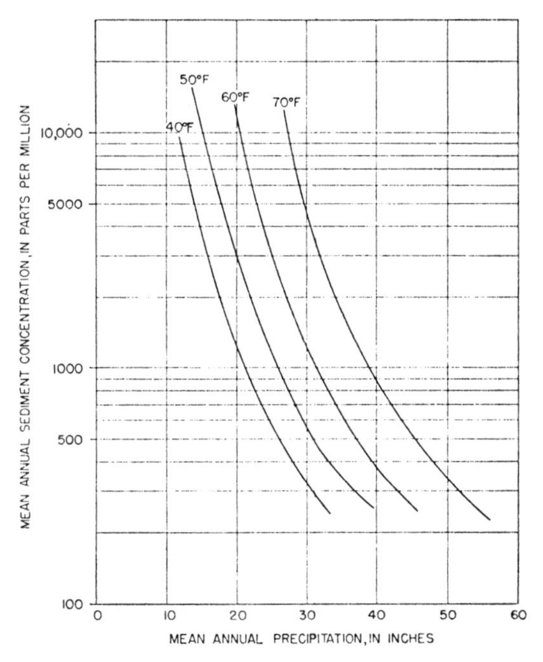

Fig. 11.II.3 Curves illustrating the effect of temperature on the relation between mean annual sediment concentration and mean annual precipitation (From Schumm, 1965).

placed laterally with respect to the 50° F curve (fig. 11.II.2). Together they indicate that, as annual temperature increases, maximum sediment yields should occur at higher amounts of annual precipitation. That is, higher annual temperatures cause higher rates of evaporation and transpiration, and less precipitation is available to support vegetation. Runoff is less, and so the maximum rate of sediment yield shifts to the right.

In addition to the amount of sediment moved, its concentration in the water by which it is moved is important. Curves were developed to show the relation between average sediment concentrations and average precipitation at different temperatures (fig. 11.II.3). For a given annual precipitation, sediment concentrations increase with annual temperature, whereas for a given annual temperature sediment concentrations decrease with an increase in annual precipitation.

One important point to be made with regard to figs. 11.II.2 and 11.II.3 is that, although more sediment is moved from a drainage system under semi-arid conditions, nevertheless sediment concentrations are greatest in arid regions. During a period of years the small number of high-concentration runoff events that occur in arid regions cannot transport the quantities of sediment that are moved by the greater number of lower concentration runoff events in semi-arid regions.

The curves of fig. 11.II.2 show that major changes in erosion rates will occur with relatively minor changes of climate if plant cover adjusts significantly to the climate change. That is, at a mean temperature of 50° F a small change of precipitation anywhere between 0 and 20 in. should elicit a significant hydrologic and geomorphic response, whereas this should not be the case in humid regions, because a major change of vegetational type and density will not accompany a small change of precipitation or temperature.

Changing type and density of vegetation exert a major control on landforms. For example, as the density of vegetation increases with annual precipitation (Langbein and Schumm, 1958, fig. 7), the rate of erosion of hill-slopes should decrease in semi-arid and sub-humid regions. However, in initially arid regions the major increase in runoff that occurs with increased precipitation (fig. 11.II.1) probably will more than compensate for any increase in vegetal cover. Therefore initially arid slopes should be subjected to more intense erosion. Although this last statement is conjectural, the peaks of the sediment–yield curves (fig. 11.II.2) demonstrate that more sediment is exported from semi-arid than from arid regions, and some of this increase must be derived from the hill-slopes.

2. Changes in the channel system

Moving down off the hill-slopes to valley floors, evidence concerning the response of river systems to climate change appears. Investigations into the relations among climate, runoff, and the character of channel systems indicate that within a given climatic region both the total length of channels per unit area (drainage density) and the number of channels per unit area (channel frequency) increases as annual runoff increases and as flood volumes increase. Therefore, increased runoff should cause lengthening of the drainage network and the addition of new tributaries to the system. This conclusion is based on data collected from within one climatic region, where, in fact, differences of soil type and geology exercise a dominant influence on both runoff and drainage density. Therefore, when the increase in runoff is accompanied by a major change in vegetational characteristics the results may be different. If a major climatic change causes a shift in vegetational type from shrubs and bunch grass to a continuous cover of grasses or from grasses to forested conditions it appears that the increased density of vegetation should prevent an accompanying increase in the length and number of drainage channels. In fact, worldwide measurements of terrain characteristics show that stream frequency is greatest in semi-arid regions, least in arid regions, and intermediate in humid regions (Peltier, 1959). It appears, then, that a major increase in precipitation can either increase or decrease the length and number of channels, and if drainage density were to be substituted for sediment yield on the ordinate of fig. 11.II.2 the curves should indicate in a very general manner the variation of drainage density with climate.

In summary, both drainage density and sediment yield should be greatest in semi-arid regions. The occurrence of maximum drainage densities and maximum sediment yields under a semi-arid climate suggests that high sediment yields are a reflection of increased channel development and a more efficiently drained system. Hence, a shift to a semi-arid climate from either an arid or a humid one should allow extension of the drainage network with increased channel and hillslope erosion.

Obviously the changing hydrology of the small upstream drainage areas and their hill-slopes will be reflected in the behaviour of the primary river channels that transport runoff and sediment to the sea. Again, the initial climate or the climate existing before a climate change is of major importance, for it determines the type of river to be considered. For example, in initially humid regions the rivers are perennial, and they remain perennial during a change to a wetter, cooler climate, although the channel will enlarge to accommodate the increased runoff.

The situation differs somewhat for an initially semi-arid region, for even the major rivers are initially either intermittent or characterized by long periods of low flow. The smaller tributaries are ephemeral, as are many in humid regions, but many more drainage channels should be present. With a shift to a wetter climate and greater runoff the flow in the major rivers and in many of the tributaries becomes perennial. Vegetation becomes denser, obliterating the smallest channels. As runoff increases, sediment yields and sediment concentrations decrease (fig. 11.II.1, 11.II.2, and 11.II.3). The result will be enlargement of the main channels. With a return to semi-arid conditions, the sediment yield increases, as both hill-slope and channel erosion increases, and runoff decreases. Deposition in and decrease in size of the main channels should result, as the tributaries again reach their maximum extent and number.

It has not always been recognized that the changes in tributary channels might not conform to changes along the larger rivers. This may not be important in humid regions, but it becomes of major importance in arid regions, where unfortunately few hydrologic data are available from which one may estimate river response to climate change. Nevertheless, increased precipitation should increase runoff in arid regions, but probably not to the extent that ephemeral rivers will be converted to perennial ones. Increased runoff should enlarge the tributary channels and extend the drainage network (arroyo cutting). Sediment will be flushed out of the tributary valleys into the main channels during local storms, and because the loss of water into the alluvium of the main channels is appreciable, aggradation of the main channels will result. In effect, increased runoff in arid regions will erode the tributary valleys, and this sediment will move downstream, where at least part of it will be deposited in the major river channels.

Without perennial flow, semi-arid and arid rivers probably are always characterized by a relative instability, and channel cutting can alternate with phases of aggradation. These events can be considered a natural part of the cycle of erosion of these ephemeral stream channels.

Deductions based on morphologic evidence have been made concerning channel adjustment to climate change, but what evidence exists that the postulated changes may be real? The evidence lies in studies of modern river behaviour in response to differing conditions of runoff and sediment yield. For example, many investigators have demonstrated that the width (w), depth (d), meander wavelength (l), and gradient (s) of rivers are related to the average quantity of water or the discharge (Qw) passing through a channel. As shown by equation (1), channel width, depth, and meander wavelength will increase with an increase in discharge, but gradient will decrease.

$$Qw \simeq \frac{w, d, l}{s} \tag{1}$$

Although the relation between these channel parameters and discharge are highly significant, nevertheless, for a given discharge a ten-fold variation in meander wavelength, width, and depth can occur. Recent work has demonstrated that with constant discharge an increase in the bedload (Qs), the quantity of sand and coarser sediment moved through a channel, will cause an increase in channel width, gradient, and meander wavelength but a decrease in channel depth, as follows:

$$Qs \simeq \frac{w, l, s}{d} \tag{2}$$

Channel shape is significantly influenced by sediment load, and as indicated by equation (2), an increase of bedload at constant discharge will cause an increase in the width–depth ratio, as a channel widens and shallows. Accompanying this change of shape is an increase in meander wavelength (l) and an increase in gradient (s). These changes are brought about by a decrease in the sinuosity, that is, a straightening of the course of the stream, which decreases the number of bends and steepens the gradient.

A glance at equations (1) and (2) reveals that changes in type of sediment load and discharge do not always reinforce one another, that is, although an increase in water discharge will increase depth, an increase in bedload will decrease it. The changes of river morphology that occur, therefore, depend on the magnitude of the changes of discharge and sediment load.

Evidence of river changes in response to climate fluctuations is difficult to obtain, because in most instances the adjustment destroys the pre-existing form, and the basis for comparison is lost. However, the Riverine Plain of New South Wales, Australia, is a unique area from which significant observations concerning river adjustment to changed climate can be obtained. Across this alluvial plain the Murrumbidgee River flows to the west, and evidence is visible on the surface of the plain of older, abandoned channels that functioned during different climatic episodes of the past (fig. 11.II.4). Oxbow lakes, which are remnants of an abandoned channel that was much larger than the modern river, are visible on the floodplain of the Murrumbidgee River. The channel shape and sinuosity of this channel is similar to that of the Murrumbidgee River, but its width, depth, and meander wavelength are larger. The abandoned channel is filled primarily with silts and clays, which must have comprised the sediment load of that channel and which is the predominant load of the modern river. Apparently, as a result of increased precipitation, much higher discharges moved out of the source area in the recent past; however, little change in the nature of the

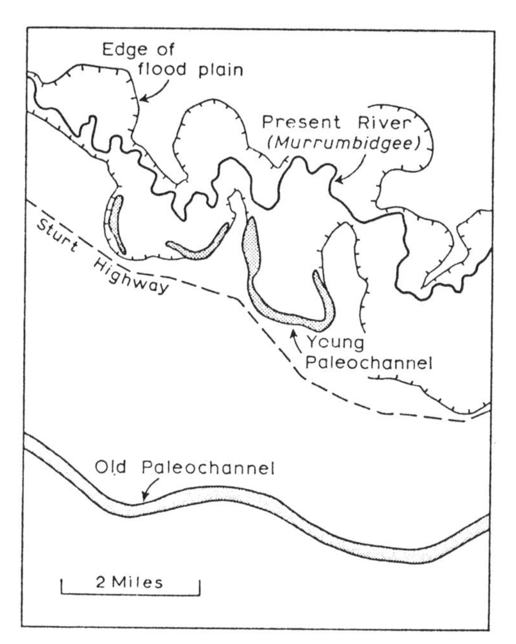

Fig. 11.II.4 Diagram made from an aerial photograph of a portion of the Riverine Plain near Darlington Point, New South Wales. The sinuous Murrumbidgee River, which is about 200 ft wide, flows to the left across the top of the figure (upper arrow). It is confined to an irregular floodplain on which a large oxbow lake (youngest paleochannel, middle arrow) is preserved. The oldest paleochannel (lower arrow) crosses the lower part of the figure.

sediment load occurred, and although the size of the channels are very different, they are in other respects morphologically similar. The contact between the Murrumbidgee River floodplain and the surface of the Riverine Plain proper (fig. 11.II.4) is a series of large meander scars, which are further evidence of the large size of this most recent palaeochannel. Another older set of abandoned channels can be detected in this area (fig. 11.II.4), and these palaeochannels are morphologically completely different from the palaeochannels, which once occupied the floodplain, and the modern river. The older palaeochannel (fig. 11.II.4) is straighter, wider, and shallower than both of the younger channels, and because of its straight course its gradient is twice that of the modern river. Its abandoned and aggraded channel is filled with sand and fine gravel, indicating that this channel was moving a very different type of sediment load from the source area, during presumably a more arid climate. The differences among these three channels can be explained by the changes in water discharge (*Qw*) and type of sediment load (*Qs*), as shown by equations (1) and (2).

Although direct evidence of the response of landforms to climate change is rare, the fact that it is the climatic effect on runoff and sediment yield that causes adjustment of river channels makes available other sources of information, as for example, downstream changes of river character with changing discharge and sediment load.

An excellent example of channel changes with changing discharge and sediment load is provided by the Kansas River system, which is tributary to the

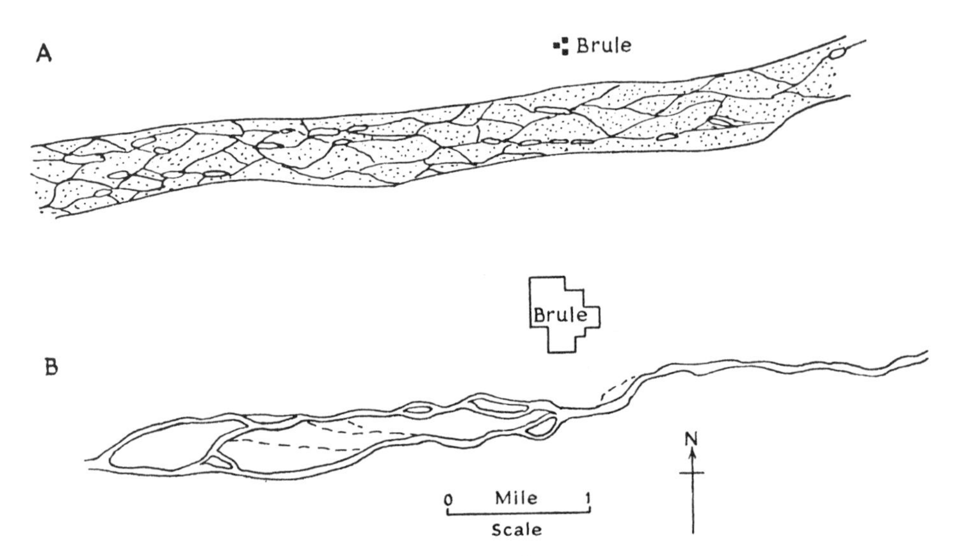

Fig. 11.11.5 South Platte River at Brule, Nebraska.

A. Sketch of channel from U.S. Geological Survey topographic map of the Ogallala Quadrangle, based on the field surveys of 1897.

B. Sketch of channel from U.S. Geological Survey topographic map of the Brule and Brule SE Quadrangles, prepared from aerial photographs taken in 1959 and field checked in 1961.

Missouri River in the western United States. Rising in eastern Colorado, the Smoky Hill River flows to the east. It drains a region of sandy sediments, and it is a relatively wide and shallow stream of steep gradient and straight course. In central Kansas two reservoirs retain much of the bedload of the Smoky Hill River, and below these reservoirs two major tributaries join the Smoky Hill River. These tributaries drain a region of fine-grained sedimentary rocks, and they introduce into the Smoky Hill River large quantities of suspended sediment (silt and clay). Although discharge is increasing in a downstream direction, the width of the channel decreases and its depth increases as a result of the influx of fine sediment and the decrease of bedload. The channel becomes more sinuous and gradient, width–depth ratio, and meander wavelength decrease. Farther downstream the Republican River, a major tributary, which drains an area about equal in size to that of the Smoky Hill River, joins the Smoky Hill to form the Kansas River. Below the junction of the two rivers, the major increase in discharge and the addition of large quantities of sand cause a major change in the channel morphology. The channel width increases markedly, as does the width–depth ratio and gradient, whereas sinuosity decreases and channel depth remains relatively unchanged.

The above changes in channel characteristics have occurred along one river as the nature of the sediment load was altered by changes in runoff and sediment yield from tributary basins. In addition, man has modified the flow patterns and sediment loads of rivers, and these changes often duplicate the effects of a climate change. Therefore, although geologic evidence is limited, the engineering literature is a fruitful source of information concerning river adjustment to man-induced changes of hydrologic variables. In fig. 11.II.5 a major change in the width of the South Platte River is illustrated. Both the North and South Platte Rivers in Nebraska were classic examples of braided rivers during the latter part of the nineteenth century, however, due to regulation of discharge, these channels have recently undergone major changes of dimensions and form. In the case of the South Platte River flood peaks have been reduced, and in response the river has changed from a wide braided channel to a narrow and somewhat more sinuous channel. Depth has probably also decreased. Similar changes have occurred along the North Platte River as a result of both a reduction in peak discharge and annual discharge. In these examples a major reduction in channel size has occurred as a result of reduced discharge, as indicated by equation (1).

Discussions of the adjustment of landforms to long-term climate changes are still highly speculative, but a more complete understanding of the possible changes that can occur is being developed. The ability to predict landscape changes resulting from a modification of the climatic or hydrologic characteristics of drainage systems has practical implications, for if man persists in his efforts to modify not only the hydrologic regime of river systems but also the climate of drainage basins, then he should be prepared to evaluate the consequences of these acts not only in terms of changed river discharge and sediment yields but also in changed channel and drainage basin morphology.

REFERENCES

FOURNIER, M. F. [1960], *Climat et erosion* (Presses Univ. France, Paris), 201 p.

LANGBEIN, W. B. *et al.* [1949], Annual runoff in the United States; *U.S. Geological Survey Circular* 52, 14 p.

LANGBEIN, W. B. and SCHUMM, S. A. [1958], Yield of sediment in relation to mean annual precipitation; *Transactions of the American Geophysical Union*, **39**, 1076–84.

24

Reprinted from *Great Britain Geographical Essays,* J. B. Mitchell, ed., Cambridge University Press, London, 1962, pp. 10-16

RELIEF

B. W. Sparks

[*Editor's Note:* In the original, material precedes this excerpt.]

Two major regions are commonly recognized in Britain, highland and lowland, with the dividing line formed by the base of the Coal Measures. Like all great generalizations this creates difficulties, especially in the diversity of landforms placed in the highlands. Probably the only really satisfactory solution is to avoid classification and to treat all areas individually, but on a broad scale a case may be made for a threefold division into what may be termed highlands, uplands, and lowlands.

Highlands are formed, essentially of Pre-Cambrian and Lower Palaeozoic rocks folded mainly in the Caledonian earth-movements. The structures are so complex and the lithology so variable yet so similar that structure is less important in the relief here than in the uplands and the lowlands. Over large areas of central Wales, the Southern Uplands and even parts of the Highlands of Scotland the relief is monotonous, uplifted, dissected plateau. The general impermeability of the rocks, allied with the heavy rainfall deriving from their westerly position, together with the general lack of limestone, results in infertile, leached, acidic soils. Having been the centres of the main ice caps of Britain the highlands bear signs of more severe glacial erosion than the other regions of Britain. Yet there are differences within and between the areas.

The most extensive of them, the Highlands and islands of Scotland, is by far the largest area of crystalline metamorphic rocks in Britain. Over the whole area occupied by the Moine and Dalradian rocks the relief is due mainly to fracturing and glaciation and not to lithology. This is the highest region in Britain, the summits usually exceeding 3,000 ft and often 4,000 ft. It is deeply dissected, often along fracture lines, by broad glacial troughs, many of them now occupied by lochs some of which, for example Ness and Morar, occupy greatly overdeepened basins. All the signs of severe highland glaciation are here: valley steps, roches moutonnées, hanging valleys, corries, recessional moraines, and streams rising either in corries or on imperceptible divides in through valleys. In this extensive area of foliated crystalline rocks few rocks have individual effects on the relief with the possible exception of the quartzites, which are held responsible for the more pointed form of some of the peaks, for example Schiehallion.

Lithological diversity increases in importance near the west coast, where, beyond the thrust planes which terminate the outcrop of the Moine series, there are tracts of Lewisian gneiss, Torridonian sandstone and very much younger Tertiary igneous rocks. On the Lewisian is developed monotonous, lowlying, bare, hummocky glaciated relief, showing in many places numerous lakes and indeterminate drainage, especially on the mainland north of Ullapool and in the Outer Hebrides, notably in Uist and Benbecula. Along the north-west coast of the mainland the Torridonian mountains rise above the Lewisian in the greatest contrast, slabs of near-horizontal, ancient, resistant sediments often weathered, as on Stac Polly and Suilven, into steep and ragged ridges. It is curious how the Lewisian and also the Moine rocks on the coast farther south around Arisaig give the appearance of having been almost ground out of existence by the strength of the glaciation, while the Torridonian rocks, both in the mountains and in lowlying outliers on the Lewisian, do not.

The Tertiary igneous rocks, although so different in age, have many features in common with the Pre-Cambrian rocks: like them they are crystalline, while the relief on some of the major intrusions, especially the gabbro of the Cuillin Hills of Skye and the granite of northern Arran, provides superb examples of mountain glaciation. Sheets of plateau basalt, especially in northern Skye and the dyke swarms of Skye, Mull and Arran are distinctive elements in the relief, though both lava-sheets, for example the Devonian andesites south of Oban, and dykes of earlier ages are not unknown in the Highlands.

None of the other highland regions is quite like the Highlands of Scotland. In general the relief is smoother, although they have the characteristic of high average elevation in common. The greatest sharpness of relief seems to be provided by a combination of igneous rocks and severe glacial erosion. Snowdonia provides a good example: a series of lavas and ashes, predominantly rhyolitic in character, and a number of intrusions, mark this region off geologically from much of the rest of Wales. Heavy glaciation has produced on these rocks the corries of the northern side of the Glyders, glacial troughs such as the Nant Ffrancon, and a number of lakes, features which make this region, in spite of its much smaller size, comparable with the Highlands of Scotland. Cader Idris shows similar corried escarpments preserved on doleritic sills, while, in the Lake District, on the andesitic Borrowdale volcanic rocks ice has produced similar glaciated highland. Outside these districts, that is in much of the Southern Uplands, the northern part of the Lake District, and central Wales, the hills are big but rounded. Glacial features occur, but they have not the angularity typical of those of the volcanic outcrops, while the more common occurrence of sheets of drift seems to have softened the landscape.

Although the highlands as described in the preceding section seem to possess certain features in common, notably the absence of a simple structural pattern and its reflexion in a neat relief pattern as well as a general lack of limestones, there is one area which is completely atypical. It is true that the Cambrian Durness Limestone of the far north-west of Scotland introduces karst features on a small scale, but it does not form an area so different from the rest as do the alternating limestones and shales of the upper part of the Silurian in the Welsh borderland. Between and around Wellington and Ludlow the comparatively gentle dips of the Wenlock and Aymestry limestones and interbedded shales have resulted in a pattern of high cuestas, ridges and vales much more characteristic of what will be described below as upland Britain than of true highland Britain. The presence of abundant limestone results in a brightness of landscape not normally found in the highlands and allies this area to the uplands, in spite of the sombre effects introduced by the extensive planting of conifers on the limestones.

The upland areas of Britain are the areas of Devonian and Carboniferous rocks folded in the Hercynian earth-movements. Sandstones and limestones, disposed in simple structures, are the principal rock types. The relief, therefore, typically has a pattern, whether it be high-level plateau as in the Pennines, low coastal plateau as in Caithness and the Orkneys, or high escarpment as in the Brecon

Beacons, the Black Mountains and the hills inland from Brora in Sutherland. Compared with the highlands the signs of glaciation are far less pronounced. Mantles of drift are common enough and seem to intensify the smoothness of many of the hillsides, but there are, for example, few indisputable corries: the soft rounded forms that occur could be spring-heads, or corries badly preserved in not particularly resistant rocks, or spring-heads modified by snow-patches. Although many parts of the Old Red Sandstone produce acidic, impermeable and infertile soils, the more marly sections form fertile lowland, while the Carboniferous Limestone adds a type of relief, soil and vegetation almost absent from highland Britain. Regional differences occur within the uplands, but the essence of the relief is comparative simplicity of structure allied to a relief of bold cuestas and plateaux.

The Pennines are typical of the high plateaux of upland Britain. Slight dips and great lithological differences both between the major divisions of the Carboniferous rocks and within the Yoredale facies of the Carboniferous Limestone ensure the prevalence of structural surfaces. Whether these surfaces are truly structural or structures approximately stripped by erosion is not fundamental in a general review of this type, for the net result is the same, the dominance of the landscape by tabular relief. In addition, the Carboniferous Limestone is unique in possessing the clearest and most complete karst features in Britain and these are nowhere better developed than in parts of the Pennines, notably around Ingleborough.

Plateau at a much lower elevation is to be found in the coastal areas of Caithness and in the Orkneys, where the rocks responsible for it are the closely bedded Devonian Caithness Flags. A more monotonous scenery than that of parts of inland Caithness would be difficult to imagine, but the coast, where marine erosion has exerted its full effect on jointed and faulted horizontal rocks to produce vertical cliffs and spectacular stacks and inlets, is magnificent.

Where a simple regional dip affects the rocks, striking escarpments and dip slopes, such as those of the Brecon Beacons and Black Mountains, occur. These are impressive features, the crests of the escarpments reaching nearly to 3,000 ft in the former and to 2,500 ft in the latter upland. In form they are magnified versions of the cuestas of lowland Britain, but, apart from the question of scale, they differ from these in the number of steps and scars caused by hard beds, and in the dull colours of their acid moorland vegetation. To the south the Pennant series of grits and sandstones is a very important relief-forming unit in the synclinal upland of the South Wales coalfield: it is responsible for the moorland plateau between the industrial valleys and for a high escarpment overlooking the dissected vale developed on Lower Coal Measures in the northern part of the region.

Like the highlands, the uplands of Britain contain areas which are not typical. North-east of the Black Mountains lies the Red Marl plain of Herefordshire, recalling much more the Triassic plains of lowland Britain than the uplands to which it belongs, except in the discontinuous ranges of bold hills which cross it in a south-west to north-east direction. Devon and Cornwall are also exceptional: superficially they are plateau and thus resemble the Pennines, but this plateau surface is an erosion form cut across a variety of rocks and complex structures, for this region was near enough to the main belt of Hercynian folding to be strongly affected and the rocks largely altered to slates. In lithology and structure it resembles highland Britain, but in elevation and surface form it is more like upland Britain. It differs from both in the absence of glaciation and the smaller extent of periglacial activity.

But the most exceptional Hercynian region by far is the Central Lowlands of Scotland, an area shattered by faults into a pavement of blocks of Devonian and Carboniferous rocks. Further, the lithology of the Carboniferous Limestone is abnormal, that bed containing almost more coal than limestone, while the area is characterized by frequent masses of igneous rocks. They cause a great variety of relief from the jointed plug on which Stirling Castle is built to the great fault-line scarp truncating the andesitic Ochils on their southern side. These and other forms, such as exhumed laccoliths, dykes, volcanic necks and plateau basalts around the lower Clyde, ensure a complex pattern of relief different from that found elsewhere in Britain.

In lowland Britain there is simplicity, a predominant dip to the south-east affecting rocks from the Permo-Trias upwards, a patterned relief of cuestas and vales reflecting the alternating beds of sandstone, limestone and clay, a relief generally below 1,000 ft in maximum elevation, and an area where features of glacial deposition are far more marked than features of glacial erosion.

The oldest beds concerned, the sandstones and marls of the Trias, wrap around the sides of the Pennines and form the great triangular lowland of the Midlands, where diversification of relief is provided principally by the upthrust horsts of Pre-Cambrian and Lower Palaeozoic rocks, such as the Malverns, Charnwood Forest and the Nuneaton ridge. Some of the sharp relief formed on these rocks is in striking contrast to the gentle character of Triassic relief.

The areas of Jurassic and Cretaceous rocks north of London provide considerable contrasts in scenery, because of lithological variations, especially within the Jurassic. In Yorkshire, a great development of sandstone is responsible for the North Yorkshire Moors, a region more akin in lithology, vegetation and elevation to upland Britain than to the rest of the lowlands. Over much of Lincolnshire and the East Midlands the relief is not great as massive limestone beds are lacking in the Jurassic. In Lincolnshire the lowlands are narrow, but, to the south, a general decrease in regional dip is responsible for the great spread of low plateaux through the East Midlands and East Anglia. In the Jurassic rocks of this area limestones are not well developed unlike clays the predominance of which in the Upper Jurassic has facilitated the excavation of the wide plain in which the Fen deposits have been laid down. Both here and in East Anglia a general spread of glacial deposits, mostly boulder clay, has added to the smooth monotony of the relief. Farther south, the main Cotswold and Chiltern section of the lowlands, with Oxford at its centre, forms much more characteristic scarpland, largely because of the strong development of limestone in the Middle Jurassic and the higher relief of the Chalk in the Chilterns, for which there is no obvious lithological cause. This area differs from the lowlands to the north not only in its relief, but also in the far smaller importance of glacial drift in the landscape.

The area south of London differs from the rest of lowland Britain in its generally east–west structural and relief trends. It has less extensive outcrops of Jurassic rocks, but much wider ones of Lower Cretaceous and Tertiary sands and clays. In addition, the finest development of chalk scenery is to be found here, for, not only does the Chalk outcrop over wide areas such as Salisbury Plain, but it is also less obscured by superficial deposits so that it exerts an immediate effect on landforms, soils and vegetation. The prevalence of minor folds causes the pattern of escarpments to be more complicated than it is to the north of London: in areas where these folds are pronounced, for example near Weymouth and in the western Weald, the escarpments are more intricate in plan than in areas such as the North Downs, where folds are far less prominent. Finally, unlike the rest of the lowlands, this region was not glaciated, although periglacial activity was responsible for some locally extensive developments of solifluxion deposits and to a certain, though arguable, degree for modification of landforms.

Although the threefold division of Great Britain used here has been made primarily for physical description, it may prove useful in the human geography of the land. The highland regions are areas of low population density, infertile soils and mainly pastoral farming; the upland regions include all the significant coalfields, apart from their concealed sections, and hence most of the major old industrial regions; the lowlands are the most fertile regions and the areas of the greatest uniformity of rural population spread.

Selected Bibliography

Department of Scientific and Industrial Research: Geological Survey and Museum. *British Regional Geology.* 18 vols. London and Edinburgh (H.M.S.O.), 1935–54.

[These are the regional accounts of British geology and are more geological than geomorphological in general approach.]

T. N. George. 'British Tertiary landscape evolution.' *Sci. Prog.*, **43**, 1955, pp. 291–307. [A comprehensive paper with an excellent list of references for further reading.]

H. J. Mackinder. *Britain and the British Seas.* London (2nd ed.), 1907.

T. G. Miller. *Geology and Scenery in Britain.* London, 1953.

L. D. Stamp. *Britain's Structure and Scenery.* London, 1946.

J. A. Steers: *The Coastline of England and Wales*, Cambridge, 1948; *The Sea Coast*, London, 1953; *The Coastline of England and Wales in Pictures*, Cambridge, 1960. [These three works offer a comprehensive survey of a part of the physical geography of Britain, which has necessarily been neglected in this chapter.]

A. E. Trueman. *The Scenery of England and Wales.* London, 1938.

The following works are considerably more detailed, but cover major parts of the country:

E. H. Brown. *The Relief and Drainage of Wales.* Cardiff, 1960.

S. E. Hollingworth. 'The recognition and correlation of high-level erosion surfaces in Britain.' *Quart. Journ. Geol. Soc. Lond.*, **94**, 1938, pp. 55–74.

S. W. Wooldridge and D. L. Linton. *Structure, Surface and Drainage in South-east England.* London (2nd ed.), 1955.

25

Reprinted from *U.S. Geol. Survey Prof. Paper 500B*, 1-10 (1962)

GEOMORPHOLOGY AND GENERAL SYSTEMS THEORY

By Richard J. Chorley

"[Nature] * * * creates ever new forms; what exists has never existed before, what has existed returns not again—everything is new and yet always old * * *. There is an eternal life, a coming into being and a movement in her; and yet she goes not forward." (Goethe: Essay on Nature).[1]

ABSTRACT

An appreciation of the value of operating within an appropriate general systematic model has emerged from the recognition that the interpretation of a given body of information depends as much upon the character of the model adopted as upon any inherent quality of the data itself. Fluvial geomorphic phenomena are examined within the two systematic models which have been found especially useful in physics and biology—closed and open systems, for which simple analogies are given. Certain qualities of classic closed systems, namely the progressive increase in entropy, the irreversible character of operation, the importance of the initial system conditions, the absence of intermediate equilibrium states and the historical bias, permit comparisons to be made with the Davisian concept of cyclic erosion. The restrictions which were inherently imposed upon Davis' interpretation of landforms thus become more obvious. It is recognized, however, that no single theoretical model can adequately encompass the whole of a natural complex, and that the open system model is imperfect in that, while embracing the concept of grade, the progressive reduction of relief cannot be conveniently included within it. The open system characteristic of a tendency toward a steady state by self-regulation is equated with the geomorphic concepts of grade and dynamic equilibrium which were developed by Gilbert and later "dynamic" workers and, despite continued relief reduction, it is suggested that certain features of landscape geometry, as well as certain phases of landscape development, can be viewed profitably as partially or completely time-independent adjustments. In this latter respect the ratios forming the bases of the laws of morphometry, the hypsometric integral, drainage density, and valley-side slopes can be so considered. The relative values of the closed and open systematic frameworks of reference are recognized to depend upon the rapidity with which landscape features can become adjusted to changing energy flow, and a contrast is made between Schumm's (1956) essentially open system treatment of weak clay badlands and the historical approach which seems most profitable in treating the apparently ancient landscapes of the dry tropics.

Finally, seven advantages are suggested as accruing from attempts to treat landforms within an open system framework:

1. The focusing of attention on the possible relationships between form and process.
2. The recognition of the multivariate character of most geomorphic phenomena.
3. The acceptance of a more liberal view of changes of form through time than was fostered by Davisian thinking.
4. The liberalizing of attitudes toward the aims and methods of geomorphology.
5. The directing of attention to the whole landscape assemblage, rather than to the often minute elements having supposed historical significance.
6. The encouragement of geomorphic studies in those many areas where unambiguous evidence for a previous protracted erosional history is lacking.
7. The introduction into geography, via geomorphology, of the open systematic model which may prove of especial relevance to students of human geography.

GEOMORPHOLOGY AND GENERAL SYSTEMS THEORY

During the past decade several valuable attempts have been made, notably by Strahler (1950, 1952A, and 1952B), by Culling (1957, p. 259–261), and by Hack (1960, p. 81, 85–86; Hack and Goodlett, 1960), to apply general systems theory to the study of geomorphology, with a view to examining in detail the fundamental basis of the subject, its aims and its methods. They come at a time when the conventional approach is in danger of subsiding into an uncritical series of conditioned reflexes, and when the more imaginative modern work in geomorphology often seems to be sacrificing breadth of vision for focus on details. In both approaches it is a common trend for workers to be increasingly critical of operating within general frameworks of thought, particularly with the examples of the Davis and Penck geormorphic systems before them, and "classical" geomorphologists have retreated into restricted historical studies of regional form elements, whereas, similarly, quantitative workers have often

[1] Goethe, Fragment über die Natur (1781–82): Translated from Goethe's sämtliche Werke, Jubiläumsausgabe, Stuttgart and Berlin, v. 39, p. 3–4, undated.

withdrawn into restricted empirical and theoretical studies based on process.

It is wrong, however, to confuse the restrictions which are rightly associated with preconceived notions in geomorphology with the advantages of operating within an appropriate general systematic framework. The first lead to the closing of vistas and the decrease of opportunity; the second, however, may increase the scope of the study, make possible correlations and associations which would otherwise be impossible, generally liberalize the whole approach to the subject and, in addition, allow an integration into a wider general conceptual framework. Essentially, it is not possible to enter into a study of the physical world without such a fundamental basis for the investigation, and even the most qualitative approaches to the subject show very strong evidence of operations of thought within a logical general framework, albeit a framework of thought which is in a sense unconscious. Hack (1960), for example, has pointed to the essential difference between the approaches to geomorphology of Gilbert and Davis, and in this respect the fundamental value of the adoption of a suitable general framework of investigation based on general systems theory becomes readily apparent.

Following the terminology used by Von Bertalanffy (1950 and 1960), it is possible to recognize in general two separate systematic frameworks wherein one may view the natural occurrence of physical phenomena; the closed system and the open system (Strahler, 1950, p. 675–676, and 1952A, p. 934–935). Hall and Fagan (1956, p. 18) have defined a system as "* * * a set of objects together with relationships between the objects and between their attributes." In the light of this definition, it is very significant that one of the fundamental purposes of Davis' approach to landforms was to study them as an assemblage, in which the various parts might be related in an areal and a time sense, such that different systems might be compared, and the same system followed through its sequence of time changes. Closed systems are those which possess clearly defined closed boundaries, across which no import or export of materials or energy occurs (Von Bertalanffy, 1951). This view of systems immediately precludes a large number, perhaps all, of the systems with which natural scientists are concerned; and certainly most geographical systems are excluded on this basis, for boundary problems and the problems of the association between areal units and their interrelationships lie very close to the core of geographical investigations.

Another characteristic of closed systems is that, with a given amount of initial free, or potential, energy within the system, they develop toward states with maximum "entropy" (Von Bertalanffy, 1951, p. 161–162). Entropy is an expression for the degree to which energy has become unable to perform work. The increase of entropy implies a trend toward minimum free energy (Von Bertalanffy, 1956, p. 3). Hence, in a closed system there is a tendency for leveling down of existing differentiation within the system; or, according to Lord Kelvin's expression, for progressive degradation of energy into its lowest form, i.e. heat as undirected molecular movement (Von Bertalanffy, 1956, p. 4). This is expressed by the second law of thermodynamics (Denbigh, 1955) which, in its classic form, is formulated for closed systems. In such systems, therefore, the change of entropy is always positive, associated with a decrease in the amount of free energy, or, to state this another way, with a tendency toward progressive destruction of existing order or differentiation.

Thus, one can see that Davis' view of landscape development contains certain elements of closed system thinking—including, for example, the idea that uplift provides initially a given amount of potential energy and that, as degradation proceeds, the energy of the system decreases until at the stage of peneplanation there is a minimum amount of free energy as a result of the leveling down of topographic differences. The Davisian peneplain, therefore, may be considered as logically homologous to the condition of maximum entropy, general energy properties being more or less uniformly distributed throughout the system and with a potential energy approaching zero. The positive change of entropy, and connected negative change of free energy, implies the irreversibility of events within closed systems. This again bears striking similarities to the general operation of the geomorphic cycle of Davis. The belief in the sequential development of landforms, involving the progressive and irreversible evolution of almost every facet of landscape geometry, in sympathy with the reduction of relief, including valley-side slopes and drainage systems, is in accord with closed system thinking. Although "complications of the geographical cycle" can, in a sense, put the clock back, nothing was considered by Davis as capable of reversing the clock. The putting back of the clock by uplift, therefore, came to be associated with a release, or an absorption into the new closed system, of an increment of free energy, subsequently to be progressively dissipated through degradation.

Also, in closed systems there is the inherent characteristic that the initial system conditions, particularly the energy conditions, are sufficient to determine its ultimate equilibrium condition. This inevitability of closed-system thinking is very much associated with the view of geomorphic change held by Davis. Not only this, but the condition of a closed system at any particular time can be considered largely as a function of the initial system conditions and the amount of time which has subsequently elasped. Thus closed systems are eminently susceptible to study on a time, or historical, basis. This again enables one to draw striking analogies between closed-system thinking and the historical approach to landform study which was proposed by Davis.

Finally, it is recognized that closed systems can reach a state of equilibrium. Generally speaking, however, this equilibrium state is associated with the condition of maximum entropy which cannot occur until the system has run through its sequential development. In addition, it is impossible to introduce the concept of equilibrium into a closed-system framework of thought without the implication that it is associated with stationary conditions. The only feature of the cyclic system of Davis which employed the general concept of equilibrium was that of the "graded" condition of stream channels and slopes which, significantly, Davis borrowed from the work of Gilbert, who had an entirely noncyclic view of landform development (Hack, 1960, p. 81). Characteristically, the concept of grade was the one feature of Davis' synthesis which seems least well at home in the cyclic framework, for it has always proved difficult to imagine how, within a closed system context, a graded or equilibrium state could exist and yet the associated forms be susceptible to continued change—namely, downcutting or reduction.

The foregoing is not meant to imply that it is unprofitable to consider any assemblage of phenomena within a closed system framework, or, as Davis did, to overstress those aspects or phases which seem to achieve most significance with reference to the closed system model. It is important, however, to recognize the sources of partiality which result, not from any inherent quality of the data itself, but from the general systematic theory under which one is operating. In reality, no systematic model can encompass the whole of a natural complex without ceasing to be a model, and the phenomena of geomorphology present problems both when they are viewed within closed and open systematic frameworks. In the former, the useful concept of dynamic equilibrium or grade rests most uncomfortably; in the latter, as will be seen, the progressive loss of a component of potential energy due to relief reduction imposes an unwelcome historical parameter.

A simple, classic example of a closed system is represented by a mass of gas within a completely sealed and insulated container. If, initially, the gas at one end of the container is at a higher temperature than that at the other, this can be viewed as a condition of maximum segregation, maximum free energy, and, consequently, of maximum ability to perform work, should this thermal gradient be harnessed within a larger closed system. This is the state of minimum entropy. It is obvious, however, that this state of affairs is of a most transient character and that immediately an irreversible heat flow will begin toward the cooler end of the container. This will progressively decrease the segregation of mass and energy within the system, together with the available free energy and the ability of this energy to perform work, bringing about a similarly progressive increase of entropy. While the system remains closed nothing can check or hinder this inevitable leveling down of differences, which is so predictable that, knowing the initial energy conditions, the thermal conductivity of the gas and the lapse of time, one could accurately calculate the thermal state of the system at any required stage. Thus the distribution of heat energy and the heat flow within the system have a progressive and sequential history, the one becoming less segregated and the other ever-decreasing. Nor is it possible to imagine any form of equilibrium until all the gas has attained the same temperature, when the motion of the gas molecules is quite random and the static condition of maximum entropy obtains.

Open systems contrast quite strikingly with closed systems. An open system needs an energy supply for its maintenance and preservation (Reiner and Spiegelman, 1945), and is in effect maintained by a constant supply and removal of material and energy (Von Bertalanffy, 1952, p. 125). Thus, direct analogies exist between the classic open systems and drainage basins, slope elements, stream segments and all the other form-assemblages of a landscape. The concept of the open system includes closed systems, however, because the latter can be considered a special case of the former when transport of matter and energy into and from the system becomes zero (Von Bertalanffy, 1951, p. 156). An open system manifests one important property which is denied to the closed system. It may attain a "steady state" (Von Bertalanffy, 1950; and 1951, p. 156–157), wherein the import and export of energy and material are equated by means of an adjustment of the form, or geometry, of the system itself. It is more difficult to present a simple mechanical analog to illustrate completely the character and operations of an open system but it may be helpful to visualize one such system as represented by the moving body of water contained in a bowl which is being constantly filled from an overhead inflow and drained by an outflow in the bottom. If the inflow is stopped, the bowl drains and the system ceases to exist; whereas, if the inflow is stopped and the outflow is blocked, the system partakes of many of the features of a closed system. In such an arrangement, changes in the supply of mass and energy from outside lead to a self-adjustment of the system to accommodate these changes. Thus, if the inflow is increased, the water level in the basin rises, the head of water above the outflow increases, and the outflow discharge will increase until it balances the increased inflow. At this time the level of water in the bowl will again become steady.

Long ago, Gilbert recognized the importance of the application of this principle of self-adjustment to landform development:

> The tendency to equilibrium of action, or to the establishment of a dynamic equilibrium, has already been pointed out in the discussion of the principles of erosion and of sculpture, but one of its most important results has not been noticed.
>
> Of the main conditions which determine the rate of erosion, namely, the quantity of running water, vegetation, texture of rock, and declivity, only the last is reciprocally determined by rate of erosion. Declivity originates in upheaval, or in the displacement of the earth's crust by which mountains and continents are formed: but it receives its distribution in detail in accordance with the laws of erosion. Wherever by reason of change in any of the conditions the erosive agents come to have locally exceptional power, that power is steadily diminished by the reaction of the rate of erosion upon declivity. Every slope is a member of a series, receiving the water and the waste of the slope above it, and discharging its own water and waste upon the slope below. If one member of the series is eroded with exceptional rapidity, two things immediately result: first, the member above has its own level of discharge lowered, and its rate of erosion is thereby increased; and second, the member below, being clogged by an exceptional load of detritus, has its rate of erosion diminished. The acceleration above and the retardation below diminish the declivity of the member in which the disturbance originated: and as the declivity is reduced, the rate of erosion is likewise reduced.
>
> But the effect does not stop here. The disturbance that has been transferred from one member of the series to the two which adjoin it, is by then transmitted to others, and does not cease until it has reached the confines of the drainage basin. For in each basin all lines of drainage unite in a main line, and a disturbance upon any line is communicated through it to the main line and thence to every tributary. And as a member of the system may influence all the others, so each member is influenced by every other. There is an interdependence throughout the system. (Gilbert, 1880, p. 117–118).

This form-adjustment is brought about by the ability of an open system for self-regulation (Von Bertalanffy, 1952, p. 132–133). Le Châtelier's Principle (originally stated for equilibrium in closed systems) can be expanded also to include the so-called "Dynamic Equilibrium" or steady states in open systems:

> Any system in * * * equilibrium undergoes, as a result of a variation in one of the factors governing the equilibrium, a compensating change in a direction such that, had this change occurred alone it would have produced a variation of the factor considered in the opposite direction. (Prigogine and Defay, 1954, p. 262.)

A geomorphic statement of this principle has been given by Mackin (1948):

> A graded stream is one in which, over a period of years, slope is delicately adjusted to provide, with available discharge and with prevailing channel characteristics, just the velocity required for the transportation of the load supplied from the drainage basin. The graded stream is a system in equilibrium; its diagnostic characteristic is that any change in any of the controlling factors will cause a displacement of the equilibrium in a direction that will tend to absorb the effect of the change.

The cyclic adaptation of the concept of grade did not give sufficient importance to the factors, other than channel slope, which a stream system can control for itself, and in this respect Davis' ignorance of the significance of the practical experiments of Gilbert (1914) is most evident. A stream system cannot greatly control its discharge, which represents the energy and mass which is externally supplied into the open system. Neither can it completely control the amount and character of the debris supplied to it, except by its action of abrasion and sorting or as the result of the rapport which seems to exist regionally between stream-channel slope and valley-side slope (Strahler, 1950, p. 689). However, besides adjusting the general slope of its channel by erosion and deposition, a stream can very effectively and almost instantaneously control its transverse channel characteristics, together with its efficiency for the transport of water and load, by changes in depth and width of the channel. As Wolman (1955, p. 47) put it:

> The downstream curves on Brandywine Creek * * * suggest that the adjustment of channel shape may be as significant as the adjustment of the longitudinal profile. There is no way in which one could predict that the effect of a change in the independent controls would be better absorbed by a change in slope rather than by a change in the form of the cross section.

It may be, therefore, that a stream or reach may be virtually always adjusted (Hack, 1960, p. 85–86), in the sense of being graded or in a steady state, without necessarily presenting the smooth longitudinal profile considered by the advocates of the geomorphic cycle as the hallmark of the "mature graded condition." The state of grade is thus analogous to the tendency for steady-state adjustment, it is perhaps always present and, therefore, this presence cannot be employed necessarily as an historical, or stage, characteristic. It is interesting that the concept of the vegetational "climax," which has often been compared to that of grade, has passed through a somewhat similar metamorphosis. The original idea of a progressive approach to a static equilibrium of the ecological assemblage (Clements, 1916, p. 98–99) has been challenged by the open system interpretation of Whittaker (1955, p. 48), with an historical link being provided by the "individualistic concept" of Gleason (1926-27; 1927), much in the same way as Mackin's concept of grade links those of Davis and Wolman.

The forms developed, together with the mutual adjustment of internal form elements and of related systems, are dependent on the flow of material and energy in the steady state. The laws of morphometry (Chorley, 1957) express one aspect of this relationship in geomorphology. In addition, adjustment of form elements implies a law of optimum size of a system and of elements within a system (Von Bertalanffy, 1956, p. 7). This is mirrored by Gilbert's (1880, p. 134–135) symmetrical migration of divides and by Schumm's constant of channel maintenance (1956, p. 607), and is illustrated by Schumm's (1956, p. 609) contrast between basin areas of differing order.

Although a steady state is in many respects a time-independent condition, it differs from the equilibrium of closed systems. A steady state means that the aspects of form are not static and unchanging, but that they are maintained in the flow of matter and energy traversing the system. An open system will, certain conditions presupposed, develop toward a steady state and therefore undergo changes in this process. Such changes imply changes in energy conditions and, connected with these, changes in the structures during the process. The trend toward, and the development of, a steady state demands not an equation of force and resistance over the landscape, but that the forms within the landscape are so regulated that the resistance presented by the surface at any point is proportionate to the stress applied to it.

> Erosion on a slope of homogeneous material with uniform vegetative cover will be most rapid where the erosional power of the runoff is greatest. This nonuniform erosional process will in time result in a more stable slope profile which would offer a uniform resistance to erosion. (Little, 1940, p. 33.)

In this way the transport of mass and energy (i.e., water and debris) is carried on in the most economical manner. With time, landscape mass is therefore being removed and progressive changes in at least some of the absolute geometrical properties of landscape, particularly relief, are inevitable. It is wrong, however, to assume, as Davis did, that all these properties are involved necessarily in this progressive, sequential change. To return briefly to the analogy of the bowl. If the rush of water through the outflow is capable of progressively enlarging the orifice, the increasing discharge at the outflow, uncompensated at the inflow, will cause the head of water in the bowl to decrease. This loss of head will itself, however, constantly tend to compensate the increasing outflow, but, if the enlargement of the outflow orifice proceeds, this is a losing battle and an important feature of the system will be the progressive and sequential loss of head. However, not all features of this system will reflect this progressive change of head, and, for example, the structure of the flow within the bowl will remain much the same while any head of water at all remains there. The dimensionless ratios between landscape forms, similarly, seem to express the steady state condition of adjusted forms from which mass is constantly being removed. The geometrical ratios which form the basis of the laws of morphometry, and the height-area ratios involved in the dimensionless, equilibrium hypsometric integral are examples of this adjustment:

> In late mature and old stages of topography, despite the attainment of low relief, the hypsometirc curve shows no significant variations from the mature form, and a low integral results only where monadnocks remain * * *. After monadnock masses are removed, the hypsometric curve may be expected to revert to a middle position with integrals in the general range of 40 to 60 percent. (Strahler, 1952B, p. 1129–1130.)

In a drainage basin composed of homogenous material, in which no monadnocks would tend to form, it seems possible, therefore, that the dimensionless percentage volume of unconsumed mass (represented by the hypsometric integral) may achieve a time-independent value. It has been suggested, however, that the construction of the hypsometric curve may be so inherently restricted as to make the hypsometric integral insensitive to variations of an order which would be necessary to recognize such an equilibrium state (Leopold, written communication, 1961). This steady state principle has been tentatively extended by Schumm (1956, p. 616–617) to certain other aspects of drainage basin form:

> * * * the form of the typical basin at Perth Amboy changes most rapidly in the earliest stage of development. Relief and stream gradient increase rapidly to a point at which about 25 percent of the mass of the basin has been removed, then remains essentially constant. Because relief ratio [the ratio between total relief of a basin and the longest dimension of the basin parallel to the principal drainage] elsewhere has shown a close positive correlation with stream gradient, drainage density, and ground-slope angles, stage of development might be expected to have little effect on any of these values once the relief ratio has become constant.

In the steady state of landscape development, therefore, force and resistance are not equated (which would imply no absolute form change), but balanced in an areal sense, such that force may still exceed resistance and cause mass to be removed. Now, as has been pointed out, removal of mass under steady-state conditions must imply some progressive changes in certain absolute geometrical properties of a landscape, notably a decrease in average relief, but by no means all such properties need respond in this simple manner to the progressive removal of mass. The existence, for example, of the optimum magnitude principle for individual systems, or subsystems, implies that if the available energy within the system is sufficient to impose the optimum magnitude on that system, this magnitude will be maintained throughout a period of time and will not always be susceptible to a progressive, sequential change. Thus, Strahler (1950) has indicated that erosional slopes which are being forced to their maximum angle of repose by aggressive basal stream action will, of necessity, retain this maximum angle despite the progressive removal of mass with time.

Total energy is made up of interchangeable potential energy and flux, or kinetic, energy (Burton, 1939, p. 328) and even if the potential energy component decreases within an open system due to its general reduction, in other words along with a continual change in one aspect of form (i.e., relief), the residual flux energy may be of such overriding importance as to effectively maintain a steady state of operation. In practice the steady state is seldom, if ever, characterized by exact equilibrium, but simply by a tendency to attain it. This is partly due to the constant energy changes which are themselves characteristic of many open system operations, but the steady state condition of tendency toward attainment of equilibrium is a necessary prerequisite, according to Von Bertalanffy (1950, p. 23; and 1952, p. 132–33), for the system to perform work at all. Now, once a steady state has been established, the influence of the initial system conditions vanishes and, with it, the evidence for a previous history of the system (Culling, 1957, p. 261) (i.e., was our bowl full or empty at the start?). Indeed, in terms of analyzing the causes of phenomena which exhibit a marked steady-state tendency, considerations regarding previous history become not only hypothetical, but largely irrelevant. This concept contrasts strikingly with the historical view of development which is fostered by closed-system thinking. Wooldridge and Linton (1955, p. 3) have gone so far as to say that:

> Any such close comprehension of the terrain can be obtained in one way only, by tracing its evolution.

An even more extreme statement of the same philosophy has been made by Wooldridge and Goldring (1953, p. 165):

> The physical landscape, including the vegetation cover, is the record of *processes* and the whole of the evidence for its evolution is contained in the landscape itself.

The whole matter hinges on the rapidity with which landscape features become adjusted to energy flow, which may itself be susceptible to rapid changes, particularly during the rather abnormal latest geologic period of earth history. Obviously, most existing features are the product of both past and reasonably contemporary energy conditions, and the degree to which these latter conditions have gained ascendancy over the former is largely a function of the ratio between the amount of present energy application and the strength (whatever this may mean) of the landscape materials. Thus, the geometry of stream channels (Leopold and Maddock, 1953) and the morphometry of weak clay badlands (Schumm, 1956) show remarkable adjustments to contemporary processes—on whatever time level the action of these processes may be defined (Wolman and Miller, 1960)—whereas, at the other end of the energy/resistance scale, erosion surfaces cut in resistant rock and exposed to the low present energy levels associated with the erosional processes of certain areas of tropical Africa can only be understood on the basis of past conditions. Between these two extremes lies the major part of the subject matter of geomorphology including considerations of slope development, and it is here where the apparent dichotomy between the two systematic approaches to the same phenomena, termed by Bucher (1941; see also Strahler, 1952A, p. 924–925) "timebound" and "timeless," is most acute. In a related context, the problem of timebound-versus-timeless phenomena becomes especially obvious when rates of change and the ability to adjust are underestimated, as when vegetational assemblages have been correlated with the assumed stages of geomorphic history in the folded Appalachians by Braun (1950, p. 241–242) and in Brazil by Cole (1960, p. 174–177).

One can appreciate that in areas where good evidence for a previous landscape history still remains, the historical approach may be extremely productive, as exemplified by the work of Wooldridge and Linton on south-eastern England. However, in many (if not most) areas the condition is one of massive removal of past evidence and of tendency toward adjustment with progressively contemporaneous conditions. It is an impossibly restricted view, therefore, to imagine a universal approach to landform study being based only upon considerations of historical development.

Another characteristic of the open system is that negative entropy, or free energy, can be imported into it—because of its very nature. Therefore, the open system is not defined by the trend toward maximum entropy. Open systems thus may maintain their organization and regularity of form, in a continual exchange of their component materials. They may even develop toward higher order, heterogeneity, hierarchical differentiation and organization (Von Bertalanffy, 1952, p. 127–129). This is mirrored in geomorphology by the characteristic development of interrelated drainage forms, and goes along with a concept of progressive segregation (Von Bertalanffy, 1951, p. 148–149). This, to a minor extent, militates against the general view of adjustment previously discussed, insofar, as, with time, rates of interactions between form elements in an open system may tend to decrease. Therefore, it is quite reasonable to assume that mutual adjustments of form within geomorphic systems might be more difficult of accomplishment and delayed where the relief, through its influence over the potential energy of the system, is low rather than where there is a higher potential energy in the system.

Steady-state conditions can be interrupted by a disturbance in the energy flow or in the resistance, leading to form adjustments allowing a new steady state to be approached. These adjustments, however, do imply a consumption of energy and there is a "cost of transition" from one steady state to another (Burton, 1939, p. 334, 348). A particular geomorphic instance of this dissipation might be presented by the phenomenon of "overshooting" where active, but sporadic, processes are operating on weak materials, as instanced when the failure of steep slopes reduces them to inclinations very much below their repose angles, and by the excessive cutting and subsequent filling of alluvial channels associated with flash floods.

The dynamic equilibrium of the steady state manifests itself in a tendency toward a mean condition of unit forms, recognizable statistically, about which variations may take place over periods of time with fluctuations in the energy flow. These periods of time may in some instances be of very short duration, and the fluctuations of transverse stream profiles are measurable in the days, or even minutes, during which changes of discharge occur. These constant adjustments to new steady-state conditions may be superimposed on a general tendency for change possibly associated with the reduction of average relief through time. This general relief change, however, does not imply a sympathetic change of all the other features of landscape geometry. As has been demonstrated by Strahler (1958) and Melton (1957), for example, drainage density is controlled by a number of factors of which relief is only one. Recent work seems to be indicating that relief (naturally including considerations of average land slope) probably has only a relatively small influence over drainage density, which may be masked or negated altogether by the other more important factors (for example, rainfall intensity and surface resistance) which are not so obviously susceptible to changes with time. Denbigh, Hicks and Page (1948, p. 491) have pointed out that:

> Quite large changes of environment may take place, without the need for more than a small internal readjustment.

Horton (1945) did not believe, as did Glock (1931), that drainage density could be employed as a measure of landscape "age," and, indeed, it is not difficult to entertain the possibility that certain features of landscape geometry may be relatively unchanging, in actual dimensional magnitude as well as in dimensionless ratio, throughout long periods of erosional history.

For many landscape units, changes on either level are slow, or in some instances nonexistent. Under steady state conditions, therefore, corresponding local morphometric units will, as regards their form and magnitude, tend to crowd around a very significant mean value, imparting to a geomorphic region its aspects of uniformity. Strahler's (1950, p. 685) "law of constancy of slopes" is an expression of one phase of this adjustment. It is interesting that the general principle of the operation of a steady state condition was intuitively recognized long ago by Playfair (1802, p. 440):

> The geological system of Dr. Hutton, resembles, in many respects, that which appears to preside over the heavenly motions. In both, we perceive continual vicissitude and change, but confined within certain limits, and never departing far from a certain mean condition, which is such, that in the lapse of time, the deviations from it on one side, must become just equal to the deviations from it on the other.

Often the achievement of exact equilibrium in nature occurs only momentarily as variations about the mean take place (Mackin, 1948), and in these instances the existence of the steady state can only be recognized statistically (Strahler, 1954). In the study of landscape, the steady state condition indicated by discrete, close and recognizable statistical groupings of similar units, is characteristic of regions of uniform ratios between process and surface resistance.

Davis' view of landscape evolution was that the passage of time, of necessity, imprinted recognizable, significant and progressive changes, on every facet of landscape geometry. The recognition, however, that landscape forms represent a steady-state adjustment with respect to a multiplicity of controlling factors obliges one to take a less rigid view of the evolutionary aspects of geomorphology. When a geometrical form is controlled by a number of factors, any change of form with the passage of time is entirely dependent upon the net result of the effect of time upon those factors. Some factors are profoundly affected by the passage of time, others are not; some factors act directly (using the term in the mathematical sense) upon the form, others inversely; some factors exercise an important control over form aspects, others a less important one. Thus, if a particular geometrical feature of landscape is primarily controlled by a factor the action of which does not change greatly with time, or if the changes of factors having direct and inverse controls tend to cancel out the net effect of the changes, then the resulting variation in geometry may itself be small—perhaps insignificant.

A last important characteristic of open systems is that they are capable of behaving "equifinally"—in other words, different initial conditions can lead to similar end results (Von Bertalanffy, 1950, p. 25; and 1952, p. 143). Davisian (closed system) thinking is instinctively opposed to this view, and the immediate and facile assumption, for example, that most breaks of stream slope are only referable to a polycyclic mechanism is an illustration of the one cause–one effect mentality. The concept of equifinality accentuates the multivariate nature of most geomorphic processes and militates against the unidirectional inevitability of the closed system cyclic approach of Davis. The approach contrasts strikingly with that of Gilbert:

> Phenomena are arranged in chains of necessary sequence. In such a chain each link is the necessary consequent of that which precedes, and the necessary antecedent of that which follows * * * If we examine any link of the chain, we find it has more than one antecedent and more than one consequent * * * Antecedent and consequent relations are therefore not merely linear, but constitute a plexus; and this plexus pervades nature. (Gilbert, 1886, p. 286–287.)

To sum up, the real value of the open system approach to geomorphology is:

Firstly, that it throws the emphasis on the recognition of the adjustment, or the universal tendency toward adjustment, between form and process. Both form and process are studied, therefore, in equal measure, so avoiding the pitfall of Davis and his more recent associates of the complete ignoring of process in geomorphology:

> In a graded drainage system the steady state manifests itself in the development of certain topographic form characteristics which achieve a time-independent condition * * * Erosional and transportational processes meanwhile produce a steady flow (averaged over a period of years or tens of years) of water and waste from and through the landform system * * * Over the long span of the erosion cycle continual adjustment of the components in the steady state is required as relief lowers and available energy diminishes. The forms will likewise show a slow evolution.
>
> Applied to erosion processes and forms, the concept of the steady state in an open system focuses attention upon the relationship between dynamics and morphology. (Strahler, 1950, p. 676.)

The relation between process and form lies close to the heart of geomorphology and, in practice, the two are often so intimately linked that the problem of cause and effect may present the features of the "hen and the egg." Approach from either direction is valuable, however, for knowledge of form aids in the understanding of process, and studies of process help in the clearer perception of the significant aspects of form.

> The study of form may be descriptive merely, or it may become analytical. We begin by describing the shape of an object in the simple words of common speech: we end by defining it in the precise language of mathematics; and the one method tends to follow the other in strict scientific order and historical continuity * * * The mathematical definition of a "form" has a quality of precision which was quite lacking in our earlier stage of mere description * * * [employing means which] are so pregnant with meaning that thought itself is economized; * * *.
>
> We are apt to think of mathematical definitions as too strict and rigid for common use, but their rigour is combined with all but endless freedom * * * we reach through mathematical analysis to mathematical synthesis. We discover homologies or identities which were not obvious before, and which our description obscured rather than revealed: * * *
>
> Once more, and this is the greatest gain of all, we pass quickly and easily from the mathematical concept of form in its statical aspect to form in its dynamical relations: we rise from the conception of form to an understanding of the forces which gave rise to it; and in the representation of form and in the comparison of kindred forms, we see in the one case a diagram of forces in equilibrium, and in the other case we discern the magnitude and the direction of the forces which have sufficed to convert the one form into the other * * *.
>
> * * * Every natural phenomenon, however simple, is really composite, and every visible action and effect is a summation of countless subordinate actions. Here mathematics shows her peculiar power, to combine and generalize * * *.
>
> A large part of the neglect and suspicion of mathematical methods in * * * morphology is due * * * to an ingrained and deep-seated belief that even when we seem to discern a regular mathematical figure in an organism * * * [the form] which we so recognise merely resembles, but is never entirely explained by, its mathematical analogue; in short, that the details in which the figure differs from its mathematical prototype are more important and more interesting than the features in which it agrees; and even that the peculiar aesthetic pleasure with which we regard a living thing is somehow bound up with the departure from mathematical regularity which it manifests as a peculiar attribute of life * * *. We may be dismayed too easily my contingencies which are nothing short of irrelevant compared to the main issue; there is a *principle of negligibility* * * *.
>
> If no chain hangs in a perfect catenary and no raindrop is a perfect sphere, this is for the reason that forces and resistances other than the main one are inevitably at work * * *, but it is for the mathematician to unravel the conflicting forces which are at work together. And this process of investigation may lead us on step by step to new phenomena, as it has done in physics, where sometimes a knowledge of form leads us to the interpretation of forces, and at other times a knowledge of the forces at work guides us towards a better insight into form. (Thompson, 1942, p. 1026–1029.)

Secondly, open-system thinking directs the investigation toward the essentially multivariate character of geomorphic phenomena (Melton, 1957; Krumbein, 1959). It is of interest to note that the physical, and the resulting psychological, inability of geographers to handle successfully the simultaneous operation of a number of causes contributing to a given effect has been one of the greatest impediments to the advancement of their discipline. This inability has prompted, at worst, a unicausal determinism and, at best, an unrealistic concentration upon one or two contributing factors at the expense of others. Davis' preoccupation with "stage" in geomorphology has been paralleled, for example, by an undue emphasis on the part of some economic geographers upon the factor of "distance" in many analyses of economic location.

Thirdly, it allows a more liberal view of changes of form with time, so as to include the possibility of nonsignificant or nonprogressive changes of certain aspects of landscape form through time.

Fourthly, while not denying the value of the historical approach to landform development in those areas to which the application of this framework of study is appropriate, open-system thinking fosters a less rigid view regarding the aims and methods of geomorphology than that which appears to be held by proponents of the historical approach. It embraces naturally within its general framework the forms possessing relict facets, those indeed which form the basis for the present studies of denudation chronology, under the general category of the "inequilibrium" forms of Strahler (1952B). There is no uniquely correct method of treatment for a given body of information, and Postan (1948, p. 406) has been at pains to demonstrate the purely subjective distinction which exists between alternative explanations of phenomena on an immediately causal or generic basis, as against an historical or biographical one:

> For the frontier they draw separates not the different compartments of the universe but merely the different mental attitudes to the universe as a whole. What makes the material fact a fit object for scientific study is that men are prepared to treat it as an instance of a generic series. What makes a social phenomenon an historical event is that men ask about it individual or, so to speak, biographical questions. But there is no reason why the process should not be reversed; why we should not ask generic questions about historical events or should not write individual biographies of physical objects. Here Spinoza's argument still holds. The fall of a brick can be treated as a mere instance of the general study of falling bricks, in which case it is a material fact, and part and parcel of a scientific enquiry. But it is equally possible to conceive a special interest in a particular brick and ask why that individual brick behaved as it did at the unique moment of its fall. And the brick will then become an historical event. Newton must have been confronted with something of the same choice on the famous day when he sat under the fabulous apple tree. Had he asked himself the obvious question, why did that particular apple choose that unrepeatable instant to fall on that unique head, he might have written the history of an apple. Instead of which he asked himself why apples fell and produced the theory of gravitation. The decision was not the apple's but Newton's.

Davis was metaphorically struck by landscape and chose to write a history of it.

Fifthly, the open-system mentality directs the study of geomorphology to the whole landscape assemblage, rather than simply to the often minute elements of landscape having supposed evolutionary significance.

Sixthly, the open-system approach encourages rigorous geomorphic studies to be carried out in those regions—and perhaps these are in the majority—where the evidence for a previous protracted erosional history is blurred, or has been removed altogether.

Lastly, open-system thinking, when applied to geomorphology, has application within the general framework of geography; for geomorphology has always influenced geographical thinking to a great, and possibly excessive, degree (as, for example, that of Whittlesey, 1929; Darby, 1953; Beaver, 1961). Open-system thinking is characteristically less rigidly deterministic in a causative and time sense than the closed-system approach. The application of this closed-system approach to problems of human geography is extremely dangerous because, of its nature, it directs the emphasis toward a narrow determinism, and encourages a concentration upon closed boundary conditions, upon the tendency toward homogeneity and upon the leveling down of differences. Open-system thinking, however, directs attention to the heterogeneity of spatial organization, to the creation of segregation, and to the increasingly hierarchical differentiation which often takes place with time. These latter features are, after all, hallmarks of social, as well as biological, evolution.

ACKNOWLEDGMENTS

The author would like to thank Professor Ludwig von Bertalanffy of the University of Alberta and Dr. Luna B. Leopold of the United States Geological Survey for critically reading this manuscript and for making many valuable suggestions both regarding the general methodology and the application of general systems theory to geomorphology.

REFERENCES

Beaver, S. H., 1961, Technology and geography: The Advancement of Science, v. 18, p. 315–327.

Braun, E. L., 1950, Deciduous forests of eastern North America: Philadelphia, Pa. Blakiston Co., 596 p.

Bucher, W. H., 1941, The nature of geological inquiry and the training required for it: Am. Inst. Mining Metall. Engineers Tech. Pub. 1377, 6 p.

Burton, A. C., 1939, The properties of the steady state compared to those of equilibrium as shown in characteristic biological behavior: Jour. Cell. Comp. Physiol., v. 14, p. 327–349.

Chorley, R. J., 1957, Illustrating the laws of morphometry: Geol. Mag., v. 94, p. 140–149.

Clements, F. E., 1916, Plant succession: an analysis of the development of vegetation: Carnegie Inst. Washington, Pub. 242, 512 p.

Cole, M. M., 1960, Cerrado, Caatinga, and Pantanal: distribution and origin of the savanna vegetation of Brazil: Geog. Jour., v. 126, p. 168–179.

Culling, W. E. H., 1957, Multicyclic streams and the equilibrium theory of grade: Jour. Geology, v. 65, p. 259–274.

Darby, H. C., 1953, On the relations of geography and history: Inst. British Geog. Trans., no. 19, p. 1–11.

Denbigh, K. G., 1955, The principles of chemical equilibrium; with applications in chemistry and chemical engineering, Cambridge, England, Cambridge University Press, 491 p.

Denbigh, K. G., Hicks, M., and Page, F. M., 1948, The kinetics of open reaction systems: Faraday Soc. Trans., v. 44, p. 479–491.

Gilbert, G. K., 1877, Report on the geology of the Henry Mountains: 2d ed. 1880, Washington, D.C., Government Printing Office, 170 p.

——— 1886, The inculcation of the scientific method by example: Am. Jour. Sci., 3d ser., v. 31, p. 284–299.

——— 1914, The transportation of debris by running water: U.S. Geol. Survey Prof. Paper 86, 263 p.

Gleason, H. A., 1926–27, The individualistic concept of the plant association: Bull. Torrey Bot. Club, v. 53, p. 7–26.

——— 1927, Further views on the succession-concept: Ecology, v. 8, p. 299–326.

Glock, W. S., 1931, The development of drainage systems: Geog. Rev., v. 21, p. 475–482.

Hack, J. T., 1960, Interpretation of erosional topography in humid temperate regions: Am. Jour. Sci., v. 258–A, p. 80–97.

Hack, J. T., and Goodlett, J. C., 1960, Geomorphology and forest ecology of a mountain region in the central Appalachians: U.S. Geol. Survey Prof. Paper 347, 66 p.

Hall, A. D., and Fagen, R. E., 1956, Definition of system: General Systems Yearbook, v. 1, Ann Arbor, Mich., p. 18–28 (mimeographed).

Horton, R. E., 1945, Erosional development of streams and their drainage basins: hydrophysical approach to quantitative morphology: Geol. Soc. America Bull., v. 56, p. 275–370.

Krumbein, W. C., 1959, The "sorting out" of geological variables, illustrated by regression analysis of factors controlling beach firmness: Jour. Sed. Petrology, v. 29, p. 575–587.

Leopold, L. B., and Maddock, T., Jr., 1953, The hydraulic geometry of stream channels and some physiographic implications: U.S. Geol. Survey Prof. Paper 252, 57 p.

Little, J. M., 1940, Erosional topography and erosion: San Francisco, Calif., A. Carlisle and Co., 104 p.

Mackin, J. H., 1948, Concept of the graded river: Geol. Soc. America Bull., v. 59, p. 463–512.

Melton, M.A., 1957, An analysis of the relation among elements of climate, surface properties, and geomorphology: Office of Naval Research Project NR 389–042, Tech. Rept. 11, Dept. Geol., Columbia Univ., 102 p.

Playfair, J., 1802, Illustrations of the Huttonian theory of the earth: Facsimile reprint, Champagne, Ill., Univ. Illinois Press, 1956, 528 p.

Postan, M., 1948, The revulsion from thought: The Cambridge Jour., v. 1, p. 395–408.

Prigogine, I., and Defay, R., 1954, Chemical thermodynamics: London, Longmans, Green and Co., 543 p.

Reiner, J. M., and Spiegelman, S., 1945, The energetics of transient and steady states, with special reference to biological systems: Phys. Chem. Jour., v. 49, p. 81–92.

Schumm, S. A., 1956, Evolution of drainage systems and slopes in badlands at Perth Amboy, New Jersey: Geol. Soc. America Bull. v. 67, p. 597–646.

Strahler, A. N., 1950, Equilibrium theory of erosional slopes, approached by frequency distribution analysis: Am. Jour. Sci., v. 248, p. 673–696, 800–814.

——— 1952A, Dynamic basis of geomorphology: Geol. Soc. America Bull., v. 63, p. 923–938.

——— 1952B, Hypsometric (area-altitude) analysis of erosional topography: Geol. Soc. America Bull., v. 63, p. 1117–1142.

——— 1954, Statistical analysis in geomorphic research: Jour. Geology, v. 62, p. 1–25.

——— 1958, Dimensional analysis applied to fluvially dissected landforms: Geol. Soc. America Bull., v. 69, p. 279–300.

Thompson, D'Arcy W., 1942, On growth and form: Cambridge, England, 1116 p.

Von Bertalanffy, L., 1950, The theory of open systems in physics and biology: Science, v. 111, p. 23–29.

——— 1951, An outline of general system theory: Jour. British Phil. Sci., v. 1, p. 134–165.

——— 1952, Problems of life: Watts and Co., London, 216 p.

——— 1956, General system theory: General Systems Yearbook, v. 1, Ann Arbor, Mich., p. 1–10 (mimeographed).

——— 1960, Principles and theory of growth; Chapter 2 *in* Fundamental aspects of normal and malignant growth: Edited by W. W. Nowinski, Amsterdam, Elsevier Pub. Co., p. 143–156.

Whittaker, R. H., 1955, A consideration of the climax theory: the climax as a population and pattern: Ecol. Monographs, v. 23, p. 41–78.

Whittlesey, D., 1929, Sequent occupance: Assoc. American Geog. Ann. v. 19, p. 162–165.

Wolman, M. G., 1955, The natural channel of Brandywine Creek, Pennsylvania: U.S. Geol. Survey Prof. Paper 271, 56 p.

Wolman, M. G. and Miller, J. P., 1960, Magnitude and frequency of forces in geomorphic processes: Jour. Geology, v. 68, p. 54–74.

Wooldridge, S. W., and Goldring, F., 1953, The Weald: London, Collins, 276 p.

Wooldridge, S. W., and Linton, D. L., 1955, Structure, surface and drainage in south-east England: London, G. Philip and Son Ltd., 176 p.

26

Reprinted from *U.S. Geol. Survey Prof. Paper 500A*, 1-7, 17, 19-20 (1962)

THE CONCEPT OF ENTROPY IN LANDSCAPE EVOLUTION

By LUNA B. LEOPOLD *and* WALTER B. LANGBEIN

ABSTRACT

The concept of entropy is expressed in terms of probability of various states. Entropy treats of the distribution of energy. The principle is introduced that the most probable condition exists when energy in a river system is as uniformly distributed as may be permitted by physical constraints. From these general considerations equations for the longitudinal profiles of rivers are derived that are mathematically comparable to those observed in the field. The most probable river profiles approach the condition in which the downstream rate of production of entropy per unit mass is constant.

Hydraulic equations are insufficient to determine the velocity, depths, and slopes of rivers that are themselves authors of their own hydraulic geometries. A solution becomes possible by introducing the concept that the distribution of energy tends toward the most probable. This solution leads to a theoretical definition of the hydraulic geometry of river channels that agrees closely with field observations.

The most probable state for certain physical systems can also be illustrated by random-walk models. Average longitudinal profiles and drainage networks were so derived and these have the properties implied by the theory. The drainage networks derived from random walks have some of the principal properties demonstrated by the Horton analysis; specifically, the logarithms of stream length and stream numbers are proportional to stream order.

GENERAL STATEMENT

In the fluvial portion of the hydrologic cycle a particle of water falls as precipitation on an uplifted land mass. Its movement to the sea gradually molds a path that will be taken by succeeding particles of water. In its movement it will do some minute portion of the grand task of reducing the land mass in average elevation by carrying ultimately to the ocean in solution and as transported sediment molecules or particles of the continental materials.

The paths taken by the various droplets that make their way, through time, from higher to lower elevations represent the drainage network of the land surface, the patterns of the channels, and the longitudinal profiles of the waterways. As these paths are being carved by erosion and solution, the features of the landscape are expressed in the topography of the constantly changing surface.

The paths possible for water and its load have a large variety. There are obviously certain constraints which can be identified. Because gravity is the main force moving the materials, and works vertically downward, each particle of water and its associated load can move downward in the fluvial portion of the hydrologic cycle. The horizontal distance of movement is governed ultimately by the relation between the uplifted land mass and the ultimate base level—the ocean. Within these two general constraints, however, a large variety of paths is still possible.

Geomorphology is concerned with these paths, with the forms assumed during the process of landscape evolution, and with the principles governing the development of paths and forms.

W. M. Davis stated (1909) that the interpretation of land forms necessitates consideration of the processes acting, the effects of lithology and geologic structure, and the stage in the total evolution of the landscape. In attempts to define governing principles, geomorphologists have expressed various aspects of process and structure in mathematical form by equations of more or less generality. Available equations have not been found useful, however, as general expressions of spatial and time relations.

The most general equations which must be satisfied are, as in all the physical sciences, the equation of continuity which states that matter is not lost, and the equation of conservation which states that energy can neither be created nor destroyed.

These laws are so obviously general that they characterize each element, or reach, in any fluvial system. Further, the equations must characterize each unit of any path and at each instant in time. These two equations, then—however necessary—are insufficient to explain the paths of particles or the relation between one part of a path to another. Therefore, they can alone tell us nothing about the surface form of the landscape. Nor can they treat completely of the progressive development or change of form with time.

The present paper is an attempt to apply another general law of physics to the subject for the purpose of obtaining some additional insight into energy distributions and their relation to changes of land forms in space and time.

We acknowledge with warm thanks the helpful suggestions and criticisms received from colleagues who read an early draft of the manuscript. Particularly we thank Brig. Ralph A. Bagnold, M. Gordon Wolman, John T. Hack, Richard Chorley, N. C. Matalas, H. W. Olsen, and Mark Melton.

Previous work with Herbert E. Skibitzke was very influential in the development of ideas included here, and we are especially grateful to him for his interest, and for the stimulation received from many fruitful discussions with him.

ASPECTS OF ENTROPY

The thesis of the present paper is that the distribution of energy in a river system tends toward the most probable state. This principle, somewhat analogous to that implied in the second law of thermodynamics in relation to thermal energy, governs ultimately the paths of movement in the fluvial process and the spatial relations between different parts of the system at any one time or stage. Further, it is suggested that this general principle also tends to govern the sequence of development of these paths from one stage in geomorphic history to the succeeding one.

The development of the landscape involves not only the total available energy, but its distribution as well, a factor that may appropriately be described as entropy, adapting that term from the comparable concept in thermodynamics.

The first law of thermodynamics is merely a restatement of the principle of conservation of energy when heat is included in the energy forms. The second law of thermodynamics stated in simplest terms is that there is an increase in entropy in every natural process providing all the system taking part in the process is considered. The increase in entropy is a measure of the decrease in availability of the energy in the sense that a certain amount of energy is no longer available for conversion to mechanical work.

The second law expressed in thermal terms is not obviously related to geomorphic systems in which mechanical rather than heat energy is of principal concern. Its applicability will be explained in the discussion that follows.

With the understanding that it is necessary to define the system to which the second law of thermodynamics is to be applied, the essential idea to be adapted is that the entropy of a system is a function of the distribution or availability of energy within the system, and not a function of the total energy within the system. Thus, entropy has come to concern order and disorder. Information theory utilizes this aspect of the entropy concept.

The order-disorder aspect of entropy may be demonstrated by looking at your desk top. Ours is in disorder. Because of this we need offer no explanation of where things are. It is equally probable that a given piece of paper is anywhere. If we should take the trouble to place things in order, it would be necessary to label the 'various piles, and the amount of labelling (i.e. information) would increase with the degree of classification. One can tolerate a certain amount of disorder on his desk, but he must continually correct this by effort expended in search. This effort is putting negative entropy into the desk system; but the work done in this effort itself involves a general increase in entropy of the environment.

The degree of order or disorder in a system may be described in terms of the probability or improbability of the observed state. This aspect might best be stated first as a simple example: In a room filled with air the individual molecules are moving at random and, because of the effects of this movement and the collision of molecules, the gas tends to become more or less uniformly distributed throughout the room. Owing to the fact that these same collisions occur at random, it is physically possible that the random motion of particles might create a situation in which all particles of the gas are for a moment concentrated in one small volume in a corner of the room. The improbability that this would happen by chance is very great indeed. The molecules of the gas tend to become distributed in a random or disorderly way. The chance that given degrees of order prevail may be described in the form of a probability statement.

It is quite possible, of course, to pump the gas in the room into a small volume, but work would have to be expended on the system in order to accomplish this result. The overall entropy including that of the external pump would thence be increased.

The distribution of the energy may be stated in terms of the probability of the given distribution occurring relative to alternative distributions possible. As expressed by Brillouin (Bell, 1956, p. 159)—

> Entropy and probability are practically synonymous for the physicist who understands the second principle of thermodynamics as a natural tendency from improbable to more probable structures.

With the statistical conception of entropy in mind, the possible application to geomorphic systems becomes recognizable. The distribution of energy in a geomorphic system is one way of expressing the relative elevation of particles of water and of sediment which gradually will, in the process of landscape evolution, move downhill toward base level. The longitudinal profile of the river, for example, is a statement of the spatial distribution of streambed materials with regard to their elevation and, thus, with regard to their potential energy.

In thermodynamics heat energy is referred to absolute temperature as a base. The absolute temperature defines an absolute limit or a base datum, the situation in which molecular motion becomes zero. It is, then, the base level or the datum against which the energy content of a thermal system can be measured.

Systems in geomorphology also have a base datum with regard to the distribution of energy. This base datum is a datum of elevation, in most cases represented by mean sea level. The longitudinal profile of the river may be described as the distribution of the potential energy of both particles of water and particles of sediment, as they are traversing the fluvial path from higher elevation toward base level. The energy distribution may be defined in terms of the probability of the occurrence of that particular distribution. This again will be developed first for the thermodynamic case.

If, in a closed thermodynamic system at absolute temperature T, E is the thermal energy per unit mass of a substance having a specific thermal energy C (energy change per unit of temperature change), then a change dE in a unit mass is equal to

$$dE=CdT. \tag{1}$$

In this situation T may be thought of as being a measure of the adverse probability p that the energy exists in the given state above absolute zero. T is not equal to p because T is defined in arbitrary units, where p is an absolute number constrained between limits of zero and 1, but T is a function of p.

Because thermodynamic entropy is defined as

$$\phi=\int\frac{dE}{T} \tag{2}$$

then, per unit of mass,

$$\phi=C\int\frac{dT}{T}$$

$$=C'\int\frac{dp}{p}$$

$$\phi=C'\log_e p+\text{constant} \tag{3}$$

where C' is the specific heat energy in appropriate units.

Thus, entropy in the abstract sense may be defined as the logarithm of a probability and may express, for example, the ratio of the probability of a given physical state to the probability of all other possible alternative states. Such a statistical definition of entropy has been used in physical chemistry, information theory, and elsewhere (Bell, 1956; Lewis and Randall, 1961).

An example of possible alternatives would be the following: If consecutive single draws are made from a deck of playing cards, replacing the drawn card and reshuffling after each operation, the probabilities of drawing an ace, a face card, or a numbered card would differ. The alternative states (aces, face cards, numbered cards) have individual and different probabilities of turning up in consecutive draws. The entropy of the system is defined not in terms of the results of any single experiment but in terms of probabilities among the alternatives and is of the form $\phi=c(\log p_A+\log p_F+\log p_N)$ where p_A is the probability of drawing an ace, p_F the probability of drawing a face card, and p_N the probability of drawing a numbered card, and c is a constant to convert to appropriate units.

To generalize, then, if a system included various alternative states 1, 2, 3, . . . n, the individual probabilities of which occurring in various examples are $p_1, p_2, p_3 \ldots p_n$, the entropy of the system is defined as the sum of the logarithms of these probabilities

$$\phi=c\Sigma\log p. \tag{4}$$

The probability of a given state, p_1, represents the fractional chance as compared with unity that any example or sample will be in the state designated "1" among the n alternatives.

To explain in another way, if two states, statistically independent, of probabilities respectively, p_1 and p_2, and corresponding entropies, ϕ_1 and ϕ_2 from an assembly of states are combined, then the probability of the combination is p_1p_2 since the probabilities are multiplicative. On the other hand, entropy being an extensive property is additive. This relation between entropy and probability is satisfied by a logarithmic relation as before, thus,

$$\phi\propto\log p+\text{constant}. \tag{5}$$

In terms of a gas in an isolated system, the Boltzman relation between entropy and probability is usually stated as

$$\phi=\frac{R}{N}\log p+\text{constant} \tag{5a}$$

where R is the gas constant, N is Avogadro's number, and p represents, as before, the probability of being in a given state. Because $\log p$ is a negative number, the sign of ϕ depends on the difference between $R/N\log p$ and the constant, representing the base level for measuring entropy.

Since p is a positive number less than 1 equation 5a can be written

$$p \propto e^{-\phi N/R}$$

or since

$$\phi_i = E_i/T$$

$$p \propto e^{-E_i/kT} \quad (5b)$$

where k is Boltzman's constant R/N. In other words, the probability of a given energy E_i of one state i is proportional to the negative exponential of its ratio to the total energy kT of all possible states.

It is now possible to specify the *most probable* distribution of energy in a system. If the system is composed of possible states, 1, 2, 3, . . . n, the individual probabilities of which are $p_1, p_2, p_3, \ldots p_n$, the entropy of the system is maximum when the sum

$$\Sigma \log p \text{ is a maximum.} \quad (6)$$

This may be written as

$$(\log p_1 + \log p_2 + \log p_3 + \ldots + \log p_n) = \text{a maximum.} \quad (7)$$

Because the states 1, 2, 3, . . . n represent various alternative states, each represented in the whole system, the sum of the probabilities is unity, or

$$p_1 + p_2 + p_3 + \ldots + p_n = 1.0. \quad (8)$$

We are now concerned with the values of the individual probabilities that would make the sum of their logarithms a maximum. It can be shown that the sum $\Sigma \log p$ is a maximum when $p_1 = p_2 = p_3 = \ldots p_n$. In other words, the most probable condition among alternatives occurs when the individual probabilities of the various alternatives are equal.

Whether a system can attain this distribution depends on the constraints on the distribution of energy. However, in this connection it must be remarked that although the quantity $[\log p_1 + \log p_2 + \log p_3 + \ldots + \log p_n]$ is a maximum when all probabilities are equal, this maximum is not a sensitive one. That is, there can be some considerable irregularity between the several values of the probabilities p, and yet the sum of their logarithms may not diverge greatly from the maximum value obtained when all probabilities are equal. For example, if there are 100 possible states and the probability of each state is 0.01, the value of $\Sigma \log_e p = -460$. If, on the other hand, 50 states have a probability of 0.015 and the remaining 50 a probability of 0.005, the value of $\Sigma \log_e p = -475$, not greatly less than that for exactly equal partition of probabilities. The geomorphic significance of this mathematical fact seems to be that one might expect considerable deviations from the theoretic "most probable" state even in the absence of marked physical constraints. Complementary to the flat nature of the peak of the curve of the sum of the logarithms, the side limbs are very steep, indicating that relatively short intervals of time are needed to reach a state approaching the maximum probable, although the rate of adjustment to the theoretic most probable state thereafter may be quite slow if ever achieved.

OPEN SYSTEMS

The classical treatment of entropy in thermodynamics deals with closed systems in which entropy continuously increases to a maximum stationary level at equilibrium. In closed systems there is no loss or addition of energy. Geomorphic processes operate, on the other hand, in open systems in which energy is being added in some places while in other places energy is being degraded to heat and is thus lost insofar as further mechanical work is concerned. A river system is an example of an open system. Let the system be defined as the water and the debris in the river channel. As water flows down the channel it gives up potential energy which is converted first to kinetic energy of the flowing water and, in the process of flow, is dissipated into heat along the channel margins.

Precipitation brings increments of energy into the system because water enters at some elevation or with some potential energy. Heat losses by convection, conduction, and radiation represent energy losses from the system. Yet, the channel may be considered in dynamic equilibrium.

The steady state possible in an open system differs from the stationary state static equilibrium of closed systems. We shall therefore equate the term steady state with dynamic equilibrium in geomorphology as defined by Hack (1960). For the general case of an open system the statement of continuity of entropy is as follows (Denbigh, 1951, p. 40):

> rate of increase of entropy in system + rate of outflow of entropy = rate of internal generation of entropy.

In an open system in dynamic equilibrium, the rate of increase of entropy in the system is zero (Prigogine, 1955, p. 82; Denbigh, 1951, p. 86). The continuity equation above then takes the form that the rate of outflow of entropy equals the rate of internal generation of entropy.

In a river system in dynamic equilibrium, therefore, between any two points along its length, the rate of outflow of entropy is represented by the dissipation of energy as heat. This equals the rate of generation of

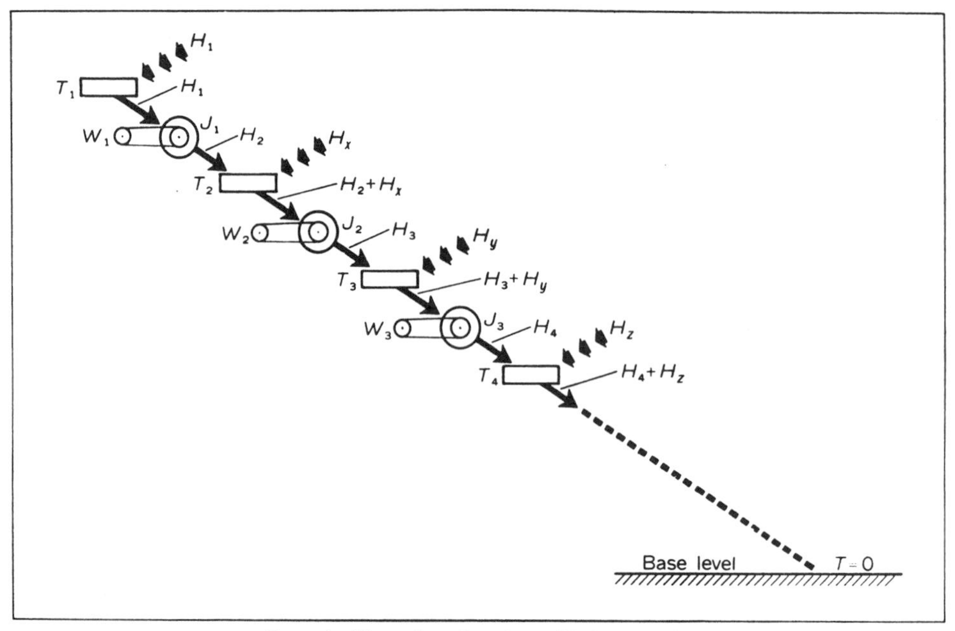

FIGURE 1.—Thermodynamic engine model of steady state.

entropy which is represented by the energy gradient toward base level. At this point it is desirable to have an equation for the generation of entropy in a river system. For this purpose we shall again consider a thermodynamic model depicted on figure 1. This model consists of a series of perfect engines, J_1, J_2, etc., operating between heat sources and sinks designated by their absolute temperatures, T_1, T_2, etc. The work produced by each engine in unit of time Δt is W_1, W_2, etc., which is delivered to a Prony brake which dissipates the work as heat. The working substance is a flow of a fluid from temperature source T_1 at a rate q_1 carrying a quantity of heat H_1 in a unit of time. The amount of work (in heat units) that is done by engine J_1 equals, in unit of time Δt,

$$\frac{W_1}{\Delta_t}=\frac{H_1}{\Delta_t}\frac{T_1-T_2}{T_1}=q_1(T_1-T_2). \quad (9)$$

It will be noted that the work done by the engine is equal to the heat transferred multiplied by the Carnot efficiency factor. Further, in thermal units, work done in unit time equals flow rate times temperature difference, so the equation is dimensionally consistent. Now the heat delivered to sink T_2 is

$$H_2=[H_1-W_1]=H_1\frac{T_2}{T_1}=q_1T_2\Delta t. \quad (10)$$

At temperature level of T_2 let us introduce an additional flow of fluid at rate q_x carrying a quantity of heat q_xT_2. The total heat therefore carried to engine J_2 is H_2+H_x and the work done in unit time by engine J_2 is

$$W_2=(H_2+H_x)\frac{T_2-T_3}{T_2}=(q_1+q_x)(T_2-T_3)\Delta t.$$

The heat delivered by J_2 to T_3 is

$$H_3=H_2+H_x-W_2=(H_2+H_x)\frac{T_3}{T_2}.$$

Hence, the heat delivered to each sink decreases in proportion to the ratio of the absolute temperatures of sink and source.

A unit of work done by engine J_1 and dissipated by the Prony brake, when divided by the absolute temperature, represents a generation of entropy or a unit export of entropy from the system. This work per unit of time is $\frac{W_1}{\Delta t}$, accomplished at temperature T_1, the quotient of which represents the change of entropy per unit of time or

$$\frac{W_1}{\Delta t}\cdot\frac{1}{T}=\left(\frac{\Delta\phi}{\Delta t}\right)_1$$

and from equation 9

$$\frac{W_1}{\Delta t T_1}=q_1\frac{T_1-T_2}{T_1}$$

and

$$\left(\frac{\Delta\phi}{\Delta t}\right)_1=q_1\frac{T_1-T_2}{T_1}.$$

Similarly at engine J_2

$$\left(\frac{\Delta\phi}{\Delta t}\right)_2=\frac{W_2}{\Delta t T_2}=(q_1+q_x)\frac{T_2-T_3}{T_2}. \qquad (11)$$

Thus, in each engine the rate of production of entropy is proportional to the rate of flow of fluid, the temperature difference between source and sink, and inversely proportional to the absolute temperature.

Hence, in more general terms, if in time dt the fluid flows a distance dx, which represents the distance apart of the adjacent engines, operating between a corresponding temperature difference dT

$$\frac{d\phi}{dt}=q\frac{dT}{dx}\frac{1}{T} \qquad (12)$$

or in terms of entropy per unit of flow

$$\frac{d\phi/dt}{q}=\frac{dT}{dx}\frac{1}{T}. \qquad (13)$$

Thus the right-hand side of equation 13 may be interpreted as describing the most probable distribution of energy in the thermodynamic model. This distribution is such that energy is in inverse proportion to temperature above base level $T=0$. Equation 13 is a thermal statement having no direct relationship to a river system. There are, nevertheless, several analogous points, which, if supported by field evidence, can lead to further inferences about river systems.

The analogy to the river system is that each engine represents a reach, Δx, of the river which receives at its upper end a flow of water at total head H_1, and delivers at its lower end the same flow at H_2. The difference in head is degenerated into heat which, like the Prony brake, exerts no influence on the transport of energy or mass in the river. Within the river reach, as within the engine, certain reversible reactions may take place, as for example transfers between kinetic and potential forms of energy as the flow expands or contracts. These, too, may be necessary adjuncts to the operation of the engine or the transport of water in a river, but being isentropic, have no effect on the equation of continuity of entropy.

If one interprets equation 13 as stating that the most probable distribution of energy in the engine model is in inverse proportion to temperature above base level $T=0$, then a corresponding statement for a river system is

$$\frac{d\phi'/dt}{Q}=\frac{dH/dx}{H} \qquad (14)$$

where ϕ' is the entropy in the special sense of the river system, H is elevation or total energy content above base level, and Q is rate of river discharge. Again note that tributary entrance does not alter the relationship because equation 14 is written in terms of rate of entropy change per unit of flow rate.

According to Prigogine (1955, p. 84), in the evolution of the stationary state of an open system the rate of production of entropy per unit volume corresponds to a minimum compatible with the conditions imposed on the system. In the case of the engine model, this means that if a stationary state prevails, the work done by each engine (the source of the entropy production) is a minimum. Hence, a stable system corresponds to one of least work, a point to which we refer in the next section, after which we shall examine the distribution of the production of entropy in a river system.

PRINCIPLE OF LEAST WORK AND ENTROPY

There is therefore considerable logic to the adage that nature follows the principle of least work. The idea is well developed in the analysis of stresses of a class of framed structures where the principles of statics are insufficient to determine the division of stresses. These structures are, therefore, called "statically indeterminate," and illustrate how inferentially the principle of maximum entropy is used to complete the solution. Consider the simple truss on figure 2. The crossed

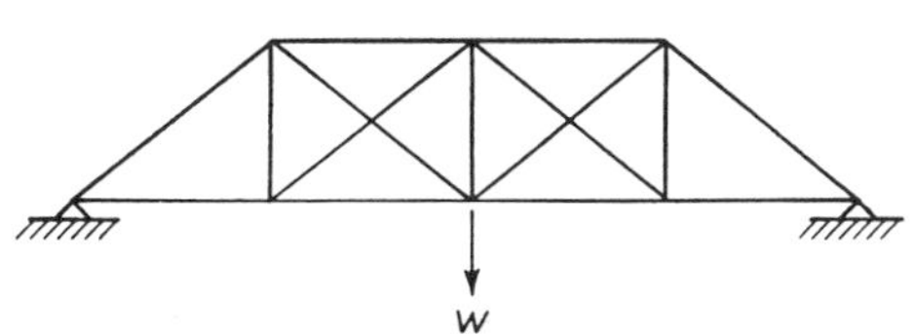

FIGURE 2.—Example of a simple statically indeterminate truss.

braces make this a common form of statically indeterminate truss. The principles of statics give 10 equations, but there are 11 unknowns. The civil engineer provides the necessary additional equation by introducing the assumption, first applied by Castigliano in 1879, that the stresses are distributed so that the total strain energy in the several members is a minimum. The strain energies are defined as $F^2L/2EA$ where F is the total stress on each member, L is its length and A its cross-sectional area, and E is the modulus of elasticity of the material. This is called the principle of least work.

In structural engineering no attention has been paid to the fact that the distribution of stresses so determined is also the most probable.

We shall demonstrate the equivalence of the principle of least work with that of maximum probability by application of the principles just discussed. As in the river profile, an articulated system such as a truss has an entropy equal to

$$\phi = \Sigma \log p + \text{constant}.$$

The most probable distribution exists when ϕ is a maximum. Let the ratios E_{n_1}/E, E_{n_2}/E, E_{n_3}/E . . ., etc., represent the proportional division of the strain energy among the several members, where E is the total strain energy in the system—the product of the weight W times deflection. (Note the analogy to the discussion leading to equation 5 and immediately following equation 5b.)

However, the joint probability that a particular combination of strain energies exists among the several members is

$$p \propto e^{-\frac{E_{n_1}}{E}} e^{-\frac{E_{n_2}}{E}} e^{-\frac{E_{n_3}}{E}} \ldots \text{etc.} \tag{15}$$

Where there are no physical limitations, this joint probability would be a maximum when all probabilities are equal; but in this case, conditions of statics set constraints on the values of the strain energies so that we can state only that the most probable combination exists when $E_{n_1}+E_{n_2}+E_{n_3}+$ . . . is a minimum. This is exactly equivalent to the statement of least work; but the principle of least work must be recognized as only one species of a larger class or genus including all of the states of maximum probability.

In the structural example the laws of statics left only one degree of freedom to be met by the principle of maximum probability. No general solution could be made based entirely on one principle—that is, entirely on the basis of statics, or entirely on the basis of maximum probability. Similarly in geomorphic problems, the set of physical factors includes many variables such as the amount of water and sediment to be carried, the fluid friction, and the river transport capacity. The equations connecting these factors leave several degrees of freedom remaining. In other words, a river system is "hydraulically indeterminate." A river can adjust its depth, width, or velocity to a given slope in several ways so that it is necessary to establish the river profile and the hydraulic geometry on the basis of maximum probability.

The general implication for rivers may then be stated as follows: The principle of least work is one of several ways in which the condition of maximum probability may be satisfied. The river channel has the possibility of internal adjustment among hydraulic variables to meet the requirement for maximum probability, and these adjustments tend also to achieve minimization of work. In systems other than rivers wherein the adjustment is actually to a condition of least work, maximum probability is achieved. The geomorphologist's intuitive inference that river equilibrium is a condition of least work (see, for example, Rubey, 1952, p. 135) is not complete. As will be shown in the section on hydraulic geometry, other factors restrain the system from attaining a state of least work.

LONGITUDINAL PROFILE OF RIVERS

A concept developed in a preceding section states that in an open system the distribution of energy tends toward the most probable. In the open system represented by the river, the unit under consideration is a unit of length along the river channel. The concept applied to the river system yields the result that the most probable distribution of energy exists when the rate of gain of entropy in each interval of length along the river is equal.

To develop the reasoning leading to that conclusion, refer back to equation 14,

$$\frac{d\phi'}{dt}\frac{1}{Q} = \frac{dH}{dx}\frac{1}{H}. \tag{14}$$

The rate of production of entropy in a river system is $\frac{d\phi'}{dt}$, and per unit of mass volume rate (discharge) is $\frac{d\phi'}{dt}\cdot\frac{1}{Q}$. This quotient is inversely proportional to the energy content above base level, H, and directly proportional to the loss of potential per unit of distance, $\frac{dH}{dx}$.

The right side of equation 14 is a statement of the proportional distribution of energy relative to base level, $H=H_0$. Then it also can be considered a statement of the probability of a given distribution of energy between $H=H_0$ and the maximum value of H in the system in question.

The probability of a particular combination of values of energy content of unit distances along the course of the open system of the river would be similar in form to equation 15, or

$$p \propto e^{-\frac{dH_1}{dx}\frac{1}{H}} e^{-\frac{dH_2}{dx}\frac{1}{H}} \ldots e^{-\frac{dH_n}{dx}\frac{1}{H}}$$

where $\frac{dH_1}{dx}$, $\frac{dH_2}{dx}$, etc., represent the loss of head in successive units of length along the river length.

[*Editor's Note:* Material has been omitted at this point.]

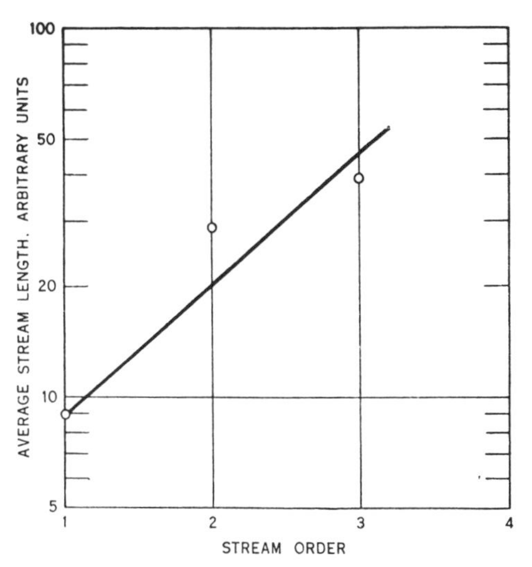

FIGURE 7.—Relation of length of streams to stream order.

DISCUSSION AND GEOMORPHIC IMPLICATIONS

In statistical mechanics the probability aspect of entropy has been demonstrated. The word entropy has, therefore, been used in the development of systems other than thermodynamic ones, specifically, in information theory and in biology. Its use is continued here.

With this understanding of the term, the present paper shows that several geomorphic forms appear to be explained in a general way as conditions of most probable distribution of energy, the basic concept in the term "entropy."

It is perhaps understandable that features such as stream profiles which occur in nature in large numbers should display, on the average, conditions that might be expected from probability considerations because of the large population from which samples may be drawn. The difficulty in accepting this proposition is that there is not one but many populations owing to the variety of local geologic and lithologic combinations that occur. Further, the geomorphic forms seen in the field often are influenced by previous conditions. Stream channels, in particular, show in many ways the effects of previous climates, of orogeny, and of structural or stratigraphic relations that existed in the past. In some instances the present streams reflect the effects of sequences of beds which have been eradicated by erosion during the geologic past.

In a sense, however, these conditions that presently control or have controlled in the past the development of geomorphic features now observed need not be viewed as preventing the application of a concept of maximum probability. Rather, the importance of these controls strengthens the usefulness and generality of the entropy concept. In the example presented here we have attempted to show that the differences between patterns derived from averaging random walks result from the constraints or controls imposed upon the system. We have, in effect, outlined the mathematical nature of a few of the controls which exist in the field. The terms in the genetic classification of streams reflect the operation of constraints. Terms like "consequent" and "subsequent" are qualitative statements concerning constraints imposed on streams which, in the absence of such constraints, would have a different drainage net and longitudinal profile.

In a sense, then, much of geomorphology has been the study of the very same constraints that we have attempted to express in a mathematical model.

The present paper is put forward as a theoretical one. It is not the purpose of a theoretical paper to compare in detail the variety of field situations with the derived theoretical relations. Rather, it is hoped merely to provide the basis for some broad generalizations about the physical principles operating in the field situations.

On the other hand, the random-walk models used here are simple demonstrations of how probability considerations enter into the problem. They are intended to exemplify how the basic equations can be tested experimentally. Thus the present paper should not be considered to deal with random walks. We hope it is concerned with the distribution of energy in real landscape problems. The random-walk models exemplify the form of the equations, but the equations describe the distribution of energy in real landscapes, simplified though the described landscape may be.

[*Editor's Note:* Material has been omitted at this point.]

Further, the relative insensitivity of the results to the lack of exactly equal probabilities among alternatives suggests that the approach to the most probable condition is at first very rapid. In the mathematical models the final elimination of minor deviations from the condition of maximum entropy requires a very large number of samples from the population of possible alternatives. In terms of the field development of forms, it seems logical that this may be equated to a long time-period required to eliminate minor variations from the theoretic most probable state.

These observed characteristics of the averages of samples used in the mathematical models make it difficult to believe that in field conditions, time periods measured in geologic epochs could elapse before fluvial systems approach the condition of quasi-equilibrium. This line of reasoning quite fails to lend support to the Davisian concept that the stage of geomorphic youth is characterized by disequilibrium whereas the stage of maturity is characterized by the achievement through time of an equilibrium state. Rather, the reasoning seems to support the concept recently restated by Hack (1960) that no important time period is necessary to achieve a quasi-equilibrium state.

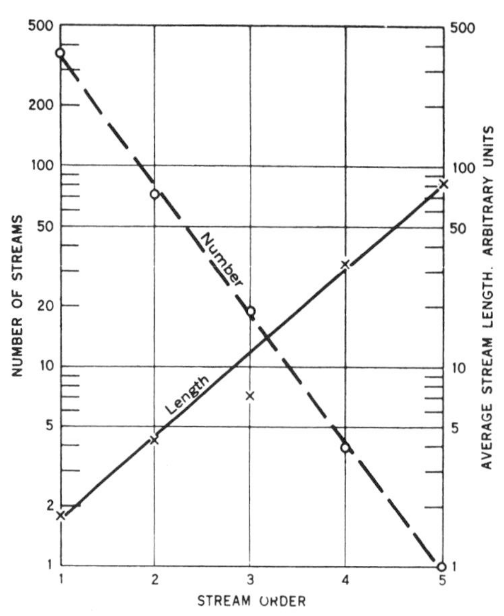

FIGURE 9.—Relation of number and average lengths of streams to stream order for model in figure 8.

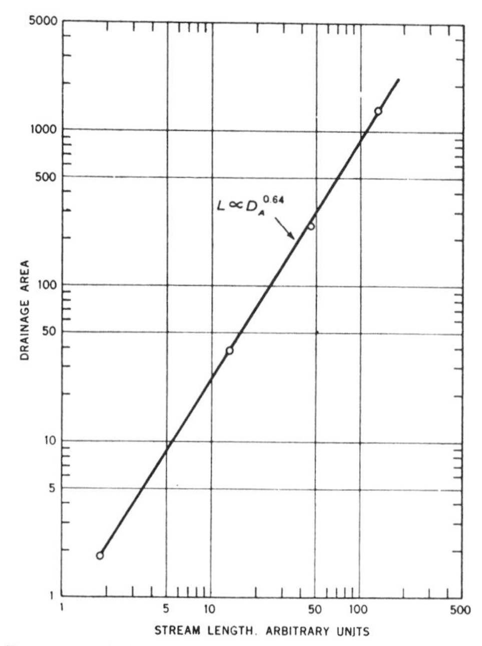

FIGURE 10.—Relation between stream length and drainage area for random-walk model.

The differing results obtained from varying the constraints imposed in the mathematical models lead us to following view as a working hypothesis: Landscape evolution is an evolution in the nature of constraints in time, maintaining meanwhile and through time essentially a dynamic equilibrium or quasi-equilibrium.

This conclusion also is in agreement with the view of Hack (1960), though arrived at by a quite different line of reasoning.

Whether or not the particular inferences stated in the present paper are sustained, we believe that the concept of entropy and the most probable state provides a basic mathematical conception which does deal with relations of time and space. Its elaboration may provide a tool by which the various philosophic premises still characterizing geomorphology may be subjected to critical test.

REFERENCES

Bagnold, R. A., 1960, Sediment discharge and stream power—a preliminary announcement: U.S. Geol. Survey Circ. 421, 23 p.

Bell, D. A., 1956, Information theory: Sir Isaac Pitman & Sons, London, 174 p.

Davis, W. M., 1909, The rivers and valleys of Pennsylvania: Geographical Essays, Dover Publications, Inc. (reprinted 1954), p. 413–484.

Denbigh, K. G., 1951, The thermodynamics of the steady state: Methuen & Co., London, 103 p.

Feller, William, 1950, An introduction to probability theory and its applications: John Wiley & Sons, New York, 419 p.

Hack, J. T., 1957, Studies of longitudinal stream profiles in Virginia and Maryland: U.S. Geol. Survey Prof. Paper 294-B, p. 46-94.

——— 1960, Interpretation of erosional topography in humid temperate regions: Am. Jour. Sci., Bradley Volume, v. 258-A, p. 80-97.

Horton, R. E., 1945, Erosional development of streams and their drainage basins; hydrophysical approach to quantitative morphology: Geol. Soc. America Bull., v. 56, p. 275-370.

Khinchin, A. I., 1957, Mathematical foundations of information theory: Dover Publications, Inc., New York, 120 p.

Langbein, W. B., 1947, Topographic characteristics of drainage basins: U.S. Geol. Survey Water-Supply Paper 968-C, p. 125-155.

Laursen, E. M., 1958, Sediment—transport mechanics in stable channel design: Am. Soc. Civil Engrs. Trans. v. 123, p. 195-206.

Leopold, L. B., 1953, Downstream change of velocity in rivers: Am. Jour. Sci., v. 251, p. 606-624.

Leopold, L. B., and Maddock, T., Jr., 1953, The hydraulic geometry of stream channels and some physiographic implications: U.S. Geol. Survey Prof. Paper 252, 56 p.

Leopold, L. B., and Miller, J. P., 1956, Ephemeral streams—Hydraulic factors and their relation to the drainage net: U.S. Geol. Survey Prof. Paper 282-A, p. 1-36.

Lewis, G. N., and Randall, Merle, 1961, Thermodynamics: McGraw-Hill Book Co., Inc., 723 p.

Prigogine, I., 1955, Introduction to thermodynamics of irreversible processes: C. C. Thomas, Springfield, Ill., 115 p.

Rubey, W. W., 1953, Geology and mineral resources of the Hardin and Brussels quadrangles: U.S. Geol. Survey Prof. Paper 217, 175 p.

27

Reprinted from *Amer. Jour. Sci.,* **263,** 305–307, 310–312 (1965)

GEOMORPHOLOGICAL SYSTEMS — EQUILIBRIUM AND DYNAMICS

Alan D. Howard

[*Editor's Note:* In the original, material precedes this excerpt.]

The concept of *equilibrium* is quite basic to system theory and is considered here to imply a complete adjustment of the internal variables to external conditions. The degree of approach of system variables to equilibrium may be measured by two methods: (a) if the external variable remains constant through time, then the parameters of a system in equilibrium should also remain constant (however, the sensitivity of specific internal variables to changes of the external variables is indeterminate by this method); (b) if the value of an external variable changes through time or space, a correlation of low variance between the value of the external variable and that of the system property indicates a close approach to equilibrium whereas a high correlation coefficient indicates a high sensitivity of the system property to changes of the external variable. Each combination of external variables defines a unique system equilibrium state. Langbein and Leopold (1964) propose that the equilibrium state represents a balance between a tendency toward equal areal distribution of energy expenditure and a tendency toward minimum total work expended upon the landforms. Thus an equilibrium state would exhibit maximum efficiency of erosion under the given external conditions.

Crucial to the study of equilibrium and dynamics of a system is the concept of the resistance to change (*inertia*) of a system variable (Chorley, 1962; Langbein and Leopold, 1964). Following a change of external variable the system will tend to adjust to the new regime. In the case of a system that is manifesting no secondary responses (discussed below) and that is changing from one equilibrium state to another, the internal parameters in many natural systems tend to approach the new equilibrium at a rate proportional to their distance from the equilibrium value (for example, Strahler, 1952, and Schumm, 1963, propose a rate of erosion proportional to the average elevation above base level). Such an exponential approach implies a time-constant characteristic of the system and of the type of change of external variable. Although a system may never achieve exact equilibrium, it will within a finite period reach any desired approximation to equilibrium.

The rate of change of an external variable compared to the capacity for adjustment of the system determines the behavior of the system. The significance of this ratio should become clear in the following discussion of the important external factors which control the dynamics of landforms.

Stratigraphy and structure act passively upon landforms, exerting influence through constraints upon the system and, in the case of stratigraphy, as a source of mass for the system. In areas where lithology and structure are of constant composition or only slowly varying in the vertical direction, a complete adjustment between landforms and geology is to be expected, and such a case is indicated by the well-defined correlation between stream and slope parameters and the parental material in areas of steep regional dip or locally homogeneous lithology (Flint, 1963; Hely and Olmsted, 1963; Hack, 1957; Wolman, 1955). On the other hand, when the lithology and structure are heterogeneous in the vertical direction, landforms must adjust as erosion exposes new parent rock and structure. For example, exposure of a more resistant layer should occasion a steepening of landform gradients and an increase of local relief for more efficient erosion. In general, items such as slope values, drainage densities, drainage basin configurations will all require adjustment to changes of geology during erosion. Associated with vertical successions of stratigraphy and structure will be transitional or non-equilibrium landforms, for example, the escarpment separating landforms on an upper resistant unit from those on a lower, weaker stratum. These transitional landforms will be zones of maximum rate of change in time and space of landform parameters, and in areas of minutely inhomogeneous geology all landforms would be transitional.

Several variables act upon the topography with great effect, but over such a geologically short time period that no equilibrium state is to be expected. Such variables are usually rare and unpredictable in time and place of action. Volcanic eruptions, glaciation, major floods, tectonics (discussed separately below), and most activities of man fall in this category. Each action of such a variable must be individually considered as an "event" in the history of the landforms. Glaciation, volcanic eruptions, and some types of tectonic movement result in landscapes that may be considered as *constructional* as opposed to erosional. The resulting landform, whether it be drumlin, volcano, or fault block, has an initial form essentially independent of erosional processes. However, when such intensive variables cease to operate, erosional landforms become superimposed upon the original structure and initial conditions become less determinate of the landforms as time proceeds. Thus in the western United States what were originally valley volcanic flows often become resistant ridges, and long-inactive faults find topographic expression only through superposition of unlike strata or through their weakness to erosion caused by fracturing. When a variable such as volcanic activity or tectonic uplift occurs intermittently, no landforms that are completely constructional will be formed and the landscape will show aspects of both.

Perhaps the greatest debate in geomorphology has been over the role of tectonics in shaping surface features. Davis constructed a theory of landform evolution around the assumption of uplift movements essentially instantaneous compared to the resulting rate of erosion, whereas Penck conceived of slowly varying uplift rates of the same magnitude as the rate of erosion. Modern studies indicate that a variety of uplift movements may occur and may be more local in extent than is commonly supposed (Chorley, 1963; Weber, 1958). Equilibrium between landforms and tectonic movement implies an equivalence of the rate of erosion to the rate of uplift, but this is probably very seldom, if ever, the case because of the slow response of landforms to tectonic movement (Schumm, 1963). Schumm (1963) and Davis, however, went a step further and proposed that no equilibrium landforms could develop in association with a uniform and long-continued uplift, although Thornbury (1954, p. 21) argued for "perpetual youth" in such cases. Davis maintained that continued uplift must produce a constant increase in relief, whereas Schumm envisions a constant extension of drainage in such a circumstance. Both authors cite the inefficiency of slope and divide erosion as causes for this disequilibrium; this position has been disputed earlier in this paper. Schumm's scheme implies either extension of some drainage basins at the expense of others (thus indirectly challenging the proposal of a tendency toward equal areal distribution of erosional intensity, advanced by Shaler, 1899, and Langbein and Leopold, 1964) or a continual increase of drainage density. Both Davis's and Schumm's viewpoints imply that landforms do not tend toward equilibrium and that geomorphic systems tend to be non-conservative. In view of the numerous manifestations of close adjustment between form and process in other aspects of geomorphic systems (for example, stream regimens), the present author allows for the possibility that the rate of uplift and the rate of erosion could be equal, even though this would rarely occur because of fluctuations in and the possible episodic nature of tectonic movement. In the case of a stable land-sea level, the only equilibrium in keeping with inactive tectonics would be peneplanation.

The external variables loosely grouped under the term *climate* are active factors of less intensity than those of volcanism, glaication, et cetera but which act with more regularity. Although we usually receive weather reports in daily lump sums, true *weather* is only the instantaneous value of temperature, precipitation, et cetera, whereas climate represents weather averaged (integrated) over an arbitrary finite time interval. Weather can have no effect upon geomorphic systems; only extended action of the external variables, expressed as climate, can be correlated with action upon the system. In illustration, an intense rainfall (severe instantaneous weather) will have little effect upon landforms if continued for only a few seconds but will have a large effect if continued for several hours (as reflected in the climatic variables of precipitation duration and average intensity). The type of climatic factor considered is arbitrary—one may define such variables as hourly precipitation averages, annual temperature, and humidity averages for a certain month. Those climatic factors to which the landforms are most sensitive are of greatest interest. Certain other definable factors exhibit the characteristics of the climatic type and are usually indirect effects of climate; in stream studies supplied load and discharge are essentially climatic in action upon stream and channel characteristics.

[*Editor's Note:* Material has been omitted at this point.]

CONCLUSIONS

High relief areas and headwater areas of large drainage basins are little affected by small-scale changes of land-sea relative levels (Thornbury, 1954, p. 106) and hence should carry little evidence of individual tectonic movements. However, areas of low relief on non-resistant rocks bordering on major streams and rivers will be affected in terms of available relief by a land-sea level change on the order of a hundred feet. Such areas may carry unmistakable evidence of a stagnation or rejuvenation of erosion. Schumm (1956) and Carter and Chorley (1961) considered erosion of the small-scale landforms of poorly consolidated rocks and show that a lowering of base level on originally muted topography results in an intensification of erosion which is progressively propagated through the area concerned in a manner similar to the initial stages of landform evolution as proposed by Davis. The main drainage channels first adjust to the lower base level, and the areas bordering the main channel because of steep gradient develop steep slopes, while smaller and more distant drainage channels incise less and the slopes are little steepened. In areas of rejuvenation of erosion, the intensification would not be reflected so much in slope profiles as in an areal variation of slope values (Strahler, 1950). Some evidence of such a pattern of drainage development may be seen in the Midwest, where postglacial drainage patterns are forming on glacial deposits; areas near main streams have rolling topography, while drainage divides remain largely flat. Areas in the eastern United States bordering on major rivers appear also to show the effects of a base-level lowering.

On the other hand, a raising of sealevel relative to the land or a prolonged period of land-sea stability would have the effect of a general decrease of relief first expressed along the main drainage channels and least affecting the headwater areas.

Despite the importance of tectonic movement and sealevel changes, their importance has been overemphasized. In many cases changes of erosional regime previously ascribed to tectonic factors must be accounted for by geologic effects upon the landforms or by climatic changes. More and more episodes of aggradation and intrenchment in the western United States are being recognized as effects of climatic change. Such would be the case especially in areas of high relief or those far removed from major drainage.

Studies in areas of homogeneous lithology show that parameters like stream and hill slopes, slope profiles, height of drainage divides, drainage density and drainage basin shape remain equal, in a statistical sense, from drainage basin to drainage basin in areas of the same tectonic unit and climatic environment. This implies that for such uniform areas an equi-partition of erosive areas is the most stable state (Langbein and Leopold, 1964; Shaler,

1899). Each type of lithology gives rise to unique landforms—limestone in humid climates supports subdued, rounded topography, while sandstone is characterized by coarse topography and straight slopes associated with a high degree of physical weathering (Hack, 1957). Likewise, areas of intense erosion on the same type of rock and in similar climates might be differentiated from areas that have received less recent uplift by coarser landforms, steeper gradients, and more predominate physical weathering. Schumm (1963) notes a correlation of intense erosion with a high ratio of basin relief to basin length. Similarly, changes of climate will call forth variations in landforms.

Conversely, lateral and vertical variations in quality of landforms in areas of the same tectonic unit and similar climate should be attributable to changes of stratigraphy or structure. For example, lateral and vertical changes of topographic parameters in areas of nearly horizontal, heterogeneous rock clearly correlate with geologic changes (Thwaites, 1960). Landforms in areas of heterogeneous rock, especially in areas of low-dipping strata, have a historical aspect not found in areas of uniform geology. In response to vertical changes of geology divides may migrate, streams may be captured, and in general slopes, drainage densities, et cetera, will change through time. In the past these geologically-induced topographic changes have often been mistaken for tectonic movement.

Because of the interrelated nature of geomorphic systems, landforms on one type of rock will be influenced by the character of the surrounding strata. Hack (1957) shows the pronounced effect of bedload transported from upstream upon stream regimen in downstream areas of differing geology. These interrelationships between rock units are most conspicuous in the transitional landforms in areas of low-dipping rocks; a resistant rock may act as a perched base level to an overlying weak unit, and conversely, escarpment retreat and overly-steep slopes on a non-resistant unit may be caused by an overlying resistant rock. Hack (1960) demonstrates that lateral planation in the eastern Appalachians is most pronounced where areas underlain by resistant rock discharge onto weaker strata, and Howard (1963) has proposed that cavern development may ultimately depend upon an inhomogeneity of geology.

References

Carter, C. S., and Chorley, R. J., 1961, Early slope development in an expanding stream system: Geol. Mag., v. 98, p. 117-130.

Chorley, R. J., 1962, Geomorphology and general systems theory: U. S. Geol. Survey Prof. Paper 500-B, 10 p.

———— 1963, Diastrophic background to twentieth-century geomorphological thought: Geol. Soc. America Bull., v. 74, p. 953-970.

Flint, R. F., 1963, Altitude, lithology, and the fall zone in Connecticut: Jour. Geology, v. 71, p. 683-697.

Hack, J. T., 1957, Studies of longitudinal stream profiles in Virginia and Maryland: U. S. Geol. Survey Prof. Paper 294-B, p. 45-97.

———— 1960, Interpretation of erosional topography in humid temperate regions: Am. Jour. Sci., Bradley v. 258A, p. 80-97.

Hely, A. G., and Olmstead, F. H., 1963, Some relations between streamflow characteristics and the environment in the Delaware River Region: U. S. Geol. Survey Prof. Paper 417-B, 25 p.

Howard, A. D., 1963, Development of karst features: Natl. Speleol. Soc. Bull., v. 25, p. 45-65.

Langbein, W. B., and Leopold, L. B., 1964, Quasi-equilibrium states in channel morphology: Am. Jour. Sci., v. 262, p. 782-794.

Schumm, S. A., 1956, Evolution of drainage systems and slopes in badlands at Perth Amboy, New Jersey: Geol. Soc. America Bull., v. 67, p. 597-646.
——— 1963, The disparity between present rates of denudation and orogeny: U. S. Geol. Survey Prof. Paper 454-H, 13 p.
Shaler, N. S., 1899, Spacing of rivers with reference to hypothesis of base-leveling: Geol. Soc. America Bull., v. 10, p. 263-276.
Strahler, A. N., 1950, Equilibrium theory of erosional slopes approached by frequency distribution analysis: Am. Jour. Sci., v. 248, p. 673-696, 800-814.
——— 1952, Dynamic basis of geomorphology: Geol. Soc. America Bull., v. 63, p. 923-938.
Thornbury, W. D., 1954, Principles of geomorphology: New York, John Wiley and Sons, Inc., 618 p.
Thwaites, F. T., 1960, Evidences of dissected erosional surfaces in the Driftless Area: Wisc. Acad. Sci., Arts, and Letters Trans., v. 49, p. 17-49.
Weber, H., 1958, Die Oberflächenformen des festen Landes: Leipzig, B. G. Teubner Verlagsgesell., p. 131-142.
Wolman, M. G., 1955, The natural channel of Brandywine Creek, Pennsylvania: U. S. Geol. Survey Prof. Paper 271, 56 p.

28

Reprinted from *Amer. Jour. Sci.*, **263**, 110, 112–115, 118–119 (Feb. 1969)

TIME, SPACE, AND CAUSALITY IN GEOMORPHOLOGY*

S. A. SCHUMM and R. W. LICHTY

U. S. Geological Survey, Denver, Colorado

ABSTRACT. The distinction between cause and effect in the development of landforms is a function of time and space (area) because the factors that determine the character of landforms can be either dependent or independent variables as the limits of time and space change. During moderately long periods of time, for example, river channel morphology is dependent on the geologic and climatic environment, but during a shorter span of time, channel morphology is an independent variable influencing the hydraulics of the channel.

During a long period of time a drainage system or its components can be considered as an open system which is progressively losing potential energy and mass (erosion cycle), but over shorter spans of time self-regulation is important, and components of the system may be graded or in dynamic equilibrium. During an even shorter time span a steady state may exist. Therefore, depending on the temporal and spacial dimensions of the system under consideration, landforms can be considered as either a stage in a cycle of erosion or as a system in dynamic equilibrium.

INTRODUCTION

Current emphasis on the operation of erosion processes and their effects on landforms (Strahler, 1950, 1952) not only has opened the way to new avenues of research but also introduces the possibility of misunderstanding the role of time in geomorphic systems. As Von Bertalanffy (1952, p. 109) put it, "In physical systems events are, in general, determined by the momentary conditions only. For example, for a falling body, it does not matter how it has arrived at its momentary position, for a chemical reaction it does not matter in what way the reacting compounds were produced. The past is, so to speak, effaced in physical systems. In contrast to this, organisms appear to be historical beings". From this point of view, although landforms are physical systems and can be studied for the information they afford during the present moment of geologic time, they are also analogous to organisms because they are systems influenced by history. Therefore, a study of process must attempt to relate causality to the evolution of the system.

It is the purpose of this discussion to demonstrate the importance of both time and space (area) to the study of geomorphic systems. We believe that distinctions between cause and effect in the molding of landforms depend on the span of time involved and on the size of the geomorphic system under consideration. Indeed, as the dimensions of time and space change, cause-effect relationships may be obscured or even reversed, and the system itself may be described differently.

[*Editor's Note:* Material has been omitted at this point.]

* Publication authorized by the Director, U. S. Geological Survey.

To resolve the controversy resulting from these two viewpoints it may be necessary to think only in terms of large and small areas or of long and short spans of time. A choice must be made whether only components of a landscape are to be considered or whether the system is to be considered as a whole. Also, a choice must be made as to whether the relations between landforms and modern erosion processes are to be considered or whether the origin and subsequent erosional history of the system is to be considered. In table 1 an attempt is made, using a hypothetical drainage basin as an example, to demonstrate that the concepts of cyclic erosion with time and timeless dynamic equilibrium are not mutually exclusive.

The variables listed in table 1 are arranged in a hierarchy we believe approximates the increasing degrees of dependence of the variables considered. For example, time, initial relief, geology, and climate are obviously the dominant independent variables that influence the cycle of erosion. Vegetational type and density depend on lithology and climate. As time passes the relief of the drainage system or mass remaining above base level is determined by the factors above it in the table, and it, in turn, strongly influences the runoff and sediment yield per unit area within the drainage basin. The runoff and sediment yield within the system establish the characteristic drainage network morphology (drainage density, channel shape, gradient, and pattern) and hillslope morphology (angle of inclination and profile form) within the constraints of relief, climate, lithology, and time. The morphologic variables, in

TABLE 1

The status of drainage basin variables during time spans of decreasing duration

Drainage basin variables	Status of variables during designated time spans		
	Cyclic	Graded	Steady
1. Time	Independent	Not relevant	Not relevant
2. Initial relief	Independent	Not relevant	Not relevant
3. Geology (lithology, structure)	Independent	Independent	Independent
4. Climate	Independent	Independent	Independent
5. Vegetation (type and density)	Dependent	Independent	Independent
6. Relief or volume of system above base level	Dependent	Independent	Independent
7. Hydrology (runoff and sediment yield per unit area within system)	Dependent	Independent	Independent
8. Drainage network morphology	Dependent	Dependent	Independent
9. Hillslope morphology	Dependent	Dependent	Independent
10. Hydrology (discharge of water and sediment from system)	Dependent	Dependent	Dependent

turn, strongly influence the volumes of runoff and sediment yield which leave the system as water and sediment discharge.

Among the variables listed on table 1, every cause appears to be an effect and every effect a cause (Mackin, 1963, p. 149); therefore, it is necessary to set limits to the system that is considered. Obviously neither the causes of geology, climate, and initial relief nor the effects of water and sediment discharge concern us here.

The three major divisions of table 1 are time spans which are termed cyclic, graded, and steady. The absolute length of these time spans is not important. Rather, the significant concept is that the system and its variables may be considered in relation to time spans of different duration.

Cyclic time, of course, represents a long span of time. It might better be referred to as geologic time, but in order to keep the terminology of the table consistent, cyclic is used because it refers to a time span encompassing an erosion cycle. Cyclic time would extend from the present back in time to the beginning of an erosion cycle.

Consider a landscape that has been tectonically stable for a long time. A certain potential energy exists in the system because of relief, and energy enters the system through the agency of climate. Over the long span of cyclic time a continual removal of material (that is, expenditure of potential energy) occurs and the characteristics of the system change. A fluvial system when viewed from this perspective is an open system undergoing continued change, and there are no specific or constant relations between the dependent and independent variables as they change with time (fig. 1a).

During this time span only time, geology, initial relief, and climate are independent variables. Time itself is perhaps the most important independent variable of a cyclic time span. It is simply the passage of time since the beginning of the erosion cycle, but it determines the accomplishments of the erosional agents and, therefore, the progressive changes in the morphology of the system. Vegetational type and density are largely dependent on climate and

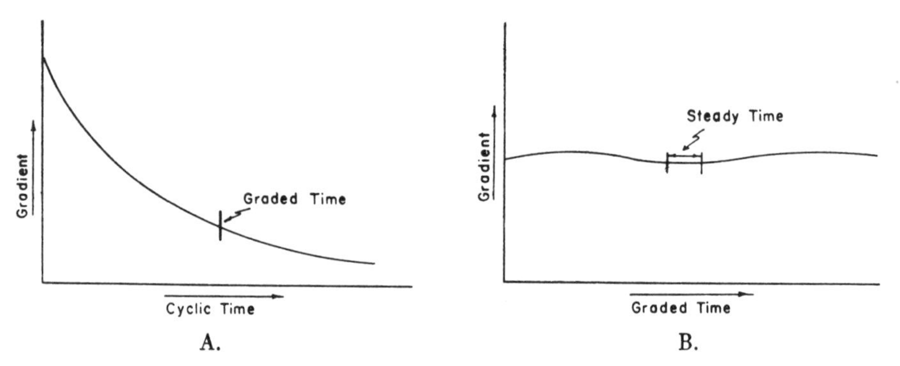

Fig. 1. Diagrams illustrating the time spans of table 1. Channel gradient is used as the dependent variable in these examples.

a. Progressive reduction of channel gradient during cyclic time. During graded time, a small fraction of cyclic time, the gradient remains relatively constant.

b. Fluctuations of gradient above and below a mean during graded time. Gradient is constant during the brief span of steady time, lithology, but they significantly influence the hydrology and erosional history of a drainage basin. If all the independent variables are constant except time, then as time passes the average relief and mass, volume of material remaining within the drainage system, will decrease. As the relief or mass of the system changes so will the other dependent morphologic and hydrologic variables.

With regard to space or the area considered, it is possible to consider an entire drainage system or any of its component parts during a cyclic time span. For example, the reduction of an entire drainage system or only the decrease in gradient of a single stream may be considered (fig. 1a) during cyclic time.

The graded time span (table 1) refers to a short span of cyclic time during which a graded condition or dynamic equilibrium exists. That is, the landforms have reached a dynamic equilibrium with respect to processes acting on them. When viewed from this perspective one sees a continual adjustment between elements of the system, for events occur in which negative feedback (self-regulation) dominates. In other words the progressive change during cyclic time is seen to be, during a shorter span of time, a series of fluctuations about or approaches to a steady state (fig. 1b). This time division is analogous to the "period of years" used by Mackin (1948, p. 470) in his definition of a graded stream by which he rules out seasonal and other short-term fluctuations, as well as the slow changes that accompany the erosion cycle.

As an erosion cycle progresses, more and more of the landscape may approach dynamic equilibrium. That is, the proportion of graded landforms may increase, and it seems likely that temporary graded conditions become more frequent as time goes on. However, it is apparent that during this time span the graded condition can apply only to components of the drainage basin. The entire system cannot be graded because of the progressive reduction of relief or volume of the system above base level, which occurs through export of sediment from the system. A hillslope profile or river reach, however, may be graded. Therefore, unlike cyclic time when no restriction was placed on space or area considered, graded time is restricted to components of the systems or to smaller areas.

During a graded time span, the status of some of the variables listed on table 1 changes. For example, time has been eliminated as an independent variable, for although the system as a whole may be undergoing a progressive change of very small magnitude, some of the components of the system will show no progressive change (that is, graded streams and hillslopes). Initial relief also has no significance because the landform components are considered with respect to their climatic, hydrologic, and geologic environment (Hack, 1960), and intial relief with time has been designated as not relevant on table 1.

In addition, some of the variables that are dependent during a long period of progressive erosion become independent during the shorter span of graded time. The newly independent hydrologic variables, runoff and sediment yield, are especially important because during a graded time span they take on a statistical significance and define the specific character of the drainage channels and hillslopes, whereas during a cyclic time span there is a progressive change in these morphologic variables.

The geomorphic variables of hillslope and drainage network morphology of graded time may be considered as "time-independent" in the meaning of Hack (1963, written communication). That is, relict features may not be present, and the landforms may be explained with regard to the independent variables without regard to time.

During a steady time span (table 1) a true steady state may exist in contrast to the dynamic equilibria of graded time (fig. 1b). These brief periods of time are referred to as a steady time span because in hydraulics steady flow occurs when none of the variables involved at a section change with time. The landforms, during this time span, are truly time independent because they do not change, and time and initial relief have again been eliminated as independent variables. During this time span only water and sediment discharge from the system are dependent variables.

Obviously the steady state condition is not applicable to the entire drainage basin. Although an entire drainage basin cannot be considered to be in a steady state over even the shortest time span, yet certain components of the basin may be. For example, a stream over short reaches may export as much water and sediment as introduced into the reach, yet the river as a whole is reducing its gradient in the headwaters (cyclic erosion). In addition, the entire drainage basin may be losing relief as hillslopes are lowered (cyclic erosion); however, segments of the hillslopes may remain at the same angle of inclination and act as slopes of transportation (steady state), or they may retreat parallel, maintaining their form (dynamic equilibrium), but the volume of the drainage basin is being reduced nevertheless. Thus over short periods of time and in small areas the steady state may be maintained. Over large areas progressive reduction of the system occurs, and this is true over long periods of time.

The preceding discussion and the relations presented in table 1 and figure 1 have the sole object of demonstrating that, depending on the time span involved, time may be either an extremely important independent variable or of relatively little significance to a study of landforms.

[*Editor's Note:* Material has been omitted at this point.]

CONCLUSIONS

The distinction of cause and effect among geomorphic variables varies with the size of a landscape and with time. Landscapes can be considered either as a whole or in terms of their components, or they can be considered either as a result of past events or as a result of modern erosive agents. Depending on one's viewpoint the landform is one stage in a cycle of erosion or a feature in dynamic equilibrium with the forces operative. These views are not mutually exclusive. It is just that the more specific we become the shorter is the time span with which we deal and the smaller is the space we can consider. Conversely when dealing with geologic time we generalize. The steady state concept can fit into the cycle of erosion when it is realized that steady states can be maintained only for fractions of the total time involved.

The time span considered also influences causality, as the sets of independent and dependent variables of tables 1 and 2 show. If the variables were not considered with respect to the time span involved, in many cases it would be difficult to determine which variables are independent. Mackin's (1963) and Kennedy and Brooks' (in press) suggestions forestall any arguments between workers in the laboratory and workers in the field. In the same manner the disparate points of view of the historically oriented geomorphologist and the student of process can be reconciled.

References

Chorley, R. J., 1962, Geomorphology and general systems theory: U. S. Geol. Survey Prof. Paper 500-B, 10 p.

Dury, G. H., 1962, Results of seismic exploration of meandering valleys: Am. Jour. Sci., v. 260, p. 691-706.

Hack, J. T., 1960, Interpretation of erosional topography in humid temperate regions: Am. Jour. Sci., v. 258-A, Bradley v., p. 80-97.

Kennedy, J. F., and Brooks, N. H., in press, Laboratory study of an alluvial stream at constant discharge: Federal Interagency Sedimentation Conf. Proc., Jackson, Miss., 1963, in press.

Mackin, J. H., 1948, Concept of the graded river: Geol. Soc. America Bull., v. 59, p. 463-511.

——— 1963, Rational and empirical methods of investigation in geology, *in* Albritton, C. C., Jr., ed., The fabric of geology: Reading, Mass., Addison-Wesley Publishing Co., p. 135-163.

Nikiforoff, C. C., 1959, Reappraisal of the soil: Science, v. 129, p. 186-196.

Schumm, S. A., 1963, Sinuosity of alluvial rivers on the Great Plains: Geol. Soc. America Bull., v. 74, p. 1089-1100.

Strahler, A. N., 1950, Equilibrium theory of erosional slopes approached by frequency distribution analysis: Am. Jour. Sci., v. 248, p. 673-696, p. 800-814.

——— 1952, Dynamic basis of geomorphology: Geol. Soc. America Bull., v. 63, p. 923-938.

Von Bertalanffy, Ludwig, 1952, Problems of life: London, Watts and Co., 216 p.

29

Reprinted from *Amer. Jour. Sci.*, **258A**, 80–81, 85–87, 96–97 (Bradley volume, 1960)

INTERPRETATION OF EROSIONAL TOPOGRAPHY IN HUMID TEMPERATE REGIONS*

JOHN T. HACK

U. S. Geological Survey, Washington, D. C.

ABSTRACT. Since the period 1890 to 1900 the theory of the geographic cycle of erosion has dominated the science of geomorphology and strongly influenced the theoretical skeleton of geology as a whole. Some of the principal assumptions in the theory are unrealistic. The concepts of the graded stream and of lateral planation, although based on reality, are misapplied in an evolutionary development, and it is unlikely that a landscape could evolve as indicated by the theory of the geographic cycle.

The concept of dynamic equilibrium provides a more reasonable basis for the interpretation of topographic forms in an erosionally graded landscape. According to this concept every slope and every channel in an erosional system is adjusted to every other. When the topography is in equilibrium and erosional energy remains the same all elements of the topography are downwasting at the same rate. Differences in relief and form may be explained in terms of spatial relations rather than in terms of an evolutionary development through time. It is recognized however that erosional energy changes in space as well as time, and that topographic forms evolve as energy changes.

Large areas of erosionally graded topography in humid regions have been considered to be "maturely dissected peneplains." According to the equilibrium theory, this topography is what we should expect as the result of long continued erosion. Its explanation does not necessarily involve changes in base level. Pediments in humid regions and some terraces are also equilibrium forms and commonly occur on a lowland area at the border of an adjacent highland.

INTRODUCTION

The part of geologic theory that deals with the interpretation of landforms and the history of landscape development has been dominated for several generations by the ideas of William Morris Davis and his followers. Davis' theory of landscape evolution was first fully presented in his essay, "The Rivers and Valleys of Pennsylvania" (Davis, 1889).[1] The important concepts that he introduced include the geographic cycle, the peneplain, and the formation of mountains by a succession of interrupted erosion cycles. Davis' theories became immensely popular among geologists in Europe as well as in America, though there were dissenters, including, for example, Tarr (1898) and Shaler (1899). His theory of the evolution of mountains as topographic features through the mechanism of multiple erosion cycles was especially influential and came to have a great influence on the theoretical skeleton of the whole science of geology. Its impact is still felt. Many of our ideas relating to the history of mountains, the internal constitution of the earth and the origin of some ore deposits are closely related to this theory. The idea that mountain ranges are vertically uplifted after they have been folded was conceived in order to explain the widespread existence of dissected peneplains (Daly, 1926). Another example is the theory of origin of bauxite and of manganese ores and other residual concentrates in the Appalachian Highlands, that are thought by some to have formed on a Tertiary peneplain surface (Hewett, 1916; Stose and Miser, 1922, p. 52-55; Bridge, 1950, p. 196.)

* Publication authorized by the Director, U. S. Geological Survey.

[1] Davis' major papers dealing with the sculpture of landscapes by streams (Davis, 1889, 1890, 1896a, 1896b, 1899a, 1899b, 1902a, 1902b, 1903, 1905a, 1905b) as well as others of his papers were collected in one volume published as "Geographical Essays" (Davis, 1909) and reprinted in 1954 (Davis, 1954). The 1954 edition of "Geographical Essays" has the same page numbers as the 1909 edition.

In the last 20 years, however, Davis' ideas have become less popular and the small but ever-present number of geologists who were skeptical of his theories has increased. Though many geologists have been dissatisfied with it, the theory of the geographic cycle and its application to the study of landforms has not generally been replaced by any other concept. Several alternative theories have been proposed, including the theory of Penck (1924, 1953) which relates the form of slopes to changes in the rate of uplift relative to the rate of erosion, and the "pediplain" theory of L. C. King (1953), an elaboration and expansion of Penck's concept of slope retreat. Both of these theories, however, are also cyclic concepts and hold that the landscape develops in stages that are closely dependent on the rate of change of position of baselevel.

During the course of my work in the Central Appalachians which began in 1952, seeking a different approach to geomorphic problems, a conscious effort was made to abandon the cyclic theory as an explanation for landforms. Instead, the assumption was made that the landforms observed and mapped in the region could be explained on the basis of processes that are acting today through the study of the relations between phenomena as they are distributed in space. The concept of dynamic equilibrium forms a philosophical basis for this kind of analysis. The landscape and the processes molding it are considered a part of an open system in a steady state of balance in which every slope and every form is adjusted to every other. Changes in topographic form take place as equilibrium conditions change, but it is not necessary to assume that the kind of evolutionary changes envisaged by Davis ever occur. The consequences and results of this kind of analysis in most cases differ from conclusions arrived at through the use of the cyclic concepts of Davis.

On rereading some of the classic American literature in geomorphology I realized that G. K. Gilbert used essentially this approach and that I have followed a way of thinking inherited either directly from him or from some of his colleagues. Even though Davis and Gilbert were contemporaries and friends, Gilbert makes little use of and few references to the theory of the geographic cycle or any of its collateral ideas. This omission is so conspicuous that it is difficult to believe Gilbert ever wholeheartedly accepted the idea. It seems to me that Gilbert's famous paper, "Geology of the Henry Mountains" (Gilbert, 1877, p. 99-150) outlines a wholly satisfactory basis for the study of landscape that does not foreshadow the developments in geomorphology that followed in the next 50 years.

[*Editor's Note:* Material has been omitted at this point.]

THE PRINCIPLE OF DYNAMIC EQUILIBRIUM IN LANDSCAPE INTERPRETATION

An alternative approach to landscape interpretation is through the application of the principle of dynamic equilibrium to spatial relations within the drainage system. It is assumed that within a single erosional system all elements of the topography are mutually adjusted so that they are downwasting at the same rate. The forms and processes are in a steady state of balance and may be considered as time independent. Differences and characteristics of form are therefore explainable in terms of spatial relations in which geologic patterns are the primary consideration rather than in terms of a particular theoretical evolutionary development such as Davis envisaged.

The principle of dynamic equilibrium was applied to the study of landforms both by Gilbert (1877, p. 123) and by Davis (1909, p. 257-261, 389-400; 1899, p. 488-491; 1902, p. 86-98). Recently Strahler has outlined the principle in more modern terms as it might be applied to landscapes (Strahler, 1950, p. 676). The concept requires a state of balance between opposing forces such that they operate at equal rates and their effects cancel each other to produce a steady state, in which energy is continually entering and leaving the system. The opposing forces might be of various kinds. For example, an alluvial fan would be in dynamic equilibrium if the debris shed from the mountain behind it were deposited on the fan at exactly the same rate as it was removed by erosion from the surface of the fan itself. Similarly a slope would be in equilibrium if the material washed down the face and removed from its summit were exactly balanced by erosion at the foot.

In the erosion cycle concept of Davis, equilibrium is achieved in some part of the drainage system when there is a balance between the waste supplied to a stream from the headwaters and the ability of the stream to move it, or in other words, when the slope of the channel is reduced just enough so that the stream can transport the material from above with the available discharge. As argued on page 9 this kind of equilibrium probably is achieved in a stream almost immediately and is not related to a particular stage in its evolution. Davis' concept would imply that some parts of a drainage system would be in equilibrium whereas at the same time other parts would not, and that the condition of equilibrium is in time gradually extended from the downstream portion to the entire drainage system.

Rather than a concept of balance between the load of a stream and the ability of the stream to move it, it is more useful in the analysis of topographic forms to consider the equilibrium of a particular landscape to involve a balance between the processes of erosion and the resistance of the rocks as they are uplifted or tilted by diastrophism. This concept is similar to Penck's concept of exogenous and endogenous forces (1924, 1953). Suppose that an area is undergoing uplift at a constant rate. If the rate of uplift is relatively rapid, the relief must be high because a greater potential energy is required in order to provide enough erosional energy to balance the uplift. The topography is in a steady state and will remain unchanged in form as long as the rates of uplift and erosion are unchanged and as long as similar rocks are exposed at the surface. If the relative rates of erosion and uplift change, however, then the state of balance or equilibrium constant must change. The topography then undergoes an evolution from one form to another. Such an evolution might occur if diastrophic forces ceased to exert their influence, in which case the relief would gradually lower; it might occur if diastrophic forces became more active, in which case the relief would increase; or it might occur if rocks of different resistance became exposed to erosion. Nevertheless as long as diastrophic forces operate gradually enough so that a balance can be maintained by erosive processes, then the topography will remain in a state of balance even though it may be evolving from one form to another. If, however, sudden diastrophic movements occur, relict landforms may be preserved in the topography until a new steady state is achieved.

The area in which a given state of balance exists and that may be considered a single dynamic system may be conceived as very small or very large. In the Appalachian region, it may be that large areas are essentially in the same state of balance. In the West, however, in an active diastrophic belt, a single dynamic system may constitute only a small area such as a single mountain range or a small part of a mountain range. Furthermore, because of sudden dislocations of the crust relict forms may be preserved in the landscape that reflect equilibrium conditions that no longer exist.

The crust of the earth is of course not isotropic and within a single erosional system, no matter how small, there is a considerable variation in the composition and structure of the crust. These variations are reflected by variations in the topography. Consider, for example, an area composed partly of quartzite and partly of shale. To comminute and transport quartzite at the same rate as shale, greater energy is required; and since the rates of removal of the two must be the same in order to preserve the balance of energy, greater relief and steeper slopes are required in the quartzite area. Similarly geometric forms differ on different rock types. An area that is underlain by mica schist or other igneous or metamorphic rock subject to rapid chemical decay, has more rounded divides than an area underlain by quartzite, if both are in equilibrium in the same dynamic system, for the schist is comminuted by weathering to silt and clay particles that are rapidly removed from hill tops on low slopes. On the other hand to remove quartzite from a divide at the same rate, steeper slopes and sharper ridges are required because the rock must be moved in the form of larger fragments.

The analysis of topography in terms of spatial or time-independent relations provides a workable basis for the interpretation of landscape. This kind of analysis is uniformitarian in its approach, for it attempts to explain landscapes in terms of processes and rates that are in existence today and therefore observable. It recognizes that processes and rates change both in space and time, and, by clarifying the relation between forms and processes, it provides a means by which the changes can be analyzed.

[*Editor's Note:* Material has been omitted at this point.]

REFERENCES CITED

Bridge, Josiah, 1950, Bauxite deposits in the southeastern United States, *in* Snyder, F. G., ed., Symposium on mineral resources of the southeastern United States: Knoxville, Tenn., Tennessee Univ. Press, p. 170-201.
Daly, R. A., 1926, Our mobile earth: New York, Charles Scribner's and Sons, 342 p.
Davis, W. M., 1889, The rivers and valleys of Pennsylvania: Natl. Geog. Mag., v. 1, p. 183-253.
——————, 1890, The rivers of northern New Jersey, with notes on the classification of rivers in general: Natl. Geog. Mag., v. 2, p. 81-110.
——————, 1896a, Plains of marine and subaerial denudation: Geol. Soc. America Bull., v. 7, p. 377-398.
——————, 1896b, The Seine, the Meuse, and the Moselle: Natl. Geog. Mag., v. 7, p. 189-202, 228-238.
——————, 1899, The geographic cycle: Geog. Jour., v. 14, p. 481-504.
——————, 1899b, The peneplain: Am. Geologist, v. 23, p. 207-239.
——————, 1902a, River terraces in New England: Harvard College, Mus. Comp. Zoology Bull., v. 38, p. 281-346.
——————, 1902b, Base level, grade, and peneplain: Jour. Geology, v. 10, p. 77-111.
——————, 1903, The mountain ranges of the Great Basin: Harvard College, Mus. Comp. Zoology Bull., v. 42, p. 129-177.
——————, 1903, The stream contest along the Blue Ridge: Geog. Soc. Philadelphia, Bull., v. 3, p. 213-244.
——————, 1905a, The geographical cycle in an arid climate: Jour. Geology, v. 13, p. 381-407.
Davis, W. M., 1905b, Complications of the geographical cycle: Internat. Geog. Cong., 8th Rept., p. 150-163.
——————, 1909, Geographical assays: Boston, Ginn and Co., 777 p.
——————, 1926, Biographical memoir of Grove Karl Gilbert, 1843-1918: Natl. Acad. Sci., 5th Mem., v. 21, 303 p.
——————, 1954, Geographical Essays: Dover Publications, Inc., 777 p.
Dietrich, R. V., 1958, Origin of the Blue Ridge escarpment directly southwest of Roanoke, Virginia: Virginia Acad. Sci. Jour., v. 9, New Series, p. 233-246.
Frye, John C., 1959, Climate and Lester King's "Uniformitarian Nature of Hillslopes": Jour. Geology, v. 67, p. 111-113.
Gilbert, G. K., 1877, Geology of the Henry Mountains (Utah): Washington, D. C., U. S. Geog. and Geol. Survey of the Rocky Mts. Region, U. S. Govt. Printing Office, 160 p.
——————, 1909, The convexity of hill tops: Jour. Geology, v. 17, p. 344-350.
Gilluly, James, 1949, Distribution of mountain building in geologic time: Geol. Soc. America Bull., v. 60, no. 4, p. 561-590.

Hack, John T., 1958a, Studies of longitudinal stream profiles in Virginia and Maryland: U. S. Geol. Survey Prof. Paper 294B, p. 45-97.
————, 1958b, Geomorphic significance of residual and alluvial deposits in the Shenandoah Valley, Virginia (abs.): Virginia Jour. Sci., v. 9, p. 425.
————, 1959, The relation of manganese to surficial deposits in the Shenandoah Valley, Virginia (abs.): Washington Acad. Sci. Jour. Proc., v. 49, p. 93.
Hack, J. T., and Goodlett, J. C., 1960, Geomorphology and forest ecology of a mountain region in the Central Appalachians: U. S. Geol. Survey Prof. Paper 347 in press.
Hewett, D. F., 1916, Some manganese mines in Virginia and Maryland: U. S. Geol. Survey Bull. 640-C, p. 37-71.
Hunt, C. B., Averitt, Paul, and Miller, R. L., 1953, Geology and Geography of the Henry Mountains region, Utah: U. S. Geol. Survey Prof. Paper 228, 234 p.
Kesseli, J. E., 1941, The concept of the graded river: Jour. Geology, v. 49, no. 6, p. 561-588.
King, L. C., 1953, Canons of landscape evolution: Geol. Soc. America, Bull., v. 64, no. 7, p. 721-752.
Leopold, L. B., 1953, Downstream change of velocity in rivers: Am. Jour. Sci., v. 251, no. 8, p. 606-624.
Leopold, L. B., and Maddock, Thos., Jr., 1953, The hydraulic geometry of stream channels and some physiographic implications: U. S. Geol. Survey Prof. Paper 252, 57 p.
Mackin, J. H., 1948, Concept of the graded river: Geol. Soc. America Bull., v. 59, no. 5, p. 463-511.
Nikiforoff, C. C., 1942, Fundamental formula of soil formation: Am. Jour. Sci., v. 240, no. 12, p. 847-866.
————, 1949, Weathering and soil evolution: Soil Sci., v. 67, p. 219-230.
————, 1955, Harpan soils of the Coastal Plain of southern Maryland: U. S. Geol. Survey Prof. Paper 267-B, p. 45-63.
————, 1959, Reappraisal of the soil: Science, v. 129, no. 3343, p. 186-196.
Penck, Walther, 1953, Morphological analysis of landforms (translation by Hella Czeck and K. C. Boswell): New York, St. Martin's Press, 429 p.
————, 1924, Die morphologische Analyse, Ein Kapitel der physikalischen Geologie: Stuttgart, Geog. Abh. 2 Reihe, heft 2, 283 p.
Rubey, W. W., 1952, Geology and mineral resources of the Hardin and Brussels quadrangles (in Illinois): U. S. Geol. Survey Prof. Paper 218, 179 p.
Shaler, N. S., 1899, Spacing of rivers with reference to hypothesis of base-leveling: Geol. Soc. America Bull., v. 10, p. 263-276.
Smith, K. G., 1958, Erosional processes and landforms of Badlands National Monument, South Dakota: Geol. Soc. America Bull., v. 69, no. 8, p. 975-1008.
Stose, G. W., and Miser, H. D., 1922, Manganese deposits of western Virginia: Virginia Geol. Survey Bull. 23, 206 p.
Strahler, A. N., 1950, Equilibrium theory of erosional slopes approached by frequency distribution analysis: Am. Jour. Sci., v. 248, no. 10, p. 673-696; no. 11, p. 800-814.
Tarr, R. S., 1898, The peneplain: Am. Geologist, v. 21, p. 341-370.
White, W. A., 1950, The Blue Ridge front—a fault scarp: Geol. Soc. America Bull., v. 61, no. 12, pt. 1, p. 1309-1346.
Wolman, M. G., 1955, The natural channel of Brandywine Creek, Pennsylvania: U. S. Geol. Survey Prof. Paper 271, 56 p.
Woodford, A. O., 1951, Stream gradients and the Monterey sea valley: Geol. Soc. America Bull., v. 62, no. 7, p. 799-851.

30

Reprinted from *Bull. Geol. Inst. Uppsala,* 25, 293-305, 442-452 (1935)

STUDIES OF THE MORPHOLOGICAL ACTIVITY OF RIVERS AS ILLUSTRATED BY THE RIVER FYRIS

F. Hjulstrøm

[*Editor's Note:* In the original, material precedes this excerpt.]

Relationship between erosion velocity and grain size.

Even though the principle of expressing the force of erosion and traction as a function of the velocity has, to a certain degree, been considered antiquated and out-of-date, investigations have been made during the last few years to make clear this relationship. In the following there are mentioned some points of view on the erosion, transportation and deposition of bedload based on old and new investigations. They are mainly caused by an endeavour to give a graphical picture of the relationship between the kind of material and the minimum erosion velocity, and would appear to be confirmed by the writer's observations in the field and in laboratory.

In order to express the relationship mentioned it is necessary first to more clearly define the variables, the velocity and kind of material. As far as the speed is concerned it would certainly, to obtain an exact result, be necessary to have a whole curve or formula stating the variation of the velocity according to the height above the bottom. As such a diagram is never obtainable it would certainly be preferable to use the bottom-velocity. But this is only stated in a limited number of cases, and is more difficult to decide than the surface- and average velocity. For these reasons the *average velocity* has been made use of, it being presumed that this is 40 % greater than the bottom velocity. This percentage depends inter alia upon the depth, but it has been presumed that this exceeds one meter. In shallower water the velocities stated here will be somewhat less, roughly about 10—20 cm./sec. less. — Greater demands as to exactitude cannot be satisfied at present.

The kind of material is, of course, characterized by the specific gravity, the shape and grain size of the particles. The last mentioned quality is undoubtedly of paramount importance, seeing that the shape has no very great effect, which is shown by experience, and the specific gravity is subject to but slight variations, 2.6—2.7. As indicated by modern investigations, for inst. GILBERT's in 1914, SCHAFFERNAK's in 1922 and KRAMER's in 1932, the composition of the material as to grain size is of very great importance. For different relative relations of quantity between the grain sizes in various materials the corresponding erosion velocities will vary, also in cases when these sizes are the same. It may be this complicated influence of the composition of the material that causes the results of all investigations of the relation between the velocity of erosion, transport and the grain size to become so inconsistent, as mentioned above. For a graphical picture the least complex case has been selected, i. e. when the material is uniform, monodispersed. But also in such cases varying results were obtained. A body is put in motion at varying velocities, dependent upon whether it is on a rough or a smooth bedding. A severely defined and practical starting point is obtained by presuming that *a uniform material moves over a bedding of loose material of the same grain size.* Table 7 gives the values stated in the literature to correspond to erosion, i. e. a spontaneous starting of quiescent material under these conditions. This velocity will in the following be called *erosion velocity.*

The difficulties encountered when making such a comparison are firstly that it is not always possible definitely to decide whether the erosion velocity in question under the conditions stated really is that concerned, and secondly that the statements of velocity, depth and grain size occasion certain questions. The information selected and contained in the table is not all equally reliable, LAPPARENT's might be questioned seeing that in his observations the eroded material was not always moved over a bedding consisting of the same material. The same lack is the most common cause for other observations having to be excluded, and it mostly occurs when studies of natural rivers have been made. On the other hand laboratory tests must also be excluded for highly dispersed systems such as clays, as the stratification may have been changed due to silting. In cases where the surface-velocity has been stated, it has been reduced 20 % to obtain the average velocity, and — as already mentioned — the bottom-velocity has been increased by 40 %.

The question of varying velocities is connected with that of varying depths. The difference between the bottom-velocity, important with regard to erosion, and the average velocity used in practice, is increased with the depth. THRUPP (1908) has made a graph of the scouring power in relation to velocity and depth, Fig. 16 being an extract showing the course of a curve. It is, however, reproduced very reluctantly as it appears to be founded upon a rather limited amount of observation material, and as it is not for uniform material. Generally speaking, it might, however, be said to give a correct idea of the conditions, at least for limited depths when the material for observation is more comprehensive. The curve in the figure states the velocity for which coarse sand is moved. — The velocities given in Table 7 and in Figures 17 and 18 are for slightly varying depths, but in most cases a correction has been inserted when the figures stated have been for such limited depths as 1 foot by adding 0.2 m./sec. The

Table 7.

Erosion velocities for a monodisperse material on a bed of loose material of the same size of particles.

Author:	Characteristics of the material (by the resp. author)	Size of particles	Erosion velocity
		mm	cm/sec.
Etcheverry (Fortier and Scobey p. 951)	Stiff clay soil	(0.0015	137
Fortier and Scobey	Stiff clay (very colloidal)	(0.0015	130
» » »	Alluvial silts, when colloidal	(0.005)	130
» » »	» » » non-colloidal	(0.005)	76
Umpfenbach (Penck, 1894, p. 283	Feiner Lehm und Schlamm	(0.05—0.1)	26
Etcheverry (Fortier and Scobey)	Very light pure sand of quick-sand character	(0.13)	27
Gilbert, 1914, p. 69	Grade B	0.38	24
Lapparent (Schoklitsch, 1914, p. 25)	Schlamm, grob	0.40	15.0
Telford (» , » , » »	Feiner sand	(0.45)	15.2
Gilbert, 1914, p. 69	Grade C	0.51	28
Lapparent (Schoklitsch, 1914, p. 25	Sand, fein	0.70	20
Gilbert, 1914, p. 69	Grade D	0.79	34.1
» , » , » 70	Grade E	1.71	34.4
Etcheverry (Fortier and Scobey, p. 951)	Coarse sand	(2)	45 à 60
Schaffernak, 1922, p. 14		2	25
Sainjon (Schoklitsch, 1914, p. 24)	Kiesel	2.50	50
Gilbert, 1914, p. 70	Grade F	3.17	54
Schaffernak, 1922, p. 14		4	49
Gilbert, 1914,	Grade G	4.94	64
Schaffernak, 1922, p. 14		6	61
Gilbert, 1914, p. 70	Grade H	7.01	85
Schaffernak, 1922, p. 14		8	81
» ,		10	104
» ,		12	120—125
» ,		14	125—150
» ,		16	130—180
» ,		20	189—197
» ,		25	203—210
» ,		30	218—221
» ,		50	238
» ,		70	266—280

velocity statements may thus be said to cover depths of at least one meter. FORTIER and SCOBEY (1926) state this correction to be suitable. But the greatest difficulties have been encountered when the size of particle should be defined. The literature often contains such very indefinite statements as for inst. »large stones». The Table has therefore been made to include both the information supplied by the writer in question and the numerical value of the size of particle stated in the Diagram. This has been put in brackets in Column 4 in case it was not given in the original. The valuation then made has, of course, occasioned a certain subjectivity due to the existing confusion in the terminology in this sphere.

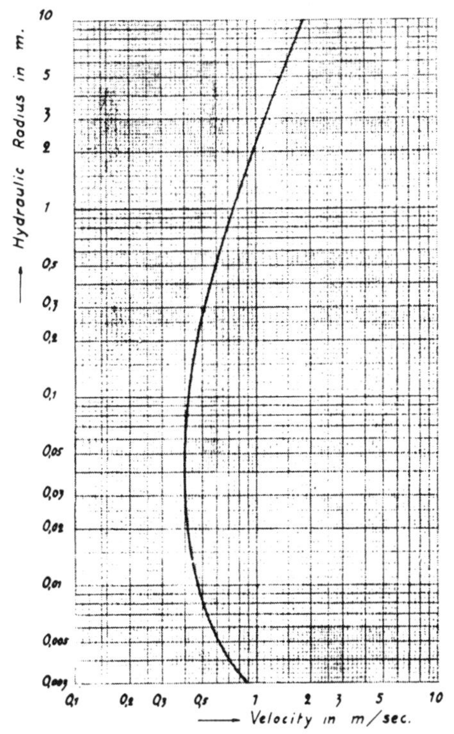

Fig. 16. The scouring power for coarse sand in relation to velocity and depth, according to THRUPP (1908).

These are the reasons why the curve in Figures 17 and 18 has not been shaped as a simple curve but as a zone. It must of course only be considered as an endeavour to make a preliminary comparison of the results obtained up to date, and may later be replaced by a more exact relation. But this will require much additional work.

In Figures 17 and 18 the values of the Table have been made the basis of a graph. (Note the upper curve.) The Figures show the same thing, but in Figure 18 the values have been dotted in a logarithmic scale in order to more clearly illustrate the interesting conditions connected with a small size of particle in a better manner than is possible in an ordinary scale.

The most noticeable deviations of the erosion curve in these illustrations from older accounts, for inst. SCHAFFERNAK's (1922, p. 14) and S. A. ANDERSEN's (1931, p. 33), is that it has a minimum and does not go down to the origin of the coordinate system. The minimum is not at the size of particle 0 but within the range 0.1—0.5 mm. This thus indicates that loose, fine sand, for inst. of quicksand character is the easiest to erode, whereas silty loam and clay as well as coarser sand and gravel, etc. demand greater velocities.

The great resistance of the clay to erosion was first strongly empha-

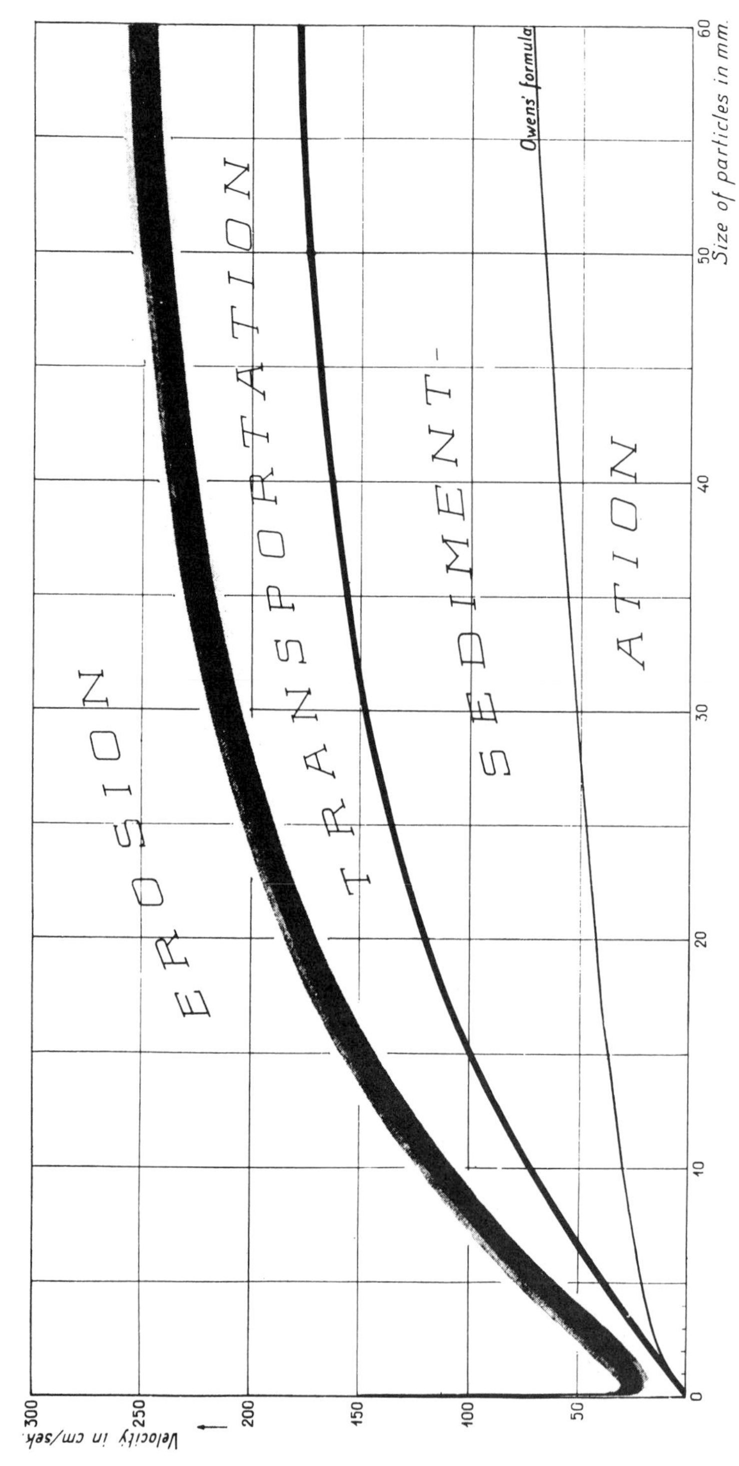

Fig. 17. The curves for erosion and deposition of a uniform material.

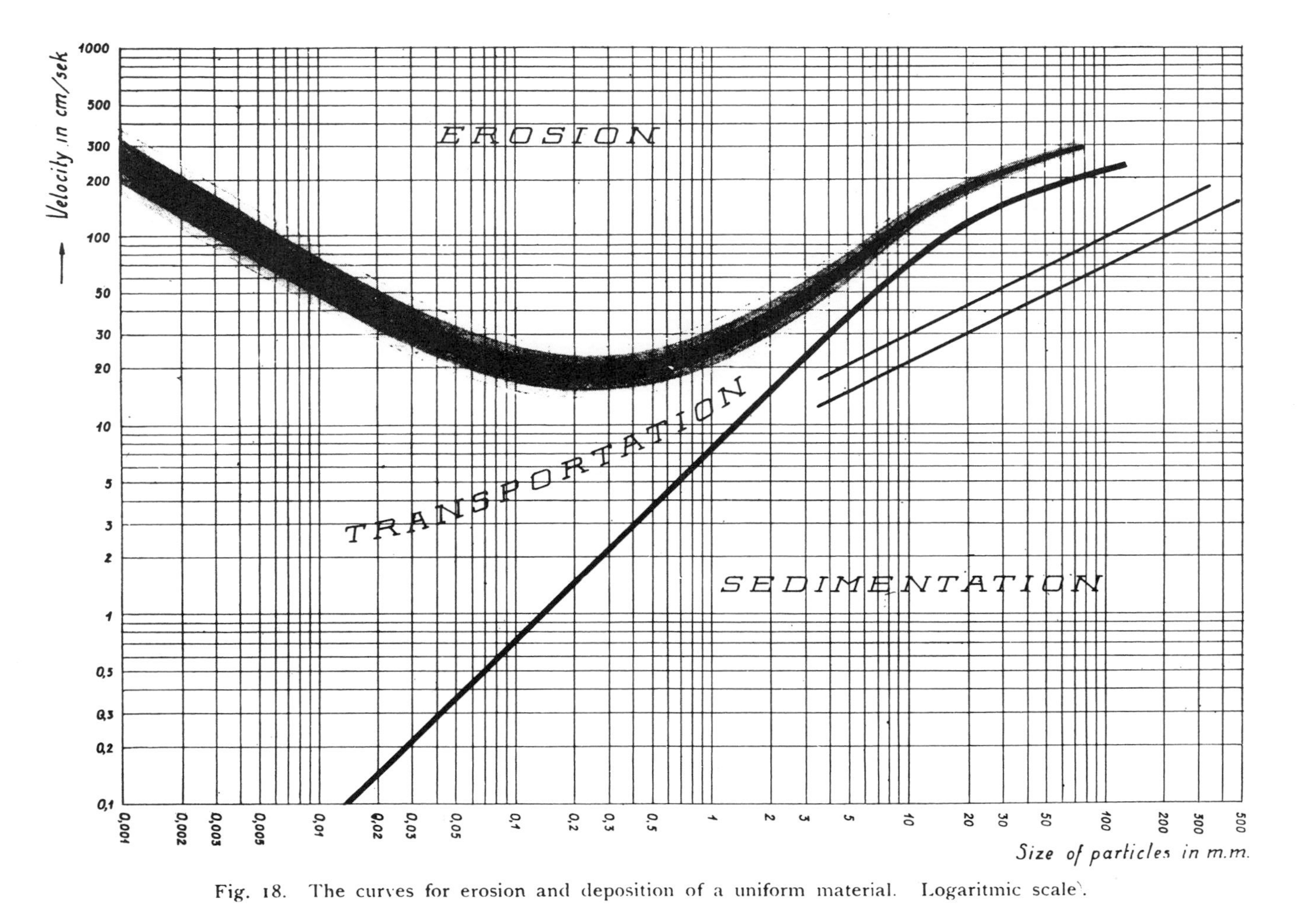

Fig. 18. The curves for erosion and deposition of a uniform material. Logaritmic scale).

sized by the American hydraulic engineers FORTIER and SCOBEY in a paper on permissible canal velocities 1926.

This quality in clay of course depends upon the influence of cohesion and adhesion, which powers tend to unite the particles. The effect increases in line with an increased degree of dispersion, inter alia due to the number of contact points between the particles of a certain weight quantity thus being increased. It is therefore only when the size of particle is small that they become noticeable in the erosion. See for inst. DENSCH in Handbuch d. Bodenlehre, Vol. VI and BRENNER 1931.

In hydraulics it must certainly be considered of great importance to be able to calculate with these great erosion velocities for clay, for consequently the cross-section of for inst. a projected irrigation canal may of course be made correspondingly smaller and the cost of construction be reduced. Even though this condition was but recently pointed out, it is an old observation. FORBES (1857, p. 475) gives the following result of experiments tried inter alia on brick-clay from Portobello: »The brick clay in its natural moist state, had a specific gravity of 2.05; and water passing over it for half an hour at a rate of 128 feet in the minute, which was the greatest velocity I could conveniently obtain, made no visible impression on the clay. When this clay was mixed with water, and allowed to settle for half an hour, it required a velocity of fifteen feet in the minute to disturb it. This mud sank in water at a rate of 0.566 feet in one minute, but the very fine particles were very much longer in subsiding.» In a work by ETCHEVERRY[1], not obtainable in Sweden, quoted by FORTIER and SCOBEY, it is also mentioned that stiff clay soil and ordinary gravel soil have the same maximum mean velocity safe against erosion, namely 4.00—5.00 feet per second (121.9—152.4 cm/sec.). When engaged on engineering work for irrigation in India already in 1874—75, also KENNEDY (quoted from GIBSON 1919, p. 345), when publishing his oft-used formula for the critical velocity at which a long canal will maintain its channel in silty equilibrium, stated that this velocity is greater for loam and silt than for light or coarse sandy soil.

Finally, also CHATLEY (1921) in his silt-studies in China arrived at the conclusion that silty and clay beds will bear very much higher velocity than sandy beds. He has made the formula

$$v = \frac{0.02}{d} \text{ centimetres per second,}$$

v being = the erosion velocity and d = the size of particle. It is valid for grains held in place only by mutual cohesion, thus for the ascending branch in Figures 17 and 18, and agrees very well with these. CHATLEY

[1] ETCHEVERRY: Irrigation Practice and Engineering. Vol. II: The Conveyance of Water. New York 1916.

states that the actual limits of velocity (200 cm/sec.) and grain-size (0.0001 cm.) in the Huangpu and Yangtze also agree with the formula. A qualitative graph of the variation of eroding velocity with size of particles also exists, in the main corresponding with the curves made by the writer.

Mixed materials.

As already mentioned these reflections are valid for uniform material. Certainly such material is not unusual, but generally, the tractional load of a natural stream includes particles with great range in size. Erosion velocity in this case means the velocity for which also the greatest size, that a *great* number of particles attain (i. e. *normal maximum* acc. to NELSON 1910, p. 21), is loosened from the bedding and removed. In his monumental work on the transportation of débris by running water (1914) GILBERT has shown how the erosion velocity[1] varies with mixed grades of the particles. He composed inter alia mixtures of two different grain-sizes with differing mutual relations of the weight quantities. The results of his experiments show that the mixture is easier eroded when an addition of fine material is made to coarser, and most easy when the mixture contains an average of 75 % of the finer sand. The tractability of the mixture then decreases to the value valid for the finer sort.

GILBERT (1914, p. 178) points out that when a finer grade of débris is added to a coarser the finer grains occupy interspaces between the coarser and thereby make the surface of the stream bed smoother. One of the coarser grains resting on a surface composed of its fellows, may sink so far into a hollow as not to be easily dislodged by the current, but when such hollows are partly filled by the smaller grains its position is higher and it can withstand less force of current. The larger particles are moved more rapidly than the smaller, a condition which the writer has always found correct when the velocity is not too violent and the particles roll or slide over the bottom. The traction then usually occurs in the shape of small stream ripples or dunes (»Transportkörper» acc. to AHLMANN, 1914a). When the velocity is increased saltation and suspension are added. In these kinds of transportation the velocity of the coarser particles is less than that of the finer ones — a condition which may thus occur occasionally but not generally as RUBEY (1933 b, p. 498) appears to have interpreted DAUBREE's and GILBERT's statements. In the Karlsruhe laboratory the writer observed in 1931 in a testing channel, whose bottom consisted of natural sand from the Rhine transported in the shape of stream ripples, the following approximate velocities for the movement of the sand particles when the velocity of the water was about 50 cm/sec.:

[1] Certainly, GILBERTS investigations concern the lowest transportation velocity, but the accordance with the results of other writers shows that they are valid also for the erosion velocity.

5	mm.'s	grain	size	30 cm/sec.
2	»	»	»	25 »
1	»	»	»	15 »

When the velocity is increased so that the water can transport also larger particles these will roll faster than the others, if once started. Nor do the larger particles stop now and then but roll incessantly. The small particles have a comparatively greater friction to surmount in the crowd of particles of the same size and will have a short moment of rest now and then, calculated in the velocity figures above.

According to GILBERT (1914, p. 173) the amount of increase in the transportation of the coarser débris appears to be greater as the contrast in fineness of components is greater, and in the extreme case the transportation of the coarser is multiplied by 3.5.

KRAMER's investigation (1932) of erosion and traction of three different kinds of sand completely confirm these observations. Sand composed of grain sizes 0.385 to 5.0 mm. is eroded at a lower velocity than sand of 0 to 5.0 mm.'s grain size. An addition of fine-grained material to a certain mixture increases the coarser material's tractability in the beginning, but at last a condition is arrived at when the added material cements the mixture and prevents the transport of the originally loose but now cemented grains. The reduction of the pore volume decreases the tractability of the material.

SCHAFFERNAK's curves (1922) also show the same thing. They clearly indicate the increased transportation due to a limited addition of fine matarial (Mischungstype II) compared with the reduction when an ample quantity of the fine material is added (Mischungstype IV) (see his Figs 10 and 12).

According to the works of various writers quoted above there is thus a tendency to decrease the tractability if a mixture of a sufficient quantity of the finer material is added, and this even if it is not so fine that cohesion and adhesion may be considered of great importance. The particles being very fine, their influence will of course be great also with a fairly limited concentration. As stated by FORTIER and SCOBEY the greatest effect is exercised by the finest particles, the colloids.

To this must be added that also the nature of the water affects the erosion velocity. Material carried along with the water has a purely mechanical effect, which may cause the bed-load to be stirred and thus make it an easy prey to erosion. FORTIER and SCOBEY report interesting experiences also in this respect. With the aid of all sources at their disposal, inter alia an inquiry to all hydraulic technicists, they have made a comparison of the maximum permissible mean canal velocities for varying nature of water, which Table is reproduced below (Table 8). The

Table 8.

Maximum permissible mean canal velocities. (From FORTIER and SCOBEY 1926).

Original material excavated for canal	Velocity in feet per second after aging of canals carrying:		
	Clear water, no detritus	Water transporting colloidal silts	Water transporting non colloidal silts, sand, gravel or rock fragments
Fine loam (non-colloidal)	1.50	2.50	1.50
Sandy » (» - »)	1.75	2.50	2.00
Silt » (» - »)	2.00	3.00	2.00
Alluvial silts when non-coll.	2.00	3.50	2.00
Ordinary firm loam	2.50	3.50	2.25
Volcanic ash	2.50	3.50	2.00
Fine gravel	2.50	5.00	3.75
Stiff clay (very colloidal)	3.75	5.00	3.00
Graded, loam to cobbles, when non-colloidal	3.75	5.00	5.00
Alluvial silts, when colloidal	3.75	5.00	3.00
Graded, silt to cobbles, coll.	4.00	5.50	5.00
Coarse gravel (non-colloidal)	4.00	6.00	6.50
Cobbles and shingles	5.00	5.50	6.50
Shales and hard-pans	6.00	6.00	5.00

velocities are valid for straight courses. »At sinous alignement a reduction of about 25 % is recommended. Likewise the figures are for depths of 3 feet or less. For greater depths a mean velocity greater by 0.5 feet per sec. may be allowed.» It is seen from the table that water transporting colloidal silts may generally be allowed to have a much greater velocity than clear water and water transporting non-colloidal silts, sand, gravel or rock fragments. The colloids »will make the bed all the more tough and tenacious, increasing its resistance to erosion.» FORTIER and SCOBEY also point out that »all experienced canal operators know the trick of holding muddy water above one chick structure after another until the mud has painted over the sides and bottom of a new canal, reducing seepage losses and making the bed of the canal less susceptible to scour.» In the discussion in the paper quoted, R. H. HART states (p. 961) that an important consideration is the position of the ground-water table with respect to the water surface of the canal. As long as the latter is higher, seepage is out of the canal, and there is a tendency for the finer materials to be carried into the interstices between coarser particles, thereby permitting a silting-up process. On the other hand, if the ground-water table is higher, as frequently happens, seepage into the canal takes place and the whole process is reversed.

It is also evident from the Table that the difference between erosion velocity for water transporting non colloidal silts, sand, gravel or rock fragments and clear water with no detritus is not so great except for coarse gravel, cobbles and shingles. The first-mentioned water in this case fills the interstices and the erosion velocity is increased. As regards finer material it has been observed that in the case that it is colloidal, it is less able to resist water with detritus than clear water. Conditions will be reversed for non-colloidal material.

These accounts show how complex natural erosion really is. It is not to be wondered at that the determinations of the erosion velocity have given such varying results. The deviations from the erosion curve in Figures 17 and 18 for non-monodispersed material may, however, be expressed in such a manner that values lying above the curve depend upon cementation with fine material, whereas values below the curve denote a less comprehensive mixture of finer components which smooth the surface.

Of the formulas that have been made and which do not agree very well, there is one by OWEN (1908, p. 418), which when re-expressed to be valid for a specific gravity of 2.7 and for cm. as a unit of length, reads

$$d = 0.0011\, v^2,$$

d being = the diameter of the particle in cm., and
v » = the erosion velocity in a special case, for the transport of coarser material over fine sand or clay.

This formula agrees remarkably well with the one obtained by JEFFREYS (1929) in a theoretical manner, see p. 268, which with the same designations as above reads

$$d = 0.0010\, v^2.$$

In the curves, Figs 17 and 18, OWEN's formula is graphically expressed as a fine line. The velocity in the formula in question is for the surface-velocity of a stream with a depth of water of 2.5 to 152 cm. It may be presumed to be approximately equal to the average velocity of a greater depth.

In this connection an interesting observation by W. W. RUBEY (1933 a) is worth mentioning, namely, that the current required to move a particle along the bottom of a stream (after OWEN's formula) is approximately the same as the settling velocity of the same particle in still water.

The description of the velocity as given here is certainly very approximate. It is not the average velocity that is decisive for the erosion but the velocity in the bottom-layer, where the increase in velocity in proportion to the height over the bottom is great, and where particles of different sizes are thus affected by different velocities. In a laboratory investigation M. WELIKANOFF (1932) aimed at a physical expansion of the erosion theory. He investigated the connection between velocity and grain size, AIRY's law, and found that the said law is not fully correct. The grain size must not be put proportional to the square of the velocity. The formula should read

$$\frac{v^2}{g} = \alpha \cdot d + \beta$$

where g = the acceleration due to gravity and α and β are constants, β being dependent upon the depth. WELIKANOFF also found that for a small grain size other conditions occur, so that from a grain size of from 0.4 or 0.5 mm. the constants α and β have other values. According to WELIKANOFF the »$\frac{1}{n}$ potential function» cannot be made the basis of a more exact theory.

Though it is thus impossible to mathematically formulate a theory explaining the particulars of erosion the process would, however, appear to be fairly well explained in its main features. The active powers are the pressure of the water in the direction of movement and further the hydrodynamic upthrust and the effect of turbulence. The latter affects the water's direction of movement which becomes greatly variable. The vertical velocities of the turbulence become also of importance. When observing the movements of the individual particles the question soon arises as to what degree of effect may be attributed to the turbulence in this respect. The grains of sand appear to be lifted; this is also seen from the film made at the Karlsruhe River Hydraulics Laboratory. See also SCHAFFERNAK's (1922, p. 12—13) expressive description.

The pulsations of the water will be of very great importance for the erosion. When the velocity fluctuates the erosion will be by fits and starts. In addition to the value for the average velocity the force and frequency of the fluctuations are also of very great importance (see page 252).

The material loosened by the erosion is easily transported when once in motion. The coarsest material which the stream can transport is tracted as bed-load and the finer particles are carried in suspension. Saltation is a transition state between these two modes of transportation.

Erosion may occur when the water with constant velocity comes across material that the stream is capable of eroding. It may also occur due to increased velocity. If the eroded material cannot be transported in suspension it will in the former case only result in an increase of the bed-load.

In the latter case, it may, on the other hand happen that part of the bed-load is put into suspension. The process of this change of new-eroded material or of bed-load to suspension has been treated above.

[*Editor's Note:* Material has been omitted at this point.]

REFERENCES

Ahlmann, Hans W:son (1914a): Beitrag zur Kenntnis der Transportmechanik des Geschiebes und der Laufentwicklung des reifen Flusses. Sveriges Geol. Undersökning, Ser. C, n:o 262 (= Årsbok 8, nr 3). Stockholm 1914.

Andersen, S. A. (1931): Om Aase of Terrasser inden for Susaa's Vandområde og deres Vidnesbyrd om Isafsmeltningens Forløb. (With an English Summary). Danmarks Geologiske Undersøgelse. II Række. Nr. 54. København 1931.

Brenner, Thord (1931): Mineraljordaternas fysiska egenskaper (Deutsches Referat). Fennia 54. Helsingfors 1931.

Chatley, Herbert (1921): Silt. Min. of proceedings of the Inst. of Civil Engineers. Vol. 212, p. 400–413. London 1921.

Forbes, J. D. (1857): Contribution to a discussion about the delta of the Irrawaddy. Roy. Soc. of Edinburgh, Proc. Vol. III, p. 475. Edinburgh 1857.

Fortier, Samuel and Fred C. Scobey (1926): Permissible canal velocities. Trans. Am. Soc. Civ. Eng. New York, Vol. 89, p. 940–984. New York 1926.

Gibson, A. H. (1909): On the Depression of the Filement of Maximum Velocity in a Stream flowing through an Open Channel. Proc. Roy. Soc., Ser. A. Vol. 82, p. 149–159, London 1909.

Gilbert, Karl Grove (1914): The transportation of débris by running water. U.S. Geol. Survey, Professional paper 86. Washington 1914.

Jeffreys, Harold (1929a): On the Transverse Circulation in Streams. Proc. Cambridge Phil. Soc. Vol. XXV, p. 20–25. Cambridge 1929.

—— (1929b): On the Transport of Sediments by Streams. Proc. Cambridge Phil. Soc. Vol. XXV, p. 272–276. Cambridge 1929.

Kramer, Hans (1932): Modellgeschiebe und Schleppkraft. Mitt. Preuss. Versuchsanstalt für Wasserbau und Schiffbau, Berlin Heft 9. Berlin 1932.-Also as dissertation.

Nelson, Helge (1910): Om randdeltan och randåsar i mellersta och södra Sverige (With an English Summary). Sveriges Geologiska Undersokning Ser. C. N:o 220 (= Årsbok 3 (1909): N:o 3.) Stockholm 1910.

Owens, John S. (1908): Experiments on the transporting power of sea currents. Geographical Journal *31* p. 415–425. London 1908.

Penck, Albrecht (1894): Morphologie der Erdoberfläche. I, II. Stuttgart 1894.

Rubey, William W. (1933a): Settling velocities of gravel, sand, and silt particles. American Journal of Science. 5th Ser., Vol. XXV, p 325–338, New Haven, Conn. 1933.

—— (1933b): Equilibrium-conditions in debris-laden streams. Transactions of the American Geophysical Union. 14th annual meeting Washington 1933. Published by the National Research Council of The Nat. Ac. of Sc. Washington 1933.

Schaffernak, F. (1922): Neue Grundlagen für die Berechnung der Geschiebeführung in Flussläufen. Leipzig und Wien 1922.

Schoklitsch, Armin (1914): Über Schleppkraft und Geschiebebewegung. Leipzig und Berlin 1914.

Thrupp, Edgar Charles (1908): Flowing Water Problems. Min. of proceedings of the Inst. of Civil Engineers. Vol. 171, p. 346-359. London 1908.
Welikanoff, M. (1932): Eine Untersuchung über erodierende Stromgeschwindigkeiten. Wasserkraft und Wasserwirtschaft. Jahrg. 1932, H. 17. München 1932.

31

Reprinted from *"Recent Development of Mountain Slopes in the Kärkevagge and Surroundings, North Scandanavia," Geog. Ann.,* **42,** 184–187 (1960)

CONCLUDING DISCUSSION ON THE TOTAL DENUDATION OF SLOPES IN KÄRKEVAGGE

Anders Rapp

[*Editor's Note:* In the original, material precedes this excerpt.]

HOW TO COMPARE THE PROCESSES QUANTITATIVELY?

It is not easy to compare and evaluate the various slope processes discussed in the previous Chapters 4–10. Some processes are rapid, others slow, some are active on steep slopes, others on gentle.

A helpful simplification to obtain fairly comparable quantitative dimensions is to calculate the "exogene mass transfer" (German "geologische Massenverlagerung") as defined by JÄCKLI (1957, p. 28) in tons × metres. It seems to be three simple possibilities of expressing mass × movement, viz. either (*a*) mass × vertical component or (*b*) mass × horizontal component or (*c*) mass × resultant or "inclined" component. Of these JÄCKLI has calculated both (*a*) and (*b*). As we restrict our discussion to processes acting on the slopes from the water divides down to the valley bottom, and exclude the further transportation by rivers, only the vertical component is calculated here.

The mass transfer expressed in ton-metres/year makes it possible to evaluate the quantitative importance of different processes within a given area. But it is not suitable for comparisons with other areas of different sizes. For this purpose it seems convenient to add what we here preliminarily call the *relative mass transfer* = the quantity of material moved within or removed from a unit area of 1 km² (tons/km²/year). The whole drainage area of Kärkevagge is about 18 km². From this we exclude section B and the top plateau of Mt. Vassitjåkko, which are in part greatly influenced by glaciers and which have not been examined continuously. The remaining area of Kärkevagge considered in Table 32 occupies 15 km². We restrict our discussion to processes acting from the water divides down to the valley bottom and exclude the further transportation by rivers.

EXAMINATION OF TABLE 32

The letters *a* and *b* in Table 32 indicate the accuracy of the calculations. *a* = Direct measurement in Kärkevagge. *b* = Extrapolation from direct measurements in Kärkevagge.

The values in the column "Average movement" refer to the "inclined" component. The pebble-falls are supposed to move 90 metres, which is half of the average height of the rockwalls. The boulders are considered to fall the same distance as the pebbles and to continue to the base of the talus slopes (90 + 135 m). In the case of dissolved salts the average transport way is estimated at 700 m, which is half of the average distance from the water divides to the valley bottom.

An examination of the mass transfer in tons/km²/year and in ton-metres/year listed in Table 32 gives the following ranking list of transporting slope processes in Kärkevagge 1952–1960.

Transportation of dissolved salts

The most important transport process is that of dissolved salts, which removes a quantity of 26 tons/km²/year. This value is based on analyses of the water in Lake Rissajaure and in streams from Mt. Kärketjårro (p. 165). Morphological evidence of the chemical weathering are rust-coloured weatering crusts on mica-schist rocks and debris, poisonous effect on vegetation (p. 95), and white crusts of lime on rockwall and talus, especially in sections K and L (p. 165). The corresponding average denudation due to chemical solution is 0.01 mm/year or 70 mm during the postglacial period.

The continuous character of this process, its wide area of activity (100 % of the valley) and its long transport increase its importance. The main quantity of salts consists of sulphates, probably essentially dissolved from the mica-schist till and other loose debris on the slopes (cf. p. 167) but also from joints in bedrock. Calcium carbonate contributes to the salt content, which is probably considerably higher than the average chemical denudation in the mountains of Lappland (p. 168). The annual supply of atmospheric salts, reduced in the calculation, is estimated at 9 tons/km² (p. 167).

Earth-slides and mudflows

The second transport process is the earth-slides, which are due to the extreme rainstorm of October, 1959. This event very likely represents a centennial, or probably even millennial maximum (pp. 150, 157), raising the figures considerably over the "normal" means.

The relative mass transfer due to sheet-slides was 23 tons/km²/year and a further 18 tons/km²/year consisted of sheet-slides which continued their movement as mudflows or stream wash. The bowl-slides comprised a transfer of 20 tons/km²/year but a very low value in ton-metres, due to their short average movement (0.5 m).

The earth-slides and mudflows etc. mainly caused a removal of till from the upper slopes of Mt. Kärketjårro and a re-deposition as alluvial fans or sheets on the lower parts of sections J, K and L. Some material was also deposited on the upper part of sheltered talus slopes. On other places the talus slopes were eroded by gullies or slides (p. 152) continuing as mudflows. This transport together with mudflows in other years gave 8.4 tons/km²/year.

Unlike the chemical solution the slides and mudflows are only local transfers that do not (or only to a minor extent) reach the valley bottom and the main stream. This kind of slope processes probably has a still greater importance in climates with more frequent and heavier torrential rains (p. 164).

Table 32. Denudation of slopes in Kärkevagge 1952—1960 given in quantities per year. The average gradient is roughly indicated in 45°, 30° or 15°. The transfer by sheet-slides has been calculated for each case (12—420 m and 70—600 m respectively). For further explanation, see comments in the text.

Process	Volume, m³	Density	Tons (t)	Tons per km²	Average movement m	Average gradient	Ton-metres (vertical)
Rockfalls							
Pebble-falls	5 b	2.6	13	1	90 a	45°	845
Small boulder-falls	10 b	2.6	26	1.7	225 a	45°	4,160
Big boulder-falls	35 a	2.6	91	6	225 a	45°	14,560
Avalanches							
Small avalanches	8 b	2.6	21	1.4	100 b	30°	1,050
Big avalanches (Slushers)	80 a	2.6	208	14	200 b	30°	20,800
Earth-slides etc.							
Bowl-slides	170 a	1.8	300	20	0.5 a	30°	75
Sheet-slides	190 a	1.8	340	23	12—420 a	30°	20,000
Sheet-slides + mudflows	150 a	1.8	270	18	70—600 a	30°	70,000
Other mudflows	70 b	1.8	126	8.4	100 b	30°	6,300
Creep							
Talus-creep	300,000 b	1.8	—	—	0.01 b	30°	2,700[d]
Solifluction	550,000 b	1.8	—	—	0.02 b	15°	5,300[e]
Running water							
Dissolved salts	150 b	2.6	390	26	700 b	30°	136,500
Slope wash	?			?			?

[d] Horizontal component of talus-creep = 4,700.
[e] Horizontal component of solifluction = 19,800.

Dirty avalanches

Dirty avalanches are separated into two types, big and small. The big ones are the three cases of slush avalanches which occurred in 1956 and in 1958 (p. 138). They are sporadic processes probably typical of mountains in high latitudes. The morphology of the avalanche tracks shows that such big events as the three slushers are rare but not at all unique. The 14 tons/km²/year removed by slush avalanches is therefore considered as a high average value. Together with the more continuous, small avalanches we get the figure of 15.4 tons.[1]

Rockfalls and frost-weathering

The mass transfer by rockfalls is 8.7 tons/km²/year, mainly in the form of big boulder-falls (p. 114). The rockfalls are of a special interest as they show the denudation of bedrock on steep slopes. The average annual retreat of the rockwalls in Kärkevagge was 0.04–0.15 mm/year, indicating a probable postglacial "continuous" retreat of about one meter (cf. pp. 115, 122).

The rockfalls are mainly released by thawing after frost-bursting (p. 105). Thus the retreat of rockwalls indicates a maximum value of frost-weathering on more gentle bedrock surfaces. There the penetration of frost-shattering is probably slower due to insulating loose debris upon the rock. The water content may in places reverse this supposed relation of weathering in rockwalls contra more gentle rock slopes. An average frost-shattering of 0.04 mm (the minimum value of rockwall retreat) corresponds to an annual production of rock waste amounting to roughly 100 tons/km² of rockwall surface in Kärkevagge.

Solifluction and talus-creep

The continuous mass-movements of solifluction and talus-creep are difficult to compare with the momentary mass-movements. Solifluction is active on an area of about 2.2 km² and the movement is estimated at 2 cm/year of a layer 25 cm thick (p. 182). Talus-creep is active on an area of 1.5 km² with an estimated movement of 1 cm/year in a layer 20 cm thick (p. 175).

As regards these slow processes, the values in ton-metres tell more than the quantities in tons/km². The mass transfer of solifluction is calculated at 20,000 ton-metres and that of talus-creep at 5,000. The talus-creep seems to decrease towards the base, indicating that it functions as a shifting process on growing talus slopes, which are later affected by momentary removing processes, such as slides, or gullying and mudflows.

The solifluction is the dominant transporting process in certain areas, marked by lobes etc. One function of solifluction is that it delivers material to runnels and slope wash for further transportation.

Other processes

Creep due to *needle-ice* has been noted by the author but it is considered to be of small importance in Kärkevagge, where a large part of the slopes are either grass-covered or consist of talus debris, too coarse for formation of needle-ice.

Wind erosion has been observed (p. 110) but its quantitative importance is believed to be very small in Kärkevagge.

Another factor not considered in Table 32 is *slope wash*. It has not been measured in Kärkevagge, only observed a few times (p. 158). The few recordings made in other periglacial areas do not permit a comparison. The opinion of the author, based on indirect evidence (p. 160), is that slope wash is a process of minor importance on the grass-covered slopes and the naked talus slopes in Kärkevagge.

Summary

The quantitative analysis summarized in Table 32 indicated the following order of the transporting processes acting on slopes in this environment (steep and moderate mountain slopes mainly in the "tundra zone", mica-schist and limestone bedrock and till, maritime, arctic climate).

1. Transportation of salts in running water
2. Earth-slides and mudflows

[1] JÄCKLI (1957, p. 126) estimated the mass transfer by avalanches in Graubünden at 450,000 tons/year, a quantity which corresponds to 105 tons/km²/year. A comparison with the 15 tons in Kärkevagge supports our view that JÄCKLI's estimation of avalanche removal is too high (cf. p. 146).

3. Dirty avalanches
4. Rockfalls
5. Solifluction
6. Talus-creep.

Where slope wash should be placed in this list is not possible to say.

In other valleys with other types of slopes, bedrock, soils etc. the processes may have quite another order.

Frost-bursting is not included in the list as it is not a transporting process and as it is very difficult to measure directly. But the annual production of rock waste by frost-bursting on the rockwalls is calculated to be 100–400 tons/km² of wall surface. If frost-bursting were included in the list above, it would possibly be the leading one.

Final remarks[1]

The figures given in Table 32 are more or less approximate and should be looked upon as an attempt to evaluate the order of magnitude of the processes acting in the selected area. The figures thus can be checked and corrected in many respects, both in Kärkevagge and by comparative studies in other mountain areas. One of the first complementary studies that should be made is measurements of slope wash, for instance by the methods used by JAHN (1961) in Spitsbergen.

Table 32 may serve as a summary of the quantitative measurements made in Kärkevagge and also as a hypothesis for future work, both in the mountains of Scandinavia and other latitudes. Even if we here in many respects have emphasized the importance of direct recordings of slope processes, the geomorphological analysis of slope forms may not be forgotten. In this connection two other methods of quantitative slope studies can be mentioned, viz. comparisons of old and new photographs, and volume measurements of rock waste in talus cones, demonstrated by the author in a previous work.

[1] See also general summary and conclusions in: RAPP, A.: Studies of the postglacial development of mountain slopes. Meddelanden från Uppsala Universitets Geografiska Institution, A: 159. 10 pp. 1961.

[*Editor's Note:* References can be found in the original volume.]

32

Reprinted from *Jour. Sed. Petrol.*, **27**, 23-26 (1957)

BRAZOS RIVER BAR: A STUDY OF THE SIGNIFICANCE OF GRAIN SIZE PARAMETERS

Robert L. Folk and William C. Ward

[*Editor's Note:* In the original. material precedes this excerpt.]

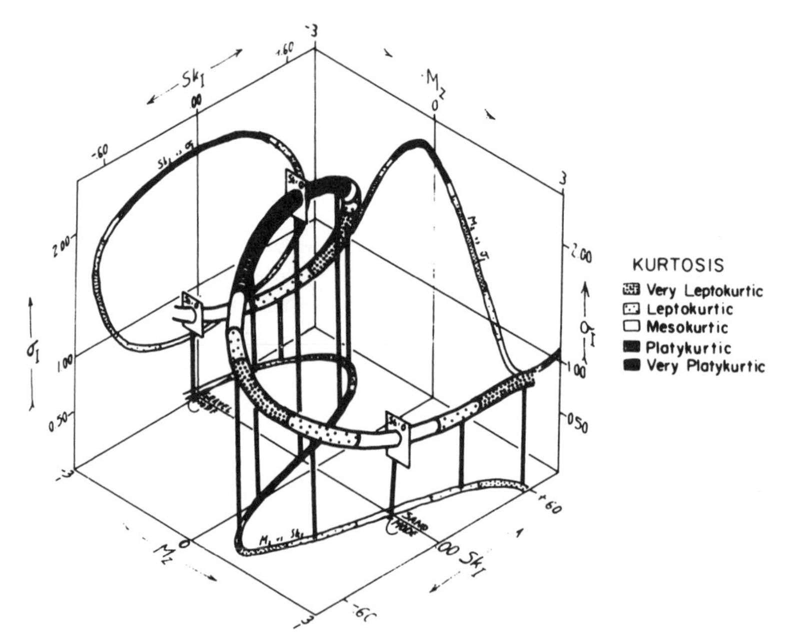

FIG. 18. Four-variate graph, showing the relation between mean size (M_z), standard deviation (σ_I), skewness (Sk_I), and kurtosis. This is an accurately-plotted isometric projection of the helix which results when mean size, standard deviation and skewness are plotted against each other. Each of the three sides of the box containing the helix represents each pair of variables plotted in turn, hence correspond to two-dimensional projections of the helical trend. Standard deviation, the vertical dimension, is shown also by the height of the "supports" to the helix: points where the helix passes through the .00 skewness plane are shown by small "signboards." Kurtosis is shown by pulsations of shading along the helix and its three projections. The following limits are used: Very Platykurtic, K_G below 0.67; Platykurtic, K_G 0.67-0.90; Mesokurtic, K_G 0.90-1.11; Leptokurtic, K_G 1.11-1.50; and Very Leptokurtic, K_G 1.50-23.90.

The equation for the helix is

$$\begin{cases} \sigma_I = 1.5\phi - 0.75 \sin [60° (M_z - 0.75\phi)] \pm 0.4\phi \\ Sk_I = -0.03\phi - 0.5 \sin [60° (M_z + 0.75\phi)] \pm 0.12\phi \end{cases}$$

where σ_I and Sk_I represent the values for standard deviation and skewness (the dependent variables), M_z is the mean size in phi units, and the last term is the standard error of estimate (two-thirds of the values will fall within the predicted value plus or minus the standard error).

For example, consider a sediment with M_z of $+0.13\phi$. In the standard deviation equation, one finds the sine of 60° (0.13 − 0.75) = sin 60° (−0.62) = sin −37.2° = −0.60. When the rest of the equation is computed, the predicted value of σ_I is 1.5 − 0.75(−0.60) = 1.5 + 0.45 = 1.95ϕ, and two-thirds of the time the actual σ_I values will lie between 1.55ϕ and 2.35ϕ. In the skewness equation, one finds the sine of 60°(0.13 + 0.75) = sin 60°(0.88) = sin 52.8° = 0.79. The predicted value of Sk_I is then 0.50(0.79) − 0.03 − 0.40 − 0.03 = −0.43, and two-thirds of the actual values will lie between −.31 and −.55.

It is possible to work out a general equation for helical trends of this type, wherein the wave length of 360° gives the phi interval between modes (in this example the distance was about 6ϕ, therefore both the equations contained the factor 360°/6 = 60°, and the skewness and standard deviation sine curves are one-quarter wave length out of phase (in this example 6ϕ/4 or 1.5ϕ, hence one equation contained the factor $M_z + 0.75\phi$ and the other $M_z - 0.75\phi$). The minimum point on the standard deviation sine curve coincides with the modal diameter, and the amplitude on the standard deviation curve is governed by the difference in σ_I between the average worst and the average best sorted samples. Actually it is not quite so simple, because seldom do both modes have equivalent σ_I values, but this type of formula may give a good approximation to the true quantities.

Preliminary work at the University of Texas and examination of previous published results (Inman, 1949) indicates that this helical trend applies in grain size distributions from many other environments. The helix probably goes through several more cycles, with each minimum of best sorting coinciding with a mode in the environment and each maximum coinciding with an inter-modal position, as shown in figure 19. These inter-modal regions can be easily identified by their platykurtic character.

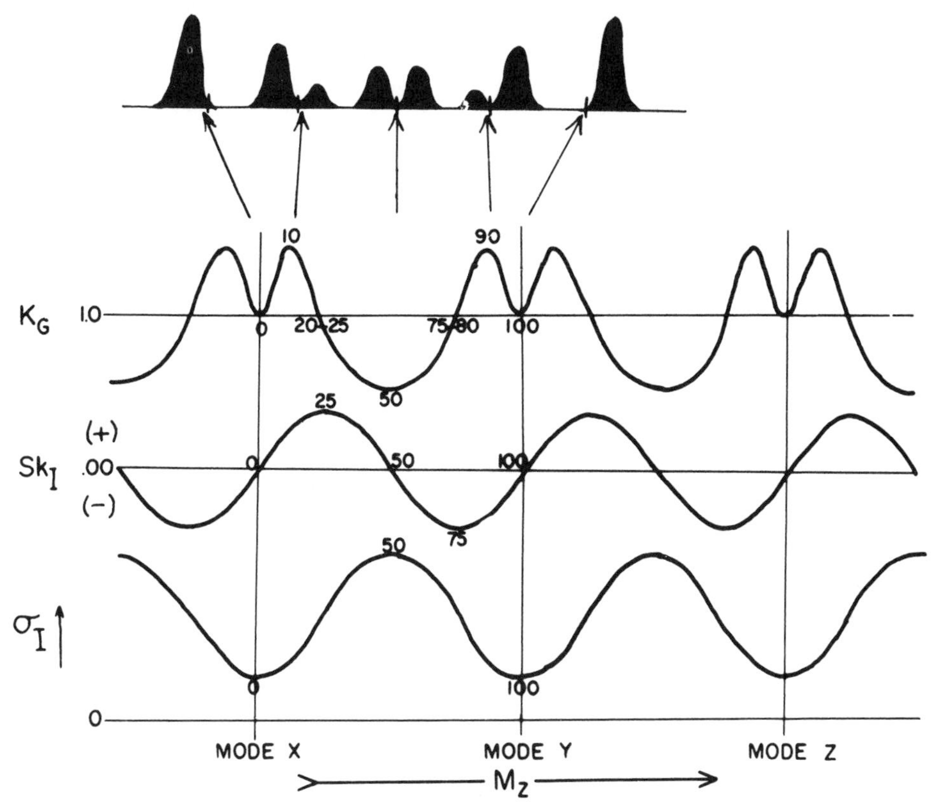

Fig. 19. Theoretical variation of standard deviation (σ_I), skewness (Sk_I) and Kurtosis (K_G) as a function of mean size (M_Z) in a hypothetical polymodal sediment. Plots of σ_I and Sk_I form sine curves one-quarter wave length out of phase (actually combining to form a helix in three dimensions), while kurtosis forms a complex rhythmic curve. For mixtures of Mode X with Mode Y, the percentages given indicate the proportion of Mode Y present at critical points on each of the curves, provided the measures described herein are used. These percentages hold for the Brazos bar, but probably are slightly different in other environments. Shaded frequency curves at the top of the diagram illustrate the appearance of grain size curves for the critical points designated.

Work in progress indicates that in some neritic environments the modes are sand (or coarse silt) plus clay, also linked by a helix which would add another cycle to the right of the one shown here for the Brazos bar. Many sedimentary environments show only a segment of this helix; to have a complete cycle one needs two distinct and fairly widely separated modes, and within the suite of samples analyzed the pure end members as well as all intermediate mixtures must be present. If, for example, the samples examined consist only of the gradation from sand to clayey sand (say at most 35 percent clay), then only one-third (120°) of a helical cycle will be completed.

Special conditions may alter the form and position of the helix. Work in progress by Todd (1956) on some Eocene sands in Texas shows that these are polymodal sediments in which there usually is a small amount of clay, regardless of the size of the sand. This has the effect of making all the samples positive-skewed and leptokurtic, but here the addition of clay has simply shifted the axis of the helix in the xz plane; instead of being parallel to x with Sk_I equal nearly to .00 (giving nearly equal frequency of positive and negative skewness values), the axis is still in the xz plane but has a Sk_I value of .00 at 1ϕ and +.30 at 3ϕ. Similarly, Miller (1955), found in the Permian Pierce Canyon siltstone of southeast New Mexico a 180° segment of a helical trend but with the axis shifted back into positive Sk_I values because of the constant presence of small "tail" of clay. These positional shifts of the helical axis also exert a strong effect on kurtosis values, tending to shift them into leptokurtic regions.

Will these helical trends show up if other measures of grain size, such as those proposed by Inman, are used? The answer appears to be affirmative, but the trends are not as distinct. The measures proposed here are based on more points, hence are more sensitive and should be expected to give a better trend. For example, addition of a secondary mode affects σ_I and Sk_I if as little as 5 percent occurs, but 16 percent is required to affect Inman's $\sigma\phi$ or $\alpha\phi$. A sample with 30 percent gravel and 70 percent sand has nearly the same mean size as one with 70 percent gravel and 30 percent sand, using Inman's measures, but M_z is considerably more sensitive because the median is included in the calculation.

POSSIBLE GEOLOGICAL SIGNIFICANCE OF SKEWNESS AND KURTOSIS

If one may be permitted to extrapolate from a small study such as this one and enter the seductive field of generalization, it appears that both skewness and kurtosis are vital clues to the bimodality of a distribution, even when the modes are not immediately apparent. For example, these modes may be hidden in an obscure sidewise kick or gentle curvature of the cumulative plot on probability paper, enough to show up as non-normal skewness and kurtosis values, but not enough to show up as a secondary mode on a frequency curve or histogram. Strictly unimodal sediments (like some beach sands) should give normal curves; non-normal values of skewness and kurtosis indicate something "wrong" with the sediment, and indicate a mixing of two or more modal fractions. As an illustration of this principle, many dune sands on the South Texas coast have slight positive skewness and leptokurtosis, caused by the presence of a very minor coarse silt mode in a size finer than the principle mode.

Extreme high or low values of kurtosis imply that part of the sediment achieved its sorting elsewhere in a high-energy environment, and that it was transported essentially with its size characteristics unmodified into another environment where it was mixed with another type of material. The new environment is one of less effective sorting energy so that the two distributions retain their individual characteristics—i.e. the mixed sediment is strongly bimodal. In the Brazos bar, extreme kurtosis values are attained because the bulk of the sand apparently received its sorting in the parent Cretaceous marine sediments, but is now being deposited in the less efficient sorting environment of a river bar, where it is rapidly dumped together with gravel or silt. In neritic sediments, extreme kurtosis values are common because the sand mode achieves good sorting in the high-energy environment of the beach, and then is transported en masse by storms to the neritic environment, where it becomes mixed with clay and hence is finally deposited in a medium of low sorting efficiency. If the sediments are near the source of the sand, they are characteristically leptokurtic and positive-skewed because the sand is in excess. The more extreme the kurtosis values, the more extreme is the sorting of the modes in their previous environment and the less effective is the sorting in the present environment. Thus one may conclude that kurtosis and skewness are very valuable clues to the "genealogy" of a sediment.

CONCLUSIONS

Once a relationship is established in an ideal case, where the changes are laid out before the observer in their most perfect form, one soon learns to recognize the same relationships in less ideal examples, where the changes are obscure. The obscure examples, hitherto unfathomable, are explained in the light provided by the ideal examples. So it has been with the Brazos bar study. We have studied a simple environment, where the changes follow an orderly helical progression because of the ideally bimodal character. Now, fortified with the knowledge of the ideal trend, we have been able to unravel many once-puzzling relationships in other sedimentary suites of more complex nature and to understand better what is going on in the sedimentary environment. The meaning of skewness and kurtosis has, we feel, been ascertained: they are vitally important distinguishing characteristics of bimodal sediments and enable us to recognize bimodality where it was previously obscure. The changes of skewness, kurtosis, and sorting with sediment transport are probably simple functions of the ratio between the two modes of the sediment. The equations tying these variables together will, we hope, be of some value in distinguishing sedimentary environments. It is not the absolute values of parameters themselves, but their four-dimensional relationships to each other which offer the best hope of further progress.

REFERENCES

CROXTON, F. E., AND COWDEN, D. J., 1939, Applied general statistics. Prentice-Hall, New York, 944 p.
DAPPLES, E. C., KRUMBEIN, W. C., AND SLOSS, L. L., 1953, Petrographic and lithologic attributes of sandstones: Jour. Geology v. 61, p. 291–317.
FOLK, R. L., 1954, The distinction between grain size and mineral composition in sedimentary rock nomenclature: Jour. Geology v. 62, p. 344–359.
— —, 1955, Student operator error in determination of roundness, sphericity, and grain size: Jour. Sedimentary Petrology, v. 25, p. 297–301.
GRIFFITHS, J. C., 1951, Size versus sorting in some Caribbean sediments: Jour. Geology v. 59, p. 211–243.
HOUGH, J. L., 1942, Sediments of Cape Cod Bay, Massachusetts: Jour. Sedimentary Petrology v. 12, p. 10–30.
INMAN, D. L., 1949, Sorting of sediments in the light of fluid mechanics: Jour. Sedimentary Petrology, v. 19, p. 51–70.
— —, 1952, Measures for describing the size distribution of sediments: Jour. Sedimentary Petrology v. 22, p. 125–145.
INMAN, D. L., AND CHAMBERLAIN, T. K., 1955, Particle-size distribution in nearshore sediments, in Finding ancient shorelines. Society of Economic Paleontologists and Mineralogists Spec. Publ. 3, p. 99–105.
KRUMBEIN, W. C., 1934, Size frequency distribution of sediments: Jour. Sedimentary Petrology, v. 4, p. 65–77.
— —, 1938, Size frequency distribution of sediments and the normal phi curve: Jour. Sedimentary Petrology, v. 8, p. 84–90.
KRUMBEIN, W. C., AND PETTIJOHN, F. J., 1938, Manual of sedimentary petrography. D. Appleton-Century, New York, 549 p.
MILLER, D. N., JR., 1955, Petrology of pierce canyon redbeds: Delaware Basin, Texas and New Mexico. Ph.D. Thesis, University of Texas.
OTTO, G. H., 1939, A modified logarithmic probability graph for the interpretation of mechanical analyses of sediments: Jour. Sedimentary Petrology, v. 9, p. 62–75.
PETTIJOHN, F. J., 1949, Sedimentary rocks. Harper and Brothers, New York, 526 p.
PLUMLEY, W. J., 1948, Black Hills terrace gravels: a study in sediment transport: Jour. Geology, v. 56, p. 526–577.
POTTER, P. E., 1955, The petrology and origin of the Lafayette Gravels. Part I; Mineralogy and Petrology: Jour. Geology, v. 63, pp. 1–38.
TODD, T. W., 1956, Comparative petrology of the Carrizo and Newby Sandstones, Bastrop County, Texas, Master's Thesis, University of Texas.

33

Reprinted from *Bull. Geol. Soc. America,* 52, 1301–1303, 1346–1351 (1941)

TILL FABRIC

C. D. Holmes

ABSTRACT

A study of the arrangement of component materials in undisturbed till, the till fabric, shows that at most localities the imbedded stones tend statistically to lie so that their long axes are parallel to the direction of glacier flow at the time of deposition. In a few localities the dominant statistical preference is for alignment at right angles to that direction. Presumably the parallel orientation was normally acquired by sliding, and the transverse orientation by rotation, and permanent deposition commonly occurred without loss of alignment. Fabric analyses indicate that stones of certain forms and degrees of roundness (enumerated in text) have a greater-than-average statistical chance for deposition either parallel to, or transverse to, the direction of transport. Such stones thus serve as guides to the direction of glacier flow and are independent of other evidence. Characteristic depositional attitudes of certain types of till stones permit inferences regarding the probable nature of the transportational environment.

INTRODUCTION

FACT OF TILL ORGANIZATION

Undisturbed till has an inherent organization. At most localities this organization manifests itself in the tendency of imbedded stones to lie so that their longest dimension or axis coincides approximately with the direction of glacier flow at the time of deposition. Such tendency is best revealed by a statistical study of the positions of at least 100 till stones from any one locality. The results can be expressed by diagrams such as the simple "rose" figure or by the contoured diagram commonly used in petrofabric studies. This makes possible the determination of ice-flow direction at places where no striae or other criteria are available. Moreover, the direction of glacier flow is known to have varied from time to time at any one place. Hence, in problems such as locating mineral deposits by tracing drift fragments to their sources (Sauramo, 1924), data from till-stone orientation may be as essential as those from striae on bedrock beneath the till.

HISTORICAL STATEMENT

Miller (1884) published probably the first critical observations on till-stone orientation. In describing "pavement boulders"[1] in the till near Edinburgh, Scotland, he stated (p. 167):

"The longer axis of the stone is often directed in the line of glaciation, and the pointed end is frequently, but not always, toward the ice."

Smaller stones may be oriented in "fluxion structures" around the larger boulders.

"It does not follow, however, that wherever we find an orientation of boulders in the till there was fluidal motion in the layer in which they lie. If the ice had a fluxion structure of its own, such boulders as were incorporated within its mass would arrange their axes conformably; and, when they lagged and came to rest and were imbedded, they might retain in many cases the arrangement that marked them when in motion" (p. 187).

[1] Stones with strongly striated upper surfaces, generally occurring in definite horizons in till and constituting a "pavement" on which the superjacent till rests.

Bell (1888b) carried the investigation further by visiting a number of Swiss glaciers to determine whether such orientation of stones existed in transit. He concluded (p. 341)

". . . that the tendency of boulders on all glaciers is to assume a longitudial position, and that this is most observable . . . on large glaciers, where the obstructions are fewer in proportion to the mass, and produce the least disturbing effect."

Unfortunately, both Miller's and Bell's data are largely qualitative and selective.

Upham (1891) described the characteristic position of "oblong" stones in subglacial till as having their long axes parallel to contiguous striae. Flat stones were said to lie parallel with the surface of deposition. However, Upham recorded no systematic investigation.

The results achieved by these early investigators failed to attract the notice they deserved. James Geikie's classic treatise (1895, p. 15, 62) probably reflects the prevailing attitude of that time. Although mentioning that "in certain regions the large and small till stones are oriented in parallel fashion," and citing Miller's work, Geikie described till as a clay containing a "confused and pell-mell mixture of stones." For many years this concept of chaotic agglomeration remained practically unchallenged. Twenhofel (1932, p. 86) summarized more recent opinion by stating that ground moraine consists of "unstratified and unorganized material."

Papers by Richter (1932, 1933, 1936) are among the most important published works of recent date. His discoveries apparently developed as a modification of the method used by the Scottish geologists for determining the direction of glacier flow from striae on the upper surfaces of "pavement boulders" in areas where striated bedrock ledges are absent. In northern Germany, Richter found that the long axes of elongate stones tend to parallel the direction of pavement-boulder striae, even though such stones themselves bore no striae. Like Miller, he reasoned that these stones had been oriented as streamlined bodies in the glacier and had been deposited with but little or no change in orientation. A statistical grouping of long-axis orientations indicates the direction in which the glacier was moving. Later studies at Engebrae and Fondalsbrae glaciers in Norway verified his conclusions. Stones imbedded in debris zones near the end of the glacier show a statistical preference for long-axis orientation parallel with the direction of glacier flow. The majority of stones in the till beside the glacier have similar orientations. Although Richter's conclusions are essentially those of Miller and Bell and appear to be a rediscovery of the earlier knowledge, his conclusions are supported by quantitative, statistical data.

Richter noted that many of the larger till stones (small boulders) at the Norwegian localities tend to an orientation transverse to glacier flow. These were thought to have been oriented originally parallel with the ice-flow direction and to have been shoved during a temporary readvance of the glacier; or to have been rolled beneath the glacier because they projected through the lowermost shear layer (Scherpakete). But the small cobbles in the same deposit are oriented essentially parallel with the ice-flow direction. This, he believed, may have resulted from the action of later meltwater streams which were competent to reorient the smaller stones, leaving the larger ones unmoved.

In describing the Pleistocene glaciation of a part of Yellowstone National Park, Miner (1937) cited the phenomenon of surface boulders with their long axes in parallel orientation, trending diagonally across a valley, but he offered no special explanation.

Krumbein (1938, p. 273; 1939) has investigated statistically the axial orientation of till stones in the western Great Lakes region. He concluded that both the mode and the arithmetic mean of long-axis directions approximate the direction of glacier flow at the locality studied.

The present writer (Holmes, 1938) published a preliminary statement on till-stone orientation in central New York and suggested till fabric as a term denoting the space relations among the component rock and mineral fragments in undisturbed till.

[*Editor's Note:* Material has been omitted at this point.]

TESTS OF CRITERIA FOR DISTINGUISHING ORIENTATION POSITIONS

General statement.—A few stations in the central New York area yielded orientation diagrams that were anomalous or ambiguous. Hence these problem stations are ideal for testing the practical value of the criteria selected for determining the direction of glacier flow.

In the following analyses, no special weight is given to those indicators or guide forms that fall in more than one guide group. Each is recorded but once.

Station 15.—Station 15 is situated 0.6 mile northwest of Carpenter Pond, in the west central part of the Cazenovia quadrangle. The exposure is a road cut in one of the many till mounds on the west slope of a shallow through-valley. The rose pattern from this station is regular and symmetrical, and evidence from striae and topography is sufficient to show the relation of this pattern to the direction of ice

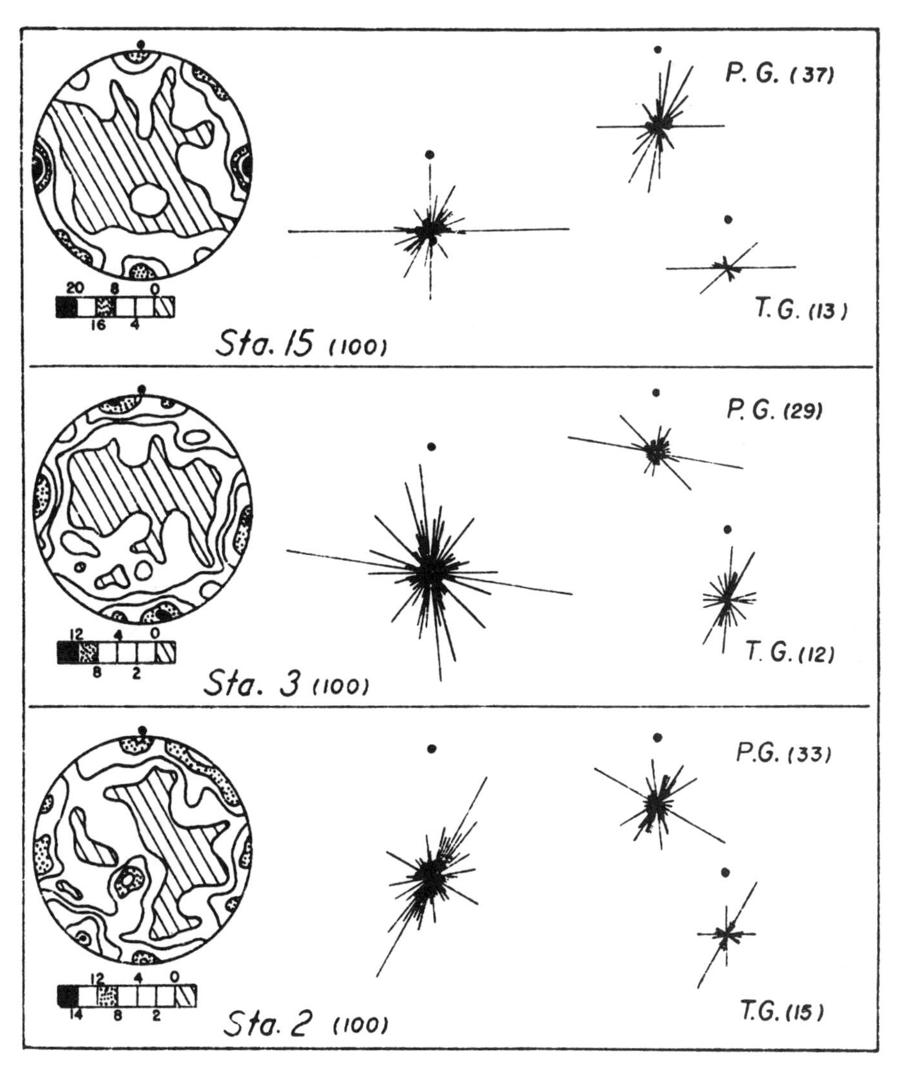

FIGURE 23.—*Analyses of anomalous patterns from three localities*

Parallel and transverse guide forms (indicated by P.G. and T.G., respectively) are shown separately. Their contribution to the complete station pattern indicates clearly the direction of glacier flow. True north is shown by a dot above each diagram.

flow, but the pattern is so different from most others that it was rejected in making up the composite group.

Figure 23 shows the various diagrams pertaining to this station. The contoured diagram shows an unusually strong east-west maximum, with almost no suggestion of a girdle across the diagram connecting the two parts of this maximum. However, an imperfect girdle is developed in a vertical plane normal to the east-west position. On the basis of these facts, the pattern is one of strong transverse orientation.

The guide forms are plotted separately. Each of the five stones oriented N. 10° E. is a guide to the parallel position, and two of them are included in more than one guide group. The average of intermediate-axis dips at this station is very low, and the five parallel guides in transverse position appear there because of their low intermediate-axis dips. Even so, the statistical effect of these forms on the entire pattern is to show clearly that the stones oriented east-west are in transverse position. Therefore the analysis indicates that the orientation of stones at this station presents a strong transverse pattern. All field evidence confirms this interpretation.

Station 3.—Station 3 is located on the south bank of Limestone Creek, one mile northeast of New Woodstock (central part of Cazenovia quadrangle). The valley has a general east-west trend. Striae about half a mile to the southwest indicate ice movement due south; and about one mile northwest, striae are aligned S. 30° E. No evidence was found within the valley to show whether the ice flowed across it or along its axis.

The rose diagram of this station (Fig. 23) is characteristic except for its slight asymmetry. Though its longest line nearly parallels the valley trend, nothing indicates whether that line represents the parallel or the transverse position. The contoured diagram shows a slight tendency to general southward dip but offers no clear solution to the problem of ice-flow direction.

The parallel indicators show that the depositing ice flowed principally up-valley (eastward) but probably shifted its direction occasionally to a more southeasterly course. Seven of the eleven stones oriented N. 85° W. are good guides to the parallel position. Likewise, four of the six stones aligned N. 45° W. are parallel guides. The transverse indicators, though few in number, are in general agreement with evidence from the parallel guides.

The lithology of the till in this part of the valley changes abruptly about half a mile farther upstream (east), marking the limit of drift brought into the valley by ice from the northwest.

Station 2.—Station 2 is located on the south bank of Middle Branch Tioughnioga Creek, one mile southeast of Sheds (southeastern part of Cazenovia quadrangle). The valley slopes to the west, and the station is less than half a mile within the limits of drift brought from the northwest. This genetic relation of the till was not fully known at the time of field study, and the singularly symmetrical pattern (Fig. 23) was then interpreted as the work of ice flowing from the northeast, which had preceded the later flow from the northwest. But after the till relation was known, the diagonal position of the pattern with respect to the valley axis left considerable uncertainty as to the direction of glacier flow.

The contoured diagram suggests the presence of girdles other than the peripheral one, but not clearly enough for safe interpretation.

The indicators of the parallel and transverse positions show clearly that the pattern is one of prevailing transverse orientation. Five of the six stones aligned N. 60° W. are good parallel indicators, and the sixth is fairly good. Therefore the depositing ice was moving toward the southeast as it reached that part of the valley.

Conclusion.—Pattern analyses of these problem stations show that the guide forms give correct results in all cases where these results can be checked by other evidence. This gives assurance that the criteria are dependable where other evidence is lacking. Parallel-position guides are more numerous in an unselected group of 100 stones and give more decisive results than transverse-position guides. Decisiveness of results generally corresponds to the degree of simplicity and symmetry of the

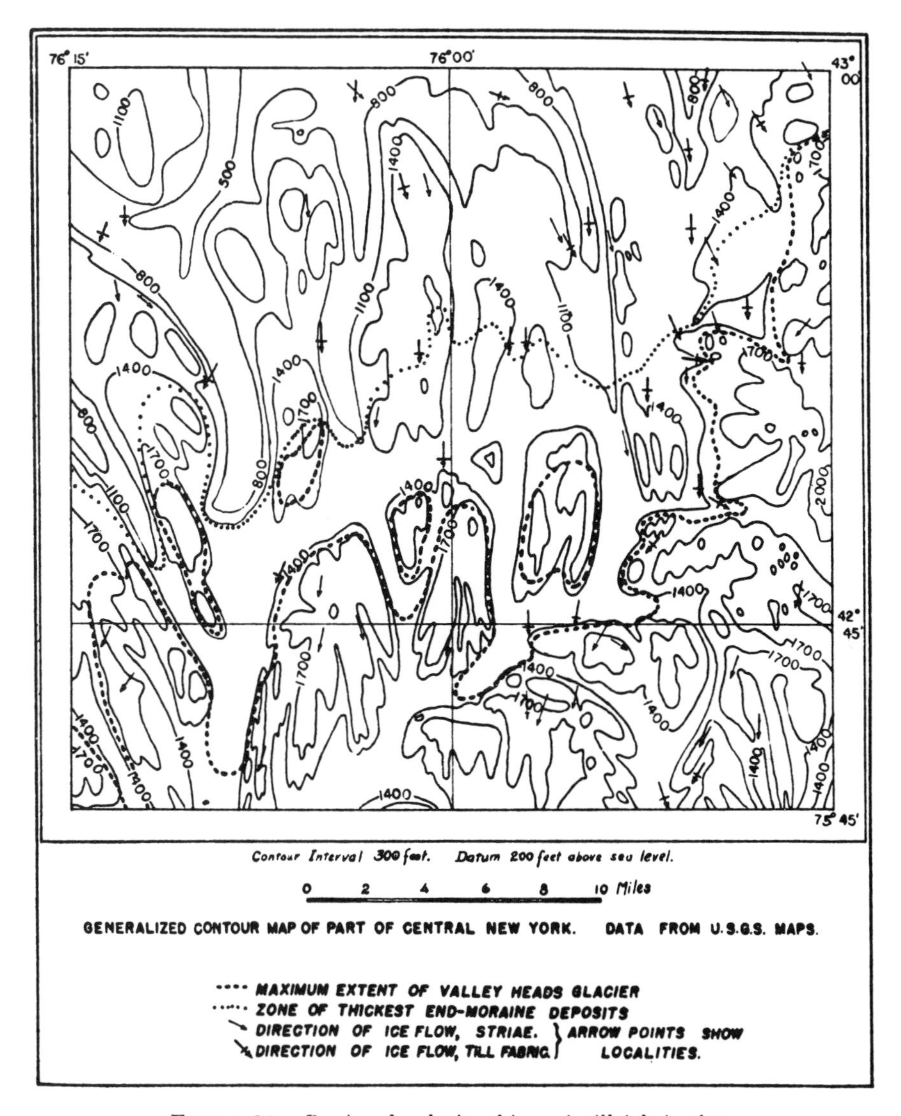

FIGURE 24.—*Regional relationships of till-fabric data*

Cazenovia quadrangle (northeast), Tully quadrangle (northwest), and parts of Pitcher quadrangle (southeast) and Cortland quadrangle (southwest). The Valley Heads glacier moved southeast, and the earlier glacier toward the southwest. Strong local topographic control of ice-flow direction is apparent.

pattern to be analyzed. This correspondence seems primarily related to the constancy in direction of glacier flow during the time of deposition.

SUMMARY OF CONCLUSIONS

The conclusions drawn from this study of till fabric are believed to be valid for till in any region where the conditions of glaciation were comparable to those of central New York. Comparative studies from other regions seem highly desirable. The principal conclusions for the central New York area may be summarized as follows:

(1) The ground-moraine till accumulated gradually beneath the moving glacier, but with occasional local and temporary intervals of erosion that gave rise to boulder-pavement phenomena. Fabric patterns from successive layers of till record occasional shifts in direction of glacier flow similar to those shown by intersecting sets of striae on rock ledges.

(2) Glacier ice in which the stones were carried moved as a plastic solid flowing in obedience to the laws of fluid mechanics. Stones carried above the floor were subject to rotation about an axis lying in a plane of uniform shear and normal to the direction of glacier flow. Ordinarily the longest axis was the axis of rotation, though under some conditions, probably following temporary orientation by sliding, rotation about the intermediate or the shortest axis probably occurred. Stones rotating on or near the floor were subject to shoving into the deposited till by thrust from higher and faster moving debris. Many of the stones thus imbedded remained permanently in that position.

(3) Stones in contact with the glacier floor (or possibly along a well-defined shear plane in the ice) moved by sliding. Normally both long and intermediate axes were parallel to the floor plane, with the long axis aligned in the direction of movement. However, prominent flat surfaces, especially on wedge-form stones, commonly exercised more control in orientation by sliding than did the trend of the long axis. The controlling surfaces were aligned parallel to the direction of movement.

(4) Inherent characters such as form, roundness, size, and relative axial lengths, predispose a stone to certain kinds of movements and thus influence the likelihood of deposition in certain diagnostic attitudes or positions. Some types of stones have a relatively high statistical probability in depositional attitude, with the long axis aligned either parallel to or transverse to the direction of glacier flow. Thus it is possible to determine the direction in which the glacier moved while the deposit was accumulating. This method of ascertaining the direction of ice flow is independent of other evidence and can be used directly in the field.

(5) Minor differences in till-stone characters, such as slight rounding, have a profound effect on the depositional orientation preference of some types of stones (for example, Fig. 18). This in turn suggests that the number of such influences has not been fully measured and that a more detailed classification of till stones will probably reveal further signifi-

cant relations. Delicate adjustment to slight degrees of change is characteristic of glacial behavior, and the enormous forces of the continental glacier seem to have responded even to very small differences in those factors commonly grouped under the term structure.

WORKS TO WHICH REFERENCE IS MADE

Bell, Dugald (1888a) *On some boulders near Arden, Lochlomond,* Geol. Soc. Glasgow, Tr., vol. 8, pt. 2, p. 254-261.

——— (1888b) *Additional note to Mr. Bell's papers (p. 237-261),* Geol. Soc. Glasgow, Tr., vol. 8, pt. 2, p. 341.

Demorest, Max (1938) *Ice flowage as revealed by glacial striae,* Jour. Geol., vol. 46, p. 700-725.

Fairchild, H. L. (1907) *Drumlins of central western New York,* N. Y. State Mus., Bull. 111, 443 pages.

——— (1932) *New York moraines,* Geol. Soc. Am., Bull., vol. 43, p. 627-662.

Geikie, James (1895) *The great ice age,* 2d ed., London, D. Appleton Co., 850 pages.

Holmes, C. D. (1938) *Till fabric* (abstract), Geol. Soc. Am., Bull., vol. 49, p. 1886-1887.

Hubbert, K. M. (1937) *Theory of scale models as applied to the study of geologic structures,* Geol. Soc. Am., Bull., vol. 48, p. 1459-1519.

Knopf, E. B. and Ingerson, Earl (1938) *Structural petrology,* Geol. Soc. Am., Mem. 6, 270 pages.

Krumbein, W. C. (1939) *Preferred orientation of pebbles in sedimentary deposits,* Jour. Geol., vol. 47, p. 673-706.

——— and Pettijohn, F. J. (1938) *Manual of sedimentary petrography,* New York, D. Appleton-Century Co., 549 pages.

Miller, Hugh (1884) *On boulder-glaciation,* Royal Physical Soc. Edinburgh, Pr., vol. 8, p. 156-189.

Miner, Neil (1937) *Evidence of multiple glaciation in the northern part of Yellowstone National Park,* Jour. Geol., vol. 45, p. 636-647.

Richter, Konrad (1932) *Die Bewegungsrichtung des Inlandeis reconstruiert aus den Kritzen und Längsachsen der Geschiebe,* Zeitschrift Geschiebeforschung, Bd. 8, p. 62-66.

——— (1933) *Gefüge und Zusammensetzung des norddeutschen Jungmoränengebietes,* Abh. geol.-paleont. Inst. Greifswald, Bd. 11, p. 1-63.

——— (1936) *Gefügestudien im Engebrae, Fondalsbrae und ihren Vorlandsedimenten,* Zeitschrift Gletscherkunde, Bd. 24, p. 22-30.

Sander, Bruno (1934) *Fortschritte der Gefügekunde der Gestein-Anwendung, Ergebnisse, Kritik,* Fortschritte der Mineralogie, Kristallographie, und Petrographie, vol. 18, p. 111-170.

Sauramo, Matti (1924) *Tracing of glacial boulders and its application in prospecting,* Comm. géol. Finlande, Bull. 67, 37 pages.

Taylor, G. I. (1923) *The motion of ellipsoidal particles in a viscous fluid,* Royal Soc. London, Pr., vol. 103 A, p. 58-61.

Tester, A. C. (1931) *The measurement of the shape of rock particles,* Jour. Sedim. Petrol., vol. 1, p. 3-11.

Twenhofel, W. H. (1932) *Treatise on sedimentation,* Baltimore, Williams and Wilkins, 926 pages.

Upham, Warren (1891) *Criteria of englacial and subglacial drift,* Am. Geol., vol. 8, p. 376-385.

34

Reprinted by permission from *Ann. Assoc. Amer. Geog.*, **55**, 516–519 (1955)

A TECHNIQUE OF MORPHOLOGICAL MAPPING

R. A. G. Savigear

[*Editor's Note:* In the original, material precedes this excerpt.]

THE TECHNIQUE

In composing the more precise technique three major problems had to be considered:

1) the definition of the forms of the ground surface that were to be recorded;
2) the definition of the nature and sizes of the lines and symbols that were to be used to represent these forms on a map;
3) the effects and use of the plotting error of the map.

The Definition of the Ground Surface

The following definitions make it possible to be more precise in defining the character of the components of the ground surface:

A *Flat* is a surface area which is horizontal or is inclined at an angle of less than two degrees.

A *Cliff* is a surface area which is vertical, or is inclined at an angle of forty degrees or more.

A *Slope* is a surface area which is inclined at two or more and less than forty degrees.

A *Facet* is a plane, horizontal, inclined, or vertical surface area.

A *Segment* is a smoothly curved concave (negative) or convex (positive) upwards surface area.

An Irregular Facet or An Irregular Segment is a facet or a segment which possesses distinct surface irregularities that are too small to be represented at the scale of the field map.

A Micro-Facet is a facet whose bounding discontinuities are too close together to be represented separately at the scale of the field map.

A Micro-Segment is a segment whose bounding discontinuities are too close together to be represented separately at the scale of the field map.

A *Morphological Unit* is either a facet, a micro-facet, a segment, or a micro-segment.

A *Break of Slope* is a discontinuity of the ground surface.

An Inflection is the point, line, or zone, of maximum slope between two adjacent concave and convex segments.

The geological concepts of true and apparent dip may also be usefully applied to the definition of surface forms.

True Slope is the direction and amount of maximum surface slope of a facet or a segment.

Apparent Slope is the direction and amount of surface slope of a facet or a segment measured in any direction other than that of the true slope.

[*Editor's Note:* Material has been omitted at this point.]

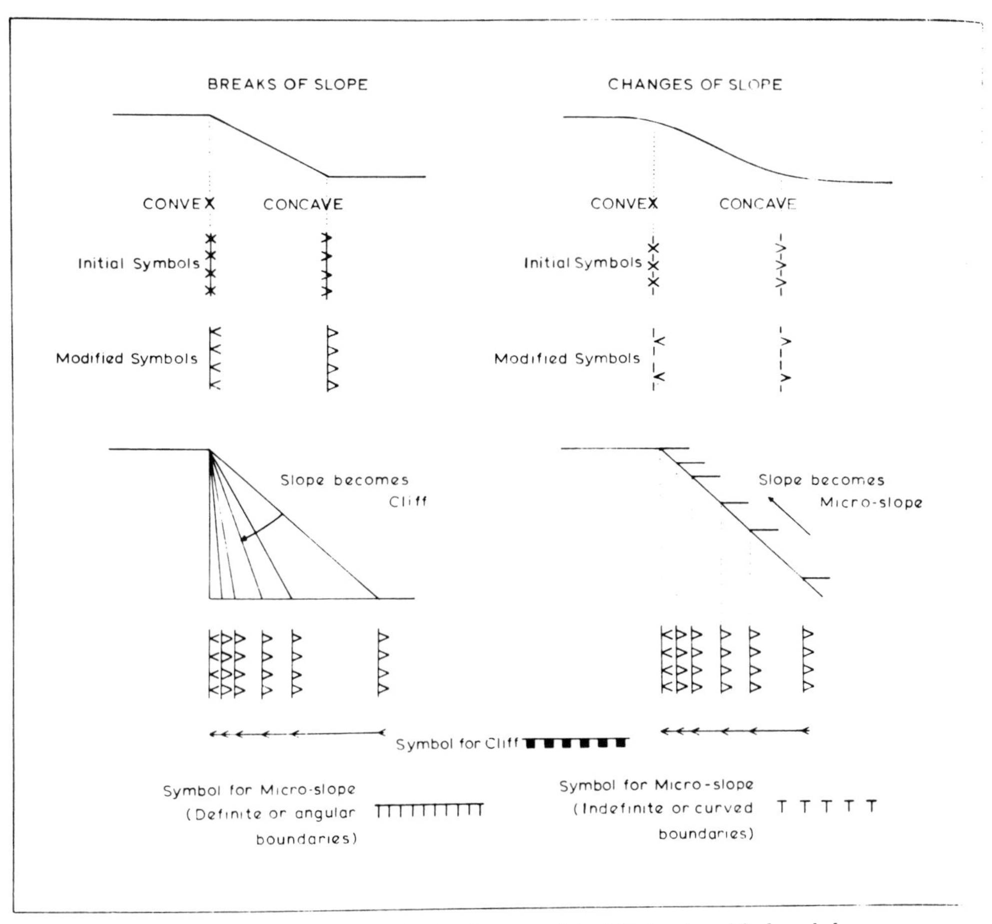

FIG. 1. Diagram illustrating the development of the initial and modified symbols.

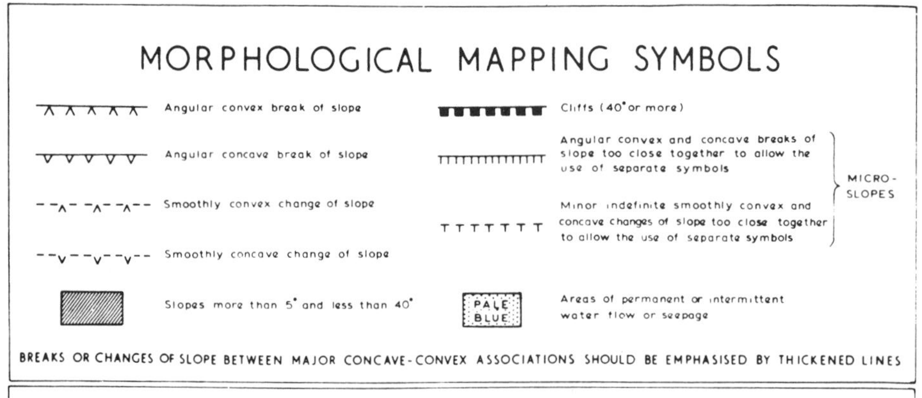

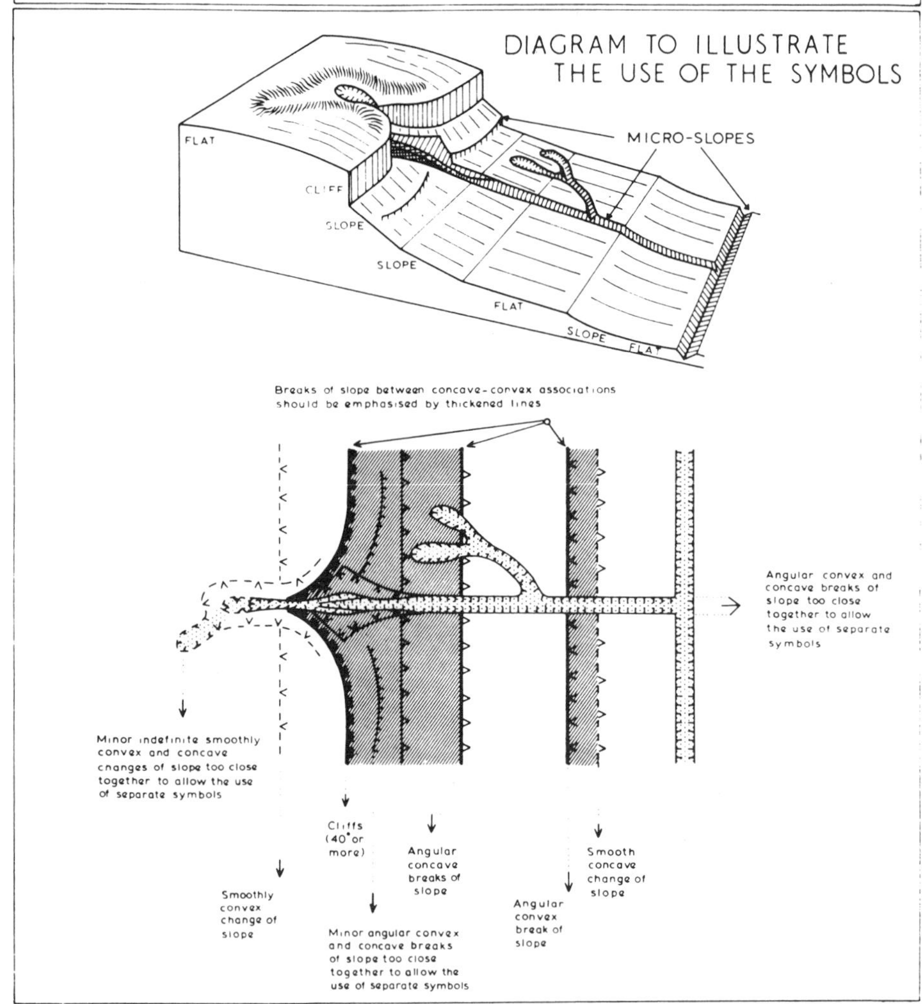

Fig. 2. An illustration of the application of the early symbols to some land form patterns typical of a part of the Southern Pennines, England.

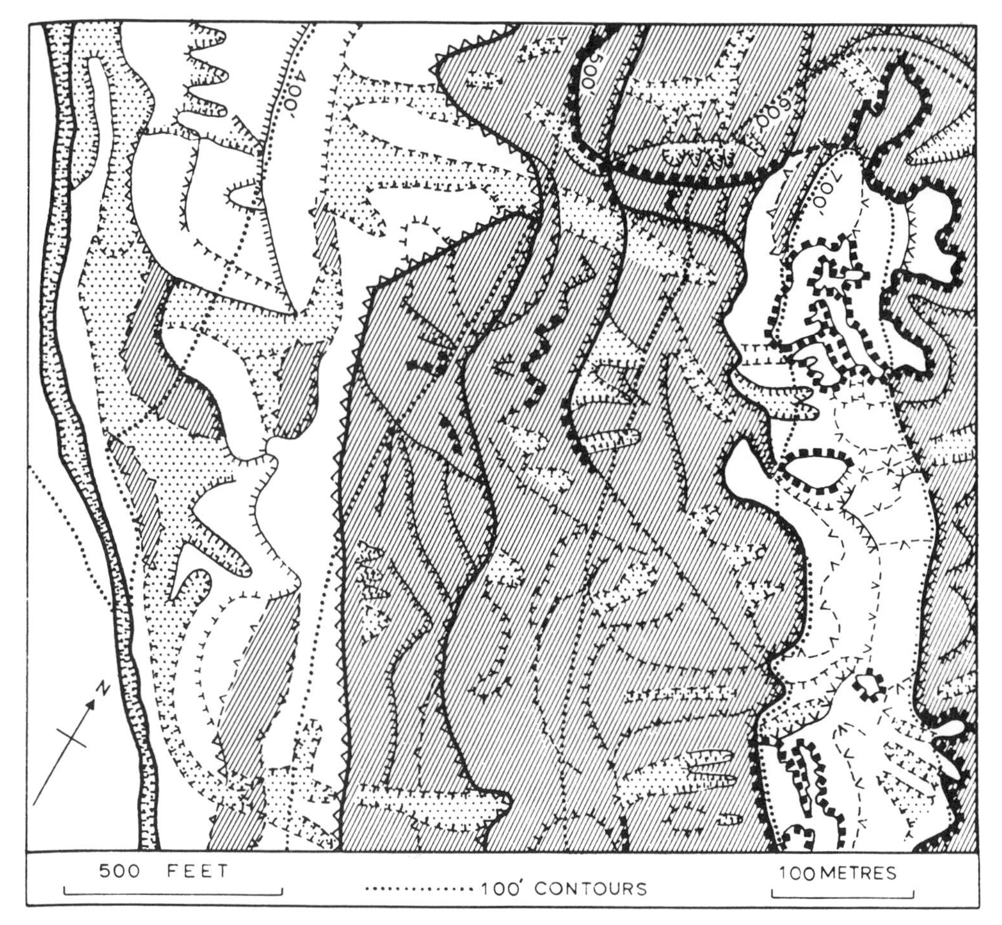

FIG. 3. A morphological map of an area in West Cornwall, England, compiled with the early symbols.

35

Reprinted from *Bull. Geol. Soc. America*, **56**, 281-284, 298-299, 366-370 (1945)

EROSIONAL DEVELOPMENT OF STREAMS AND THEIR DRAINAGE BASINS: HYDROPHYSICAL APPLICATIONS OF QUANTITATIVE MORPHOLOGY

R. E. Horton

[*Editor's Note:* In the original, material precedes this excerpt.]

QUANTITATIVE PHYSIOGRAPHIC FACTORS

GENERAL CONSIDERATIONS

In spite of the general renaissance of science in the present century, physiography as related in particular to the development of land forms by erosional and gradational processes still remains largely qualitative. Stream basins and their drainage basins are described as "youthful," "mature," "old," "poorly drained," or "well drained," without specific information as to how, how much, or why. This is probably the result largely of lack of adequate tools with which to work, and these tools must be of two kinds: measuring tools and operating tools.

One purpose of this paper is to describe two sets of tools which permit an attack on the problems of the development of land forms, particularly drainage basins and their stream nets, along quantitative lines.

An effort will be made to show how the problem of erosional morphology may be approached quantitatively, and even in this respect only the effects of surface runoff will be considered in detail. Drainage-basin development by ground-water erosion, highly important as it is, will not be considered, and the discussion of drainage development by surface runoff will mainly be confined to processes occurring outside of stream channels. The equally important phase of the subject, channel development—including such problems as those of the growth of channel dimensions with increase of size of drainage basin, stream profiles, and stream bends—will not be considered in detail.

STREAM ORDERS

In continental Europe attempts have been made to classify stream systems on the basis of branching or bifurcation. In this system of stream orders, the largest, most branched, main or stem stream is usually designated as of order 1 and smaller tributary streams of increasingly higher orders (Gravelius, 1914). The smallest unbranched fingertip tributaries are given the highest order, and, although these streams are similar in characteristics in different drainage basins, they are designated as of different orders.

Feeling that the main or stem stream should be of the highest order, and that unbranched fingertip tributaries should always be designated by the same ordinal, the author has used a system of stream orders which is the inverse of the European system. In this system, unbranched fingertip tributaries are always designated as of order 1, tributaries or streams of the 2d order receive branches or tributaries of the 1st order, but these only; a 3d order stream must receive one or more tributaries of the 2d order but may also receive 1st order tributaries. A 4th order stream receives branches of the 3d and usually also of lower orders, and so on. Using this system the order of the main stream is the highest.

To determine which is the parent and which the tributary stream upstream from the last bifurcation, the following rules may be used:

(1) Starting below the junction, extend the parent stream upstream from the bifurcation in the same direction. The stream joining the parent stream at

the greatest angle is of the lower order. Exceptions may occur where geologic controls have affected the stream courses.

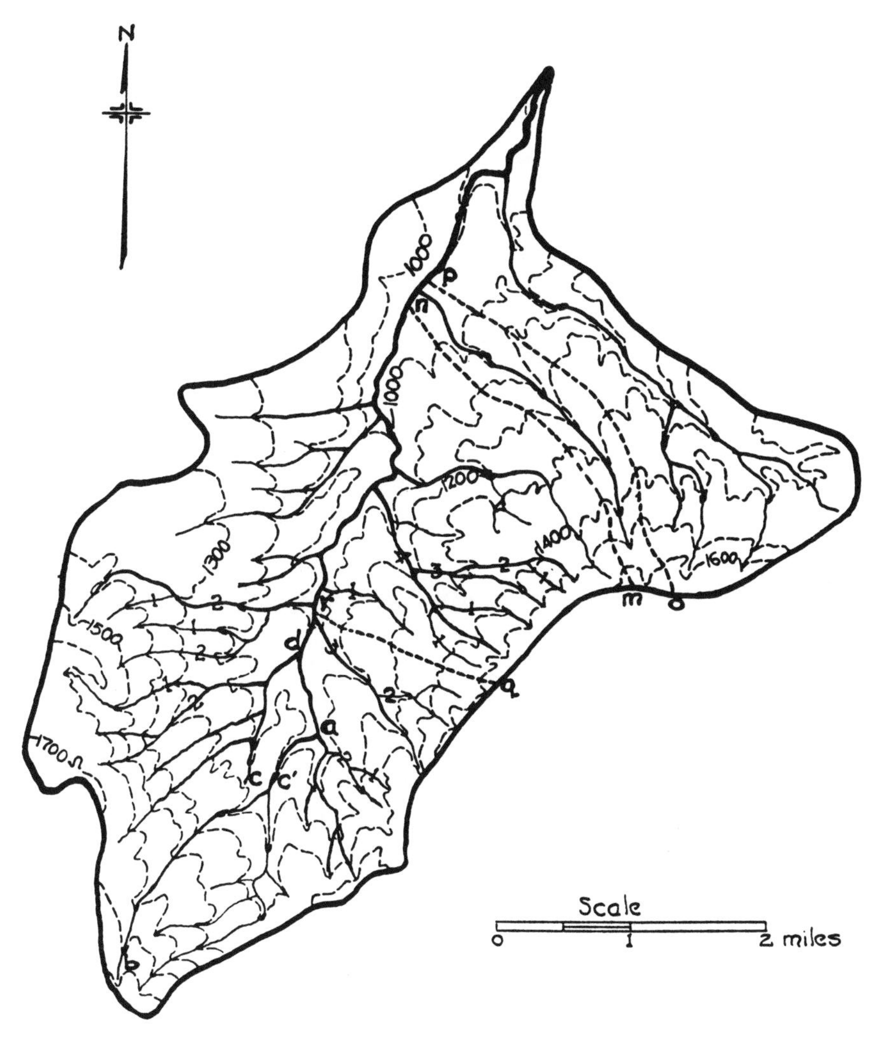

FIGURE 1.—*Well-drained basin*
(Cherry Creek, N. Y., quad., U. S. G. S.)

(2) If both streams are at about the same angle to the parent stream at the junction, the shorter is usually taken as of the lower order.

On Figure 1 several streams are numbered 1, and these are 1st order tributaries. Streams numbered 2 are of the 2nd order throughout their length, both below and above the junctions of their 1st order tributaries. The main stream is apparently *ac′b* although it joins *ad* at nearly a right angle. It is probable that the original course of the stream was *dcb*, but the portion above *dc* was diverted by headwater erosion into stream *ac′*. The well-drained basin (Fig. 1) is of the 5th order, while the poorly drained basin (Fig. 2) is of the 2d order. Stream order therefore affords a simple quantitative basis for comparison of the degree of development in the drainage nets of basins of comparable size. Its usefulness as a basis for such comparisons is limited by the fact that, other things equal, the order of a drainage basin or its stream system generally increases with size of the drainage area.

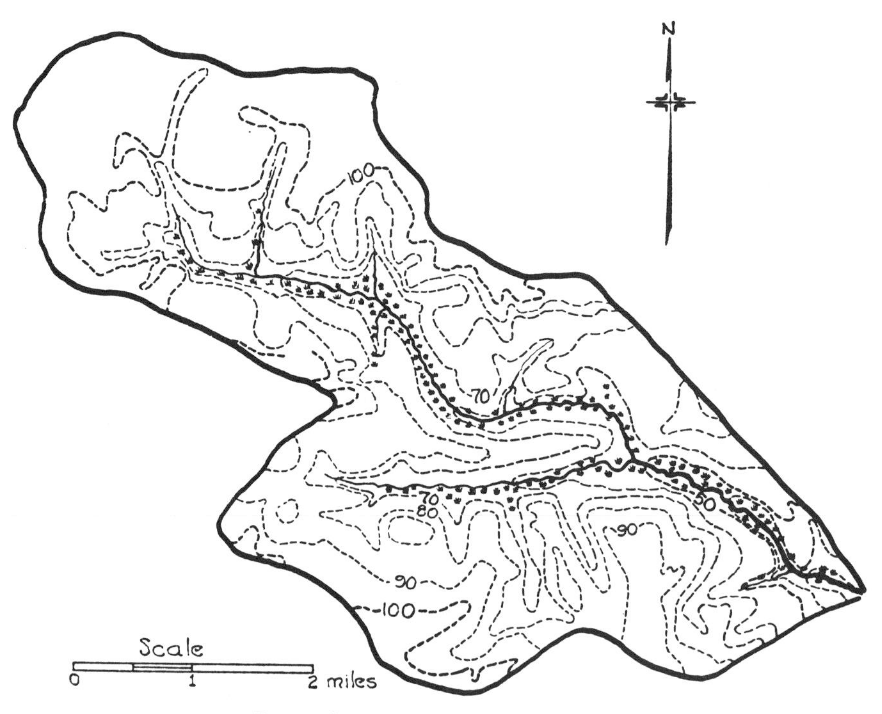

FIGURE 2.—*Flat sandy area, poorly drained*
(Bridgeton, N. J., quad. U. S. G. S.)

DRAINAGE DENSITY

Figures 1 and 2 show two small drainage basins, both on the same scale; one well drained, the other poorly drained. These terms, well drained and poorly drained, while in common use in textbooks on physiography, are purely qualitative, and something better is needed to characterize the degree of drainage development within a basin. The simplest and most convenient tool for this purpose is drainage density or average length of streams within the basin per unit of area (Horton, 1932). Expressed as an equation

$$\text{Drainage density, } D_d = \frac{\Sigma L}{A} \qquad (1)$$

where ΣL is the total length of streams and A is the area, both in units of the same system. The poorly drained basin has a drainage density 2.74, the well-drained one, 0.73, or one fourth as great.

For accuracy, drainage density must, if measured directly from maps, be determined from maps on a sufficiently large scale to show all permanent natural stream channels, as do the U. S. Geological Survey topographic maps. On these maps perennial streams are usually shown by solid blue lines, intermittent streams by dotted blue lines. Both should be included. If only perennial streams were included, a drainage basin containing only intermittent streams would, in accordance with equation (1), have zero drainage density, although it may have a considerable degree of basin development. Most of the work of valley and stream development by running water is performed during floods. Intermittent and ephemeral streams carry flood waters, hence should be included in determining drainage density. Most streams which are perennial in their lower reaches or throughout most of their courses have an intermittent or ephemeral reach or both, near their headwaters, where the stream channel has not cut down to the water table. These reaches should also be included in drainage-density determinations.

In textbooks on physiography, differences of drainage density are commonly attributed to differences of rainfall or relief, and these differences in drainage density are largely used to characterize physiographic age in the sense used by Davis (Davis, 1909; Wooldridge and Morgan, 1937). In the poorly drained area (Fig. 2) the mean annual rainfall is about 30 per cent greater than in the well-drained area (Fig. 1). Therefore some other factor or factors are far more important than either rainfall or relief in determining drainage density. These other factors are infiltration-capacity of the soil or terrain and initial resistivity of the terrain to erosion.

[*Editor's Note:* Material has been omitted at this point.]

If it is assumed that the main stream is of the

4th order, then $l_1 = 1.38$ miles;
5th order, then $l_1 = 0.50$ mile;
6th order, then $l_1 = 0.20$ mile.

Since l_1 is not far from half a mile, the main stream is of the 5th order. From line B (Fig. 8) the number of 2d order streams is 3.15. This is the bifurcation ratio. From line A the lengths of 2d and 1st order streams are, respectively, 1.38 and 0.52 miles. This gives the stream length ratio:

$$r_l = \frac{1.38}{0.52} = 2.70.$$

Data for at least four stream orders are required to determine the order of the main stream from incomplete data by this method. Care must also be used in determining the lines A and B accurately to secure correct results.

The values of the stream lengths as far as known are then plotted on semilog paper (Fig. 8A), in terms of inverse stream orders, a line of best fit drawn to represent the plotted points and this line extended downward to stream length unity or less.

To determine the order of the main stream it is necessary to know the order of magnitude but not the exact value of the average length of streams of the 1st order. The length l_1 of streams of the 1st order is rarely less than a third of a mile, a value which is approached as a minimum limit in mountain regions with heavy rainfall, as in the southern Appalachians. Also it is rarely greater than 2 or 3 miles, values

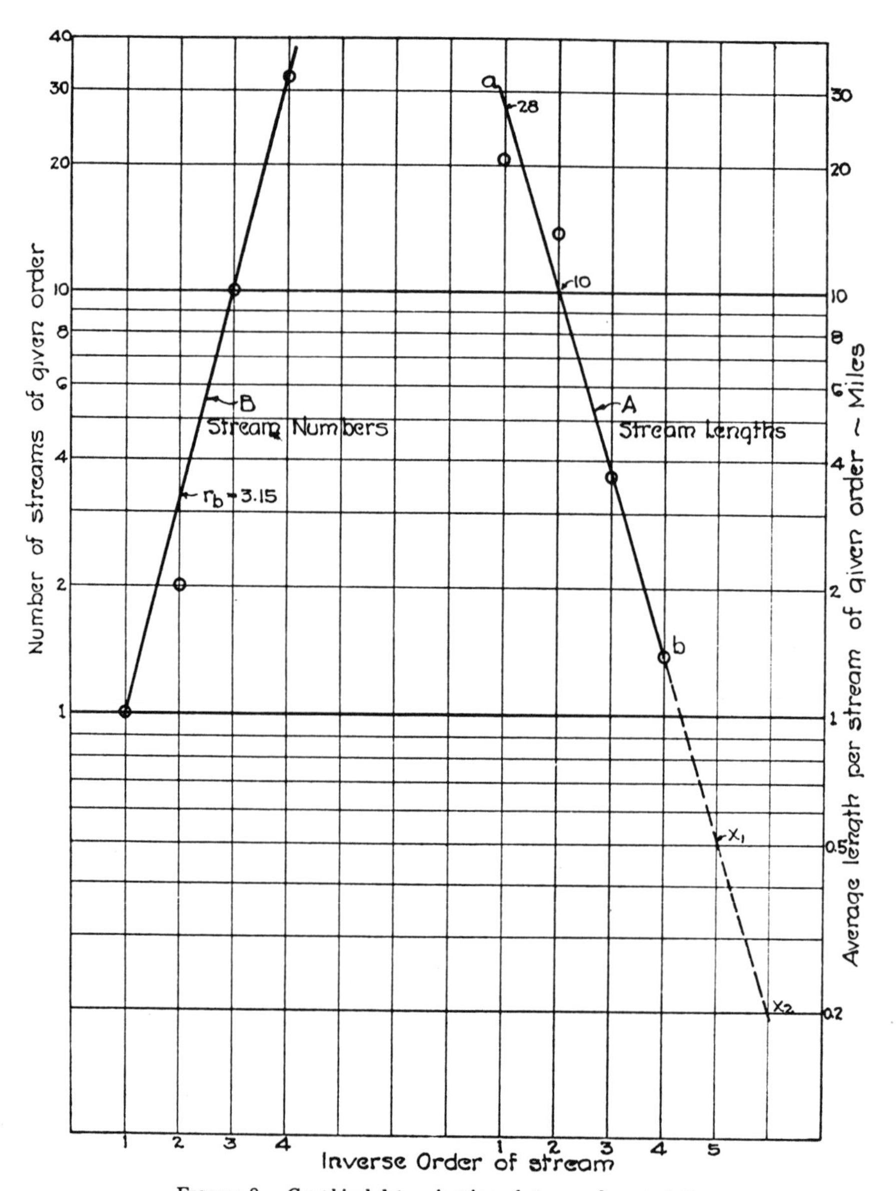

FIGURE 8.—*Graphical determination of stream characteristics*

which are approached as maximum limits under some conditions in arid and semiarid regions. Data from which the order of magnitude of l_1 can be determined are always available from some source. In general all that is required is to know whether l_1 is of the order of half a mile, 1 mile, or 2 miles or more. The point at which the stream length shown by the line *ab* (Fig. 8A) extended downward has a value about the same as the known value of l_1 for the given order indicates the order of the main stream.

This method for determining the order of the main stream is of limited value in some drainage basins, particularly large drainage basins, such as that of the Mississippi River, which are not homogeneous, and where there may be large variations in the length of 1st order streams in different portions of the drainage basins, so that the order of magnitude of l_1 may be difficult to determine. A small portion of a drainage basin, with suitable conditions of high rainfall, steep slopes, etc., may add several units to the value of s for the main stream, although it has little effect on the weighted average value of l_1 for the drainage basin as a whole. For basins which are reasonably homogeneous the method is accurate. Proof of its validity is readily obtained by applying this method to a drainage basin where the values of l_1 and the drainage density D_d have been determined from measurements on a map showing streams of all orders, but using in the determination only the data for streams of higher orders. This was done in preparing Figure 8, which is of the 5th order, although only data for the first four stream orders were used in the computation, it being assumed that l_1 was of an order of magnitude between 1 and 1.5.

This determination of s gives also the average length l_1 of 1st order streams. The bifurcation ratio r_b and the stream-length ratio r_l are determined by the slopes of the lines A and B on Figure 8. It is not necessary to know the order of the main stream to determine these quantities. When r_b, r_l, A, s, and l_1 are known, the drainage density can be determined by means of equation (17).

[*Editor's Note:* Material has been omitted at this point.]

REFERENCES CITED

Davis, W. M. (1909) *Geographical essays*, Ginn & Co., Boston, 777 pages.

Gravelius, H. (1914) *Flusskunde*, Goschen'sche Verlagshandlung, Berlin, 176 pages.

Horton, Robert E. (1932) *Drainage basin characteristics*, Am. Geophys. Union, Tr., p. 350–361.

Wooldridge, S. W., and Morgan, R. S. (1937) *The physical basis of geography*, Longmans, Green & Co., London, 435 pages.

36

Reprinted from *Jour. Geol.*, 77(4), 397, 407–412, 414 (1969)

STREAM LENGTHS AND BASIN AREAS IN TOPOLOGICALLY RANDOM CHANNEL NETWORKS[1, 2]

RONALD L. SHREVE
University of California, Los Angeles, California 90024

ABSTRACT

In order to comprehend the geometry of drainage basins and channel networks, which is prerequisite to explaining their mechanics, it is necessary to understand the close connection between network topology and such planimetric elements as stream lengths and basin areas. Topologically random channel networks constitute an important theoretical case. In an infinite topologically random network, (1) the expected magnitude of a randomly drawn link of order ω is $(2^{2\omega-1} + 1)/3$, (2) $\frac{2}{3}$ of all links and $\frac{1}{6}$ of the interior links head streams, (3) complete subnetworks of any given order have the same distribution of magnitudes as all networks of that order, which explains why it seems to make little difference whether or not basins chosen for investigation are complete, (4) the probability that a randomly drawn stream of order ω will consist of λ links is $2^{-(\omega-1)}(1 - 2^{-(\omega-1)})^{\lambda-1}$, and (5) the average stream of order Ω will have $2^{\Omega-1} - 1$ tributaries entering from the sides, of which $2^{\Omega-\omega-1}$ will on the average be of order ω. Link lengths measured on 1:24,000-scale maps of eastern Kentucky are approximated more closely by a gamma density with parameter $\nu = 2$ than by either a log normal or an exponential. The mean length of the exterior links is almost twice that of the interior links, in agreement with the findings of others. In an infinite topologically random channel network whose interior link lengths are gamma distributed with $\nu = 2$, (1) the densities of the logarithms of the stream lengths are slightly left-skewed and increase in dispersion with increasing order, and (2) the densities of the logarithms of the Schumm total length of all streams in a subbasin of given order are highly symmetrical, decrease slightly in dispersion with increasing order, and for orders 1 and 2 agree well with Schumm's observations at Perth Amboy, New Jersey. If the link lengths are exponentially distributed, then the stream lengths will also be exponentially distributed. In finite topologically random channel networks with specified stream numbers, the expected number of links per stream and per subnetwork of given order are given by closed but complicated formulas, from which the expected stream lengths and Schumm lengths can be calculated. As observed in natural networks, in a topologically random population, (1) the expected stream lengths do not satisfy Horton's geometric-series law as well as the expected stream numbers, whereas the expected Schumm lengths do, (2) on Horton diagrams the curves are straightest for the most probable networks, and are concave upward or downward for networks of order lower or higher, respectively, than the most probable, and (3) the geometric-mean stream-length and Schumm-length ratios are about 2 and 4.5, respectively. The corresponding results for basin areas are exactly the same as for Schumm lengths. In the Perth Amboy basin studied by Schumm, the average area draining directly overland into unit length of channel is not a constant independent of the particular position of the channel in the basin as he proposed, but is less for exterior links than for interior links. Substantial differences in the relative areas draining into sources and in other geomorphic characteristics are not reflected in Horton diagrams because the Horton variables are averages that are mainly determined by network topology. Of most fundamental significance, therefore, are not the Horton variables, but the more elementary quantities, such as link lengths and source areas.

INTRODUCTION

The way in which the planimetric elements of a drainage basin are put together is intimately connected with the topology of the channel network. Indeed, certain geomorphological relationships, such as the laws of drainage composition proposed by Horton (1945, p. 286–291), apparently are in large part the result of randomness in network topology (Shreve 1966, 1967; Smart 1968). Thus, in order to comprehend the geometry of drainage basins and channel networks, which is prerequisite to explaining their mechanics, it is necessary to understand the perhaps dominant topological effects. The purpose of this paper is to investigate these effects in the important theoretical case of stream lengths and basin areas in topologically random channel networks.

[1] Manuscript received September 30, 1968; revised January 16, 1969.

[2] Publication 709, Institute of Geophysics and Planetary Physics, University of California, Los Angeles, California 90024.

[JOURNAL OF GEOLOGY, 1969, Vol. 77, p. 397–414]

[*Editor's Note:* Certain tables and other material have been omitted at this point.]

HORTON'S LAW

Horton's law of stream lengths states that "the average lengths of streams of each of the different orders in a drainage basin tend closely to approximate a direct geometric series in which the first term is the average length of streams of the 1st order" (Horton 1945, p. 291). The "streams of each of the different orders" are those defined by Horton's system of ordering, which involves subjective classification of one of the streams entering each fork as *trunk* and the other as *tributary* (Shreve 1966, p. 21–22). In order to avoid the ambiguity inherent in Horton's system, Strahler (1952, p. 1120) proposed the slightly different system used in this paper, which is now the one most commonly used. Unfortunately, stream lengths defined by Strahler's system do not fit Horton's law very well (Strahler 1952, p. 1137; 1957, p. 915; Broscoe 1959, p. 5; Maxwell 1960, p. 23; Bowden and Wallis 1964, p. 769–770); hence, modification of the content of the law or of the definition of its terms has been proposed by a number of investigators. Strahler (1957, p. 915), for example, suggested changing the law to state that the total length of channel of each order varies inversely as some power of the order. Broscoe (1959, p. 5) and Bowden and Wallis (1964, p. 770) suggested that the geometric-series progression with order could be preserved by substituting the sum of the average stream lengths from the first through a given order, which they termed *cumulative mean length*, for the average stream length of that order. The cumulative lengths are approximations to the Horton lengths when the Strahler system of ordering is used. In a similar vein, Schumm (1956, p. 604) suggested using the average total length of channels in the subbasins of the given order.

At first sight it is surprising that such modifications should seem necessary, because (Mood and Graybill 1963, p. 147) the average stream lengths should cluster around the expected lengths (in the sense that, as the number of measurements increases, the mean observed lengths should approach arbitrarily close to the expected lengths with probability 1), and in infinite topologically random networks, according to (9c), the expected lengths will increase with order as a geometric series just as required by Horton's law. The explanation is that the population whose expectation is given by (9c) is not the one to which Horton's law refers. First, natural networks probably are not strictly topologically random. Such evidence as exists, however, suggests that in many, if not most, areas free of geologic controls they are not far from it (Shreve 1966, p. 31–36; 1967, p. 184–185; Smart, personal communication). Second, as already mentioned, the mean length of the exterior links in a basin generally is significantly greater than that of the interior links. This fact probably explains some of the concave-upward curvature in Horton diagrams of logarithm of stream length versus order reported by Broscoe (1959, p. 5) and Maxwell (1960, p. 62, fig. 11; note that text on p. 23 contradicts the figure). Third, and most important, regardless of which system of ordering is used, Horton's law applies to average stream lengths in finite networks, whereas (9c) refers to average stream lengths in infinite networks. Thus, comparison with (9c) is not necessarily proper, except as an indicator of general behavior (as in Shreve 1967, p. 184).

Expected average Strahler stream lengths in finite topologically random channel networks with given order and magnitude can be calculated in terms of the expected interior and exterior link lengths $\mathbf{E}(l_i)$ and $\mathbf{E}(l_e)$ independently of the specific distributions of the link lengths, assuming as before that the distribution of interior link lengths is independent of order, magnitude, or any other characteristic. With this assumption, the expected average length of streams of order 1 is $\mathbf{E}(l_e)$, and that of streams of order $\omega > 1$ is equal to the product of $\mathbf{E}(l_i)$ and the expected average number of links per stream of order ω. The expected average number of links per stream in turn is the average of $\mathbf{E}(\boldsymbol{\nu}_\omega)$ over all of the possible sets of stream numbers, weighted according to the probability of occurrence of each set, where

$$\mathbf{E}(\boldsymbol{\nu}_\omega) = \prod_{\beta=2}^{\omega}(n_{\beta-1} - 1) \Big/ (2n_\beta - 1)\,, \qquad (16)$$

$$\omega = 2, 3, \ldots, \Omega\,,$$

(Smart 1968, p. 1007) is the expected number of links per stream of order ω in networks with stream numbers $n_1, n_2, \ldots, n_{\Omega-1}, 1$. For a topologically random population of networks of given magnitude and order, the sets of stream numbers and their probabilities can be computed by means of the algorithm and formulas given in a previous paper (Shreve 1966, p. 29, 31).

The expected average total channel lengths of Schumm can be calculated in similar fashion from the weighted average of $\mathbf{E}(\xi_\omega)$ over the possible sets of stream numbers, where

$$\mathbf{E}(\xi_\omega) = 1 + \sum_{\beta=2}^{\omega}\prod_{\alpha=2}^{\beta} 2(n_{\alpha-1} - 1) \Big/ (2n_\alpha - 1)\,, \qquad (17)$$

$$\omega = 2, 3, \ldots, \Omega\,,$$

is the expected number of links per subnetwork of order ω in networks with stream numbers $n_1, n_2, \ldots, n_{\Omega-1}, 1$. The expected total length per subnetwork of order ω is then $\frac{1}{2}\mathbf{E}(l_e)[\mathbf{E}(\xi_\omega) + 1] + \frac{1}{2}\mathbf{E}(l_i)[\mathbf{E}(\xi_\omega) - 1]$ from Melton's relationships (Melton 1959, p. 345; Shreve 1966, p. 27).

The cumulative mean lengths of Broscoe

can be computed simply by summation of the expected average Strahler stream lengths.

To derive (16) and (17), consider the process of constructing topologically random networks with given stream numbers $n_1, n_2, \ldots, n_{\Omega-1}$, 1 by starting with the single main stream and in cycles adding the streams of successively lower order as done by Shreve [1966, p. 29] and in the original derivation of (16) by Smart [1968, p. 1005–1007]. If after the αth cycle the expected number of links in streams of some given order ω is $\mathbf{E}_\alpha(\mathbf{v}_\omega)$, then after the next cycle it will be

$$\mathbf{E}_{\alpha+1}(\mathbf{v}_\omega) = \mathbf{E}_\alpha(\mathbf{v}_\omega)R_\alpha \,, \quad R_\alpha = (n_{\Omega-\alpha} - 1)/(2n_{\Omega-\alpha+1} - 1) \,, \quad \alpha = \Omega - \omega + 1 \,, \; \Omega - \omega + 2, \ldots, \Omega - 1 \,, \quad \omega = 2, 3, \ldots, \Omega \,, \tag{18a}$$

counting addition of the streams of order $\Omega - 1$ as the second cycle. These equations express the fact that in the construction of topologically random networks, in which all topologically distinct arrangements are equally likely, the expected number of links in individual streams of a given order increases in direct proportion to the total number of links of that order and higher. Streams of order ω will consist of a single link when $\alpha = \Omega - \omega + 1$; hence, by induction,

$$\mathbf{E}(\mathbf{v}_\omega) = \prod_{\alpha=\Omega-\omega+1}^{\Omega-1} R_\alpha \,, \quad \omega = 2, 3, \ldots, \Omega \,, \tag{18b}$$

from which (16) follows by letting $\beta = \Omega - \alpha + 1$ and reversing the order of multiplication.

Similarly, if after the αth cycle the expected number of links in subnetworks of order ω is $\mathbf{E}_\alpha(\xi_\omega)$, then after the next cycle it will be

$$\mathbf{E}_{\alpha+1}(\xi_\omega) = \mathbf{E}_\alpha(\xi_\omega) + 2\mathbf{E}_\alpha(\xi_\omega) \times (n_{\Omega-\alpha} - 2n_{\Omega-\alpha+1})/(2n_{\Omega-\alpha+1} - 1) + \mathbf{E}_\alpha(\xi_\omega) + 1 \,, \quad \alpha = \Omega - \omega + 1 \,, \; \Omega - \omega + 2, \ldots, \Omega - 1 \,, \quad \omega = 2, 3, \ldots, \Omega \,. \tag{19a}$$

The second term on the right-hand side is the increase in the expected number of links due to addition of new tributary links along the sides of the already existing streams; and the sum of the last two terms is, from Melton's relationships (Melton 1959, p. 345; Shreve 1966, p. 27), the increase due to addition of two new tributary links at the head of each stream. As before, subnetworks that will ultimately be of order ω will consist of a single link when $\alpha = \Omega - \omega + 1$; hence, by induction, combining the terms in (19a),

$$\mathbf{E}(\xi_\omega) = 1 + 2R_{\Omega-1}\{1 + 2R_{\Omega-2} \times [1 + \ldots + 2R_{\Omega-\omega+2} \times (1 + 2R_{\Omega-\omega+1}) \ldots]\} \,, \quad \omega = 2, 3, \ldots, \Omega \,, \tag{19b}$$

from which (17) follows by successively eliminating the parenthetical expressions.

Figures 12, 13, and 14 show Horton diagrams of the expected stream lengths and numbers for topologically random populations of networks of various magnitudes and orders. Certain generalizations are immediately apparent. First, as in natural networks, the Strahler lengths do not satisfy Horton's law nearly so well as do the Strahler numbers, whereas the Broscoe and Schumm lengths do. Second, the curves are straightest for the most probable networks, as observed in natural networks by Smart (1968, p. 1007), and, except for the first-order lengths, are concave upward for networks whose order is less than the most probable, that is, whose geometric-mean bifurcation ratio is less than about 4 (Shreve 1966, p. 31), and the converse. Finally, for μ given and $\mathbf{E}(l_e)/\mathbf{E}(l_i)$ in the range from 1 to 2, the geometric-mean Strahler length ratio, which, incidentally, is preferable to the arithmetic-mean ratio sometimes used, is close to 2, and the Broscoe and Schumm ratios are close to 2.5 and 4.5, respectively, in approximate agreement with observed values (Schumm 1956, p. 604–605).

BASIN AREAS

Although Horton did not specifically include basin areas in his laws of drainage composition, he implied that they should satisfy a geometric-series law like stream numbers and lengths (Horton 1945, p. 294; Schumm 1956, p. 606). This *law of basin areas*, as it was subsequently formulated by Schumm (p. 606) in the style of Horton, states that "the mean drainage-basin areas of streams of each order tend to approximate closely a direct geometric series in which the first-order term is the mean area of the first-order basins." This law not only is identical to the law of stream lengths but also can be treated theoretically in exactly the same way.

Let $\boldsymbol{a}_f$ and $\boldsymbol{a}_i$ be, respectively, the first-order areas and the double-triangular areas draining directly overland into individual interior links. Like link lengths, these areas are random variables. Judging from the data of Schumm (p. 607, 609), they are distributed according to right-skewed densities very similar to those for link lengths. In the absence of better in-

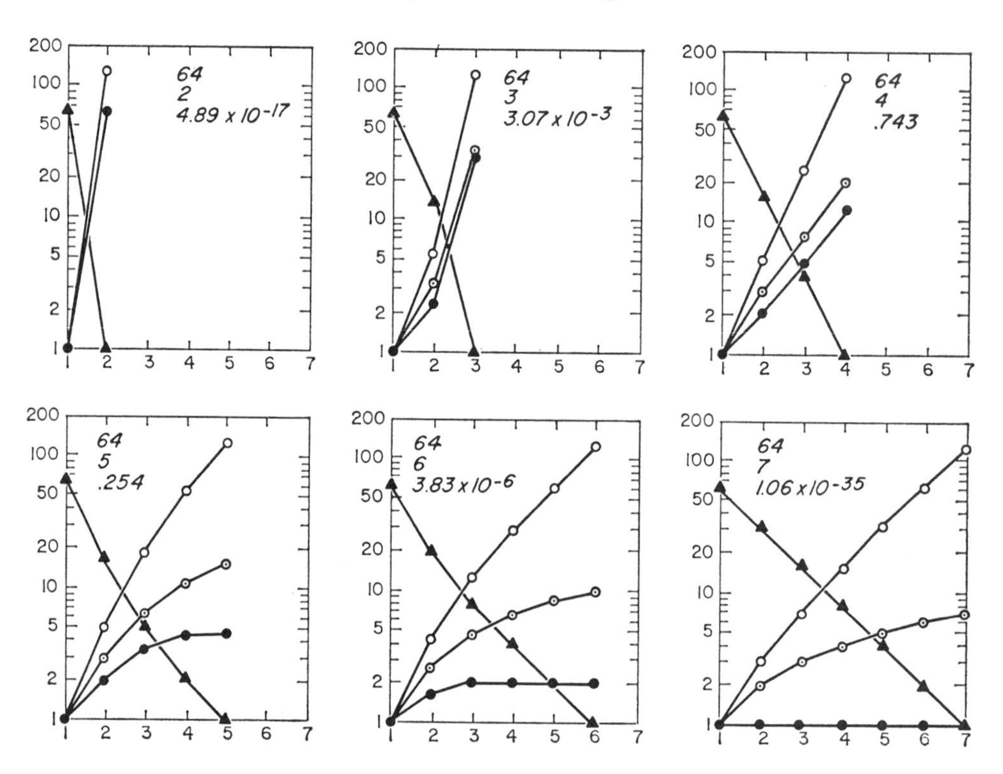

FIG. 12.—Horton diagrams of expected stream lengths and numbers for topologically random populations of networks of magnitude 64 in which expected interior and exterior link lengths are equal. Strahler stream lengths shown by solid circles, Broscoe lengths by circles with dots, Schumm lengths by open circles, and stream numbers by triangles. Unit of stream length is expected interior link length. Horizontal coordinate is stream order. The three numbers in each diagram are, from top to bottom, network magnitude, network order, and probability that a network of the given magnitude will have the specified order.

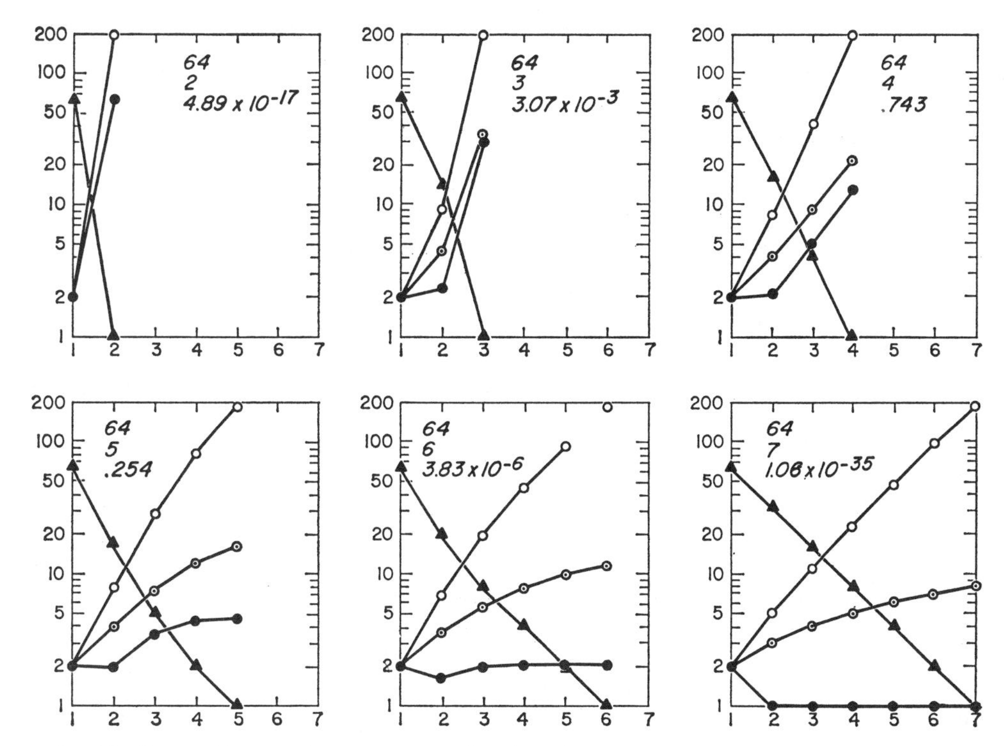

FIG. 13.—Horton diagrams of expected stream lengths and numbers for topologically random populations of networks of magnitude 64 in which expected exterior link lengths are twice expected interior link lengths. Coordinates and symbols same as in figure 12.

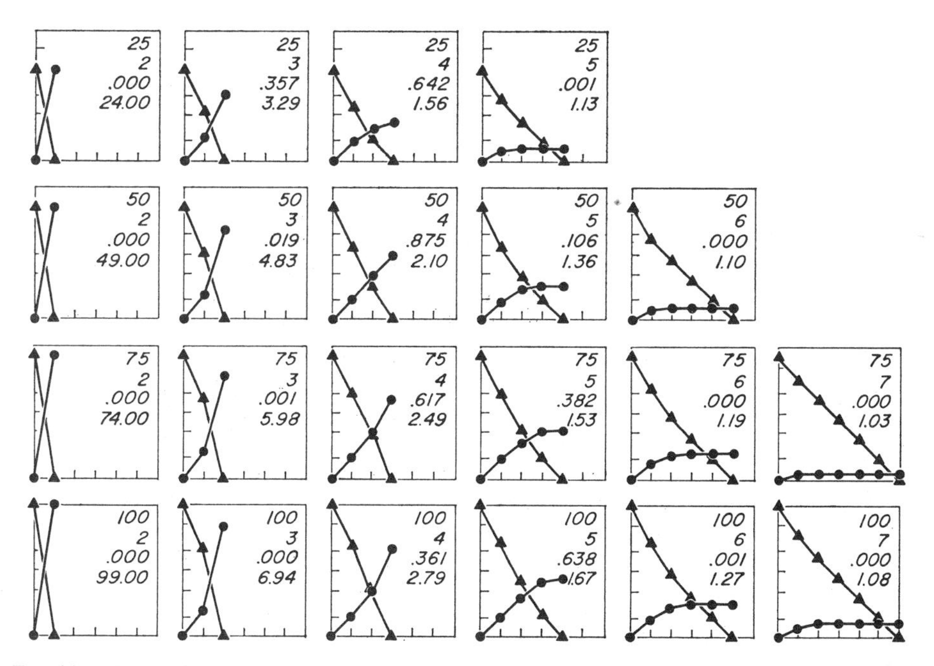

FIG. 14.—Horton diagrams of expected Strahler stream lengths and numbers for topologically random populations of networks of magnitude 25, 50, 75, and 100 in which expected interior and exterior link lengths are equal. Coordinates and symbols same as in figure 12. Fourth number in each diagram is the geometric-mean length ratio.

formation, therefore, it seems justifiable as a first approximation to assume the same type of densities, namely, gamma densities with $\nu = 2$ and expectations $\mathbf{E}(\boldsymbol{a}_f) = 2/\alpha_f$ and $\mathbf{E}(\boldsymbol{a}_i) = 2/\alpha_i$. With this assumption all of the previous results for Schumm total stream lengths are directly applicable to basin areas.

The theoretical densities of basin areas in infinite topologically random channel networks are shown in figures 9 and 10, and are compared with Schumm's histograms in figure 15. Again, the agreement between

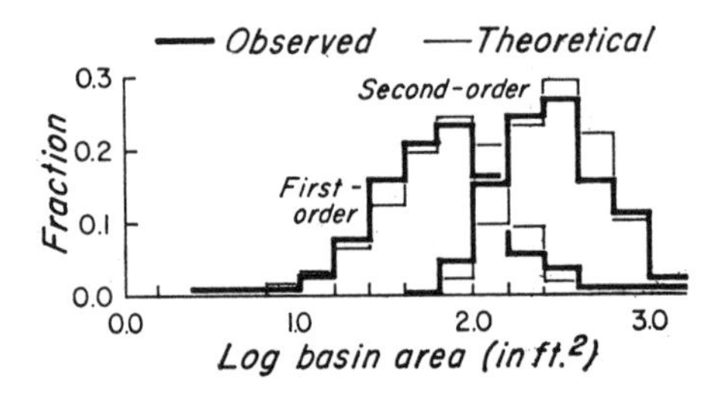

FIG. 15.—Comparison of theoretical and observed histograms of subbasin area for the Perth Amboy, New Jersey, badlands basin studied by Schumm (1956, p. 607). Assumptions same as for figure 9.

the calculated and the observed histograms is highly encouraging.

Incidentally, in the table given by Schumm (p. 608, table 4), the frequencies for the first-order areas are one column to to the left of their correct positions.

Just as in the case of link lengths, the observed mean interior area $\bar{a}_i$ needed as the estimator of $\mathbf{E}(\boldsymbol{a}_i)$ was calculated using the equation

$$A = \mu\bar{a}_f + (\mu - 1)\bar{a}_i\,, \qquad (20)$$

where $\bar{a}_f$ and A are the mean first-order and total areas in a basin of magnitude μ (tables 1 and 2). Using these estimators in the equation analogous to (13i) gives an expected second-order area of 376 ft^2, which may be compared with the average (for forty-five subbasins) of 343 ft^2 found by Schumm (p. 606, table 2).

Horton diagrams of the expected areas in finite topologically random channel networks are shown in figures 12, 13, and 16. Like expected lengths, expected areas do not depend upon the specific distributions of the first-order and interior areas, but only upon their expectations, provided as before that the distribution of interior areas is independent of order, magnitude, or any other characteristic. The diagrams show that the expected areas satisfy Horton's law rather well. The curves are straightest for the most probable networks, that is, for those whose geometric-mean bifurcation ratio is about 4 (Shreve 1966, p. 31), and are concave upward for networks whose order is less than the most probable, and the converse. Finally, the geometric-mean area ratio for the expected areas in the most probable networks is about 4.5, in approximate agreement with observed values (Schumm 1956, p. 604–605).

[*Editor's Note:* Certain tables and other material have been omitted at this point.]

REFERENCES CITED

Aitchison, J., and Brown, J. A. C., 1957, The lognormal distribution: New York, Cambridge Univ. Press, 176 p.

Bowden, K. L., and Wallis, J. R., 1964, Effect of stream-ordering technique on Horton's laws of drainage composition: Geol. Soc. America Bull., v. 75, p. 767–774.

Broscoe, A. J., 1959, Quantitative analysis of longitudinal stream profiles of small watersheds: New York, Columbia Univ. Dept. Geology, Office of Naval Research Project NR 389-042, Tech. Rept. no. 18, 73 p.

Chapman, D. G., 1956, Estimating the parameters of a truncated gamma distribution: Annals of Mathematical Statistics, v. 27, p. 498–506.

Dwight, H. B., 1961, Tables of integrals and other mathematical data: New York, Macmillan Co., 336 p.

Feller, W., 1957, An introduction to probability theory and its applications: New York, John Wiley & Sons, v. 1, 461 p.

——— 1966, An introduction to probability theory and its applications: New York, John Wiley & Sons, v. 2, 626 p.

Horton, R. E., 1945, Erosional development of streams and their drainage basins; hydrophysical approach to quantitative morphology: Geol. Soc. America Bull., v. 56, p. 275–370.

Leopold, L. B., and Langbein, W. B., 1962, The concept of entropy in landscape evolution: U.S. Geol. Survey Prof. Paper 500–A, p. A1–A20.

Maxwell, J. C., 1960, Quantitative geomorphology of the San Dimas Experimental Forest, California: New York, Columbia Univ. Dept. Geology, Office of Naval Research Project NR 389-042, Tech. Rept. no. 19, 95 p.

Melton, M. A., 1957, An analysis of the relations among elements of climate, surface properties, and geomorphology: New York, Columbia Univ. Dept. Geology, Office of Naval Research Project NR 389-042, Tech. Rept. no. 11, 102 p.

——— 1959, A derivation of Strahler's channel-ordering system: Jour. Geology, v. 67, p. 345–346.

Mood, A. M., and Graybill, F. A., 1963, Introduction to the theory of statistics (2d ed.): New York, McGraw-Hill Book Co., 443 p.

Schumm, S. A., 1956, Evolution of drainage systems and slopes in badlands at Perth Amboy, New Jersey: Geol. Soc. America Bull., v. 67, p. 597–646.

Shreve, R. L., 1966, Statistical law of stream numbers: Jour. Geology, v. 74, p. 17–37.

——— 1967, Infinite topologically random channel networks: Jour. Geology, v. 75, p. 178–186.

Smart, J. S., 1968, Statistical properties of stream lengths: Water Resources Research, v. 4, p. 1001–1014.

———; Surkan, A. J.; and Considine, J. P., 1967, Digital simulation of channel networks, *in* Symposium on river morphology: Internat. Assoc. Sci. Hydrology Pub. 75, p. 87–98.

Strahler, A. N., 1952, Hypsometric (area-altitude) analysis of erosional topography: Geol. Soc. America Bull., v. 63, p. 1117–1142.

——— 1957, Quantitative analysis of watershed geomorphology: Am. Geophys. Union Trans., v. 38, p. 913–920.

37

Reprinted from *Bull. Geol. Soc. America,* 63, 923–925, 934–937 (1952)

DYNAMIC BASIS OF GEOMORPHOLOGY

By Arthur N. Strahler

Abstract

To place geomorphology upon sound foundations for quantitative research into fundamental principles, it is proposed that geomorphic processes be treated as gravitational or molecular shear stresses acting upon elastic, plastic, or fluid earth materials to produce the characteristic varieties of strain, or failure, that constitute weathering, erosion, transportation and deposition.

Shear stresses affecting earth materials are here divided into two major categories: gravitational and molecular. Gravitational stresses activate all downslope movements of matter, hence include all mass movements, all fluvial and glacial processes. Indirect gravitational stresses activate wave- and tide- induced currents and winds. Phenomena of gravitational shear stresses are subdivided according to behavior of rock, soil, ice, water, and air as elastic or plastic solids and viscous fluids. The order of classification is generally that of decreasing internal resistance to shear and, secondarily, of laminar to turbulent flow.

Molecular stresses are those induced by temperature changes, crystallization and melting, absorption and desiccation, or osmosis. These stresses act in random or unrelated directions with respect to gravity. Surficial creep results from combination of gravitational and molecular stresses on a slope. Chemical processes of solution and acid reaction are considered separately.

A fully dynamic approach requires analysis of geomorphic processes in terms of clearly defined open systems which tend to achieve steady states of operation and are self-regulatory to a large degree. Formulation of mathematical models, both by rational deduction and empirical analysis of observational data, to relate energy, mass, and time is the ultimate goal of the dynamic approach.

CONTENTS

ILLUSTRATIONS

Introduction

The aim of this paper is to outline a system of geomorphology grounded in basic principles of mechanics and fluid dynamics, that will enable geomorphic processes to be treated as manifestations of various types of shear stresses, both gravitational and molecular, acting upon any type of earth material to produce the varieties of strain, or failure, which we recognize as the manifold processes of weathering, erosion, transportation and deposition. The concepts set forth in this paper have been established as guiding principles underlying the quantitative investigation of erosional landforms by the writer and his associates under Contract N6 ONR 271, Task Order 30, Project No. NR. 089–042, Office of Naval Research, Geography Branch. The writer is grateful to Professor W. C. Krumbein of Northwestern University and Professors Sidney Paige and Donald Burmister of Columbia University for their kindness in reading the manuscript and suggesting various improvements, and to Mr. Samuel Katz of the Lamont Geological Observatory for developing the mathematical analysis of river profiles.

For more than half a century, the study of landforms in North America was dominated by an explanatory-descriptive method of study used by W. M. Davis and his students. Davis himself maintained that the aims of his method were geographic; that the consideration of process was introduced merely to permit an orderly genetic system of landform classification. The weakness in understanding of geomorphic processes (and hence also a weakness in the understanding of the origin of landforms) has not been confined to the American continent. The geomorphologists of France and England, closely attached to schools and departments of geography, have also tended to give much attention to descriptive, deductive studies of landform development and to regional geomorphological treatments. Even in so distinctively different a treatment as Walther Penck's morphological analysis, processes and forms are analyzed by a deductive method safely removed from the reality of existing landforms and without mention of basic principles of soil mechanics and fluid dynamics.

If geomorphology is to achieve full stature as a branch of geology operating upon the frontier of research into fundamental principles and laws of earth science, it must turn to the physical and engineering sciences and mathematics for vitality which it now lacks. The geomorphic processes that we observe are, after all, basically the various forms of shear, or failure, of materials which may be classified as fluid, plastic, or elastic substances, responding to stresses which are most commonly gravitational, but may also be molecular.

Unless the fundamental nature of materials is understood, we are in a poor position to add anything worthwhile to what is already largely self-evident concerning the behavior of streams, landslides, glaciers, or wave-induced currents. We cannot hope for anything better than a superficial knowledge of the form and motion of a sand dune unless we can interpret the dune in terms of aerodynamic principles; or of a stream profile unless we understand the principles of fluid dynamics and the transportation of sediment; or of the moulding of a drumlin unless we study the flowage and fracturing of ice; or of the production of an off-shore bar unless we know something of the dynamics of waves and the transport of sediment by oscillating or pulsating wave-induced currents; or of the causes of a great earth-flow unless we can appreciate the principles of plastic flowage.

To delegate to the civil engineer all fundamental research on geomorphic processes and forms has certain disadvantages. With his attention focused upon problems dealing with man-imposed modifications of the natural landscape, the engineer may have neither the time nor the inclination to investigate a broad range of natural phenomena where they are best displayed. Furthermore he is likely to have only a limited acquaintance with geologic materials and forms, whereas the geomorphologist, trained as a geologist, has built up a life-time store of information and experience, much of it relating to theoretical and historical aspects of geology.

Although the study of fundamental principles of geomorphology by engineers is to be welcomed and encouraged, there is a real danger that the engineer will find it necessary to take over an increasingly greater proportion of geomorphic research and thereby cut it off from its most logical parent, the field of theoretical geology. A specific example is the field of landslides and related gravity movements. The geomorphologists now owe their most penetrating analysis of the fundamental principles of these phenomena to research by specialists in soil physics and soil mechanics. Karl Terzaghi's (1950) work is outstanding in this respect. Far from being a supplier of basic theoretical knowledge to civil engineers in this field, as he should be, the geomorphologist has been receiving this information from them.

It is appropriate in this paper, which is philosophical in nature, to clarify the function of time in geomorphology. Two quite different viewpoints are used in dynamic (analytical) geomorphology and in historical (regional) geomorphology. The student of processes and forms *per se* is continually asking "What happens?"; the historical student keeps raising the question "What happened?". Bucher (1941) has aptly labeled the two types of geological information as *timeless* and *timebound* knowledge respectively. It is largely with the timeless knowledge that the field of dynamic geomorphology deals. The principles are usually most easily discovered by a study of contemporary processes and existing forms, but the dynamic geomorphologist will refer to any part of the geologic record for evidence which will increase his understanding. He is not, however, primarily concerned with the actual series of events within a particular geologic period and in a particular geographical location, as is the historical geomorphologist. Although many historical-regional geomorphic investigations have been ably conducted with minimum reference to geomorphic processes, it is only reasonable to suppose that a better knowledge of how processes operate and normal forms evolve will increase the effectiveness of historical studies and reduce the likelihood of drawing erroneous inferences of past events.

On the other hand, studies of dynamic geomorphology based on existing landforms cannot be prosecuted in ignorance or disregard of past changes of climate, and hence, of relative rates and importance of various processes dependent on climate, during the long period required to develop the forms. One cannot, for example, extrapolate rainfall intensity-frequency-duration statistics of the past 50 years back over a period of 50,000 years in correlating drainage basin forms with rainfall characteristics. The dynamic geomorphologist must be alert to evidences of important differences in processes operating during earlier stages of development of the forms he is studying. Thus historical-regional geomorphic treatment cannot be divorced from dynamic investigations. The difference in the two types of study lies in proportion of each one involved, for neither can successfully be pursued independently of the other.

Dynamic Open Systems and the Steady State

Geomorphology will achieve its fullest development only when the forms and processes are related in terms of dynamic systems and the transformations of mass and energy are considered as functions of time (von Bertalanffy, 1950a; 1950b). Walling himself off in sacrosanct confines of his geological societies and journals, the geomorphologist has paid little attention to the development of thermodynamic principles and their steady infiltration from pure physics and chemistry into sciences of biology, economics, psychology, and political science. True, the Davisian concept of cycle related changes of form to changes of time, but the treatment is not based upon mathematical law, and no thought is given to energy relationships, despite the inviting opportunities for such treatment.

Many of the geomorphic processes operate in clearly defined systems that can be isolated for analysis. A drainage system—whether of water or ice—within the geographical confines of a watershed represents such a dynamic system. A cross-sectional belt of unit width across a shore line or sand dune, or down a given slope from divide to stream channel, would constitute another, more limited, type of dynamic system.

Two major types of thermodynamic systems may be recognized (von Bertalanffy, 1950a): (1) the *closed system* which has a clearly defined boundary through which neither materials nor energy are exchanged, and (2) the *open system* which exchanges either material or energy (or both) with outside environments. The closed system tends to establish an equilibrium in which entropy attains the maximum, available free energy the minimum. An example may be found in the state of water vapor in the air standing above a water surface in a sealed jar. If no heat flows into or out of the jar, the water vapor will attain a certain concentration, maintained without further change of temperature or pressure. The only activity in this system will be exchange of a few molecules between the gas and water. Such a system obviously does not describe a stream or glacier where motion continues with time and material is continually entering and leaving the system.

Form and composition of the open system depend upon the continuous import and export of materials and energy. Normally a time-independent steady state is achieved in which the form remains unchanged but the activity continues. In the case of a segment of a stream in uniform flow, a steady state ensues when the energy developed through the descent of the water is entirely dissipated in overcoming resistance to shear within the fluid and against the channel boundary and to the movement of bed load. The discharge is constant throughout this stretch of stream, and there is no acceleration of motion except what is accountable to changes of channel form, roughness, or slope. In this steady state, the form of the stream is unchanging with time, but should the import of water be cut off the form will be destroyed at once. Open systems require energy from outside to maintain a steady state; equilibrium of the closed system requires none.

Open systems such as streams or glaciers, or cells of living matter, are able to adjust internally to changes in supplies of material or energy from outside. The open system is, in other words, a self-regulating mechanism (von Bertalanffy, 1950a). When a stream is graded, it is in a steady state. If the bed-load supply is reduced or increased, the stream changes its slope so as to readjust to a new steady state. A shore segment whose beach possesses the "profile of equilibrium" is in a steady state with respect to the energy supplied by breaking waves. When the wave characteristics change, the beach profile is altered in slope until a new profile, independent of time, is established.

Mathematical Models in Geomorphology

In attempting to quantify his statements of geomorphic process and form, the geomorphologist has, in general, two types of mathematical procedure open to him, both fruitful. He may, by statistical analysis of experimental and sample field data, derive empirical equations that best state the observed interrelationships between two variables. In its first form, this sort of empirical equation states the degree of correlation between two form elements, both of which may be products of a third and unknown independent variable. For example one might relate drainage density to length of overland flow in a statistical correlation. Although neither quality of terrain is a cause of the other, the degree of correlation is very close because both qualities are controlled by a third, independent factor. Aside from predicting the magnitude of one form element when the other is known, this mathematical statement is of limited value for it does not improve understanding of the genesis of the landform.

In a higher form, the empirical equation may describe a regression in which the independent variable is a force, or time itself, whereas the dependent variable is a landform element. For example, if drainage density were plotted against surface resistivity, a close but inverse relationship would be found. Here resistivity, or resistive force, is a cause; drainage density an effect. Analysis of this type is found in general engineering practice; it is, in fact, the only way in which carefully observed sets of values can be impartially and objectively related.

As a second general procedure, the geomorphologist may formulate, through a type of invention or intuition based upon the sum total of his experience, a relatively simple mathematical model (Rafferty, 1950) which is a quantitative statement of some point of important general theory otherwise definable only in words, qualitatively. The establishment of such mathematical models may be regarded as the highest form of scientific achievement because the models are precise statements of fundamental truths. The two methods—empirical and rational—would tend to converge as time goes on and the fund of information grows. The statistical analyst cannot hope to derive quantitative relationships of general application from small samples because of their inherent variability, but, as his sample data increase and the influences of variables are isolated, his empirical equations tend to approach the status of general laws. New knowledge of the observed influences of variables in turn results in keener deduction on the part of the analyst who is formulating his general mathematical laws by intuitive, deductive mental processes.

To illustrate the formulation of mathematical models in geomorphology, a specific example is offered on the subject of the longitudinal profile of the graded stream. Mr. Samuel Katz of the Lamont Geological Observatory of Columbia University has very kindly worked out the following steps at the suggestion of the author:

We are interested in deriving a time relationship between the elevation of a given point on a graded stream and the horizontal distance of the point from the head of the stream, making only the minimum number of *a priori* assumptions. If y is the elevation, x the horizontal distance measured from the head of the stream, and t the time, we shall take y to be a function of the two independent variables x and t, and shall look for a quantitative expression of this function.

From an analysis of many graded stream profiles (Shulits, 1941; Krumbein, 1937), we feel confident that the relationship between the elevation y and the horizontal distance x of a graded stream is given at a particular time t_0 by an exponential function of the form

$$y_0 = A_0 e^{-k_1 x} \tag{1}$$

(The constant A is determined by the value of y when $x = 0$, and the constant k_1 by the value of y when $x = 1/k$.)

The overall reduction of relief and valley-wall slope steepness will cause a steady reduction of load and a diminishing supply of potential energy in a drainage system. Consequently, the regrading of the master stream becomes increasingly slower. We therefore postulate that, at any given time, the rate at which the stream profile is lowered at a given point is proportional to the slope at that point. Expressed analytically,

$$\left(\frac{\partial y}{\partial t}\right)_x = k_2 \left(\frac{\partial y}{\partial x}\right)_t . \tag{2}$$

From the basic definition of the differential of a function of two variables $y(x, t)$,

$$dy = \left(\frac{\partial y}{\partial x}\right)_t dx + \left(\frac{\partial y}{\partial t}\right)_x dt . \tag{3}$$

Substituting (2) into (3), we eliminate the partial derivative with respect to the time and obtain

$$dy = (dx + k_2 dt)\left(\frac{\partial y}{\partial x}\right)_t . \tag{4}$$

Equation (4) holds for all values of t, in particular for $t = t_0$. Differentiating (1),

$$\left(\frac{\partial y}{\partial x}\right)_{t_0} = -k_1 y_{t_0} . \tag{5}$$

Substituting (5) in (4),

$$dy = (dx + k_2 dt)(-k_1 y) \quad \text{or}$$

$$\frac{dy}{y} = -k_1 dx - k_1 k_2 dt. \tag{6}$$

Integrating and using the initial conditions that for $x = 0$ (the head of the stream) and

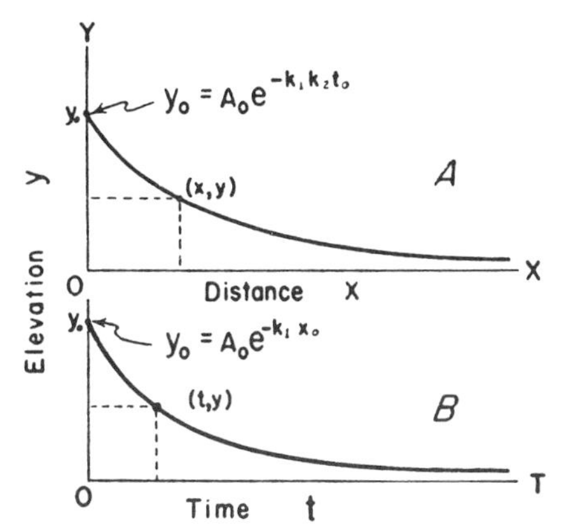

FIGURE 2.—STREAM PROFILE AS A FUNCTION OF DISTANCE AND TIME

A. Profile at a given time, t.
B. Variation in elevation at a point, x, with time.

$t = t_0$, $y = A_0$ (the elevation at the head of the stream), we obtain

$$\ln y - \ln A_0 = -k_1 x - k_1 k_2 t \quad \text{or}$$

$$y = A_0 e^{-k_1(x+k_2 t)}. \tag{7}$$

Equation (7) gives the longitudinal profile of the graded stream both as a function of the time and of the distance from stream head. It has been obtained with no explicit assumption about the variation of slope profile with time.

Figure 2A is a schematic plot of equation (7) for a given value of $t = t_0$, and thus shows the stream profile at a given moment. Figure 2B is a similar plot for a given value of $x = x_0$, and shows the reduction of elevation with time at a point (x_0) on the profile. The exponential relationship of distance to elevation is readily checked by existing profile data, but the rate of reduction of elevation with time will be extremely difficult to verify with field evidence extending back into geologic time. Perhaps controlled model experiments would be required in the latter case.

In summary, the proposed program for future development of geomorphology on a dynamic-quantitative basis requires the following steps: (1) study of geomorphic processes and landforms as various kinds of responses to gravitational and molecular shear stresses acting upon materials behaving characteristically as elastic or plastic solids, or viscous fluids; (2) quantitative determinations of landform characteristics and causative factors; (3) formulation of empirical equations by methods of mathematical statistics, (4) building of the concept of open dynamic systems and steady states for all phases of geomorphic processes, and finally (5) the deduction of general mathematical models to serve as quantitative natural laws. The program is vast and qualified investigators few, but we are already a half-century behind if development is to be measured against chemistry, physics, and the biological sciences. The need for rapid dynamic-quantitative advances is, therefore, all the more pressing.

REFERENCES CITED

Bucher, Walter H. (1941) *The nature of geological inquiry and the training required for it*, Am. Inst. Min. Metall. Engs., Tech. Pub. 1377, 6 pages.

Krumbein, W. C. (1937) *Sediments and exponential curves*, Jour. Geol., vol. 45, p. 577–601.

Rafferty, J. A. (1950) *Mathematical models in biological theory*, Am. Scientist, vol. 38, p. 549–567.

Shulits, Samuel (1941) *Rational equation of river-bed profile*, Am. Geophys. Union, Tr., vol. 22, p. 622–630.

Terzaghi, Karl (1943) *Theoretical soil mechanics*, John Wiley, New York, 510 pages.

von Bertalanffy, Ludwig (1950a) *The theory of open systems in physics and biology*, Science, vol. 111, p. 23–29.

—— (1950b) *An outline of general system theory*, British Jour. Philos. Sci., vol. 1, p. 134–165.

38

Reprinted from *Trans. Inst. Civil Engr. Paper 6699,* 473 (1963)

FIELD AND MODEL INVESTIGATION INTO THE REASONS FOR SILTATION IN THE MERSEY ESTUARY

by

William Alan Price, B.Sc.(Eng.), A.M.I.C.E.
Principal Scientific Officer, Hydraulics Research Station

and

Mary Patricia Kendrick, B.A.
Experimental Officer, Hydraulics Research Station

For discussion at an Ordinary Meeting on Tuesday, 14 May, 1963 at 5.30 p.m., and for subsequent written discussion

SYNOPSIS

The Paper gives an account of an investigation carried out by the Hydraulics Research Station, Wallingford, into the reasons for siltation in the estuary of the River Mersey. A short description of the estuary and its tidal characteristics is followed by a statement of the main problems and an account of the methods adopted in tackling them, viz., field work, a historical chart analysis and scale model tests.

The source of the material contributing to upper estuary siltation being Liverpool Bay, there follows a discussion of prototype observations of currents, salinities, and concentrations of suspended solids establishing the existence of a net landward drift of water near the bed of the estuary which carries sand upstream. The importance of salinity/density currents in a well-mixed estuary is emphasized, confirmation being provided by experiments on a large tidal model with a movable bed in which the natural salinity distribution was reproduced.

A second tidal model showed how the construction of training walls in Liverpool Bay had altered the circulation pattern thereby increasing the supply of material arriving at the mouth, and a field experiment using radioactive tracers confirmed the view that dredged material should be dumped where it will not return to the estuary. A study of the upper estuary established that there has been some restriction in the movement of the low-water channel above Eastham—a further factor likely to have contributed to upper estuary siltation.

After a discussion of the general problem of the falling capacity of the estuary, attention is directed to the more immediate, local problems associated with shoaling in the Eastham and Garston channels. Current-velocity measurements and fluorescent tracer experiments conducted on the large movable-bed model demonstrated the circulation of water and material in the upper estuary, and in conjunction with a historical analysis of siltation in the Eastham channel suggested that a reduction in the quantity of material dredged might well lead to increased depths. Support was given to a suggestion put forward by the authorities concerned that dredging methods be modified.

39

Reprinted from *Nach. Akad. Wiss. Göttingen, II. Math.-Phys. Klasse,* 5, 53-60, 65-66 (1963)

Deductive Models of Slope Evolution

By *Anthony Young*

University of Sussex

[*Editor's Note:* In the original, material precedes this excerpt.]

The results of the slope models: *1. Retreat of a rectilinear slope with no basal erosion*

In the first group of models the evolution of a rectilinear 35° slope, with level ground above, will be considered. Slope forms approximating to this are frequently observed in the field where an ungraded river is rapidly incising its course below an erosion surface remnant. It will be assumed that no further erosion takes place, but that a river or other agency continues to remove all weathered material carried to the foot of the slope. This approximates to field conditions following the attainment of grade by a river. In models 1—6 the effects of the types of action of denudational processes, as distinguished above, upon the evolution of this initial rectilinear slope will be compared.

The quantitative assumptions used in computing each of the models are given in table III. The results of the models are shown in figs. 2, 3, and 4. In addition, the profile form at one selected stage of evolution is given in table IV, which shows both the angles of each section and the curvature, as indicated by the difference in angle between adjacent sections.

Models 1—6 are shown in fig. 2a. In model 1 it is assumed that denudation is caused by downslope soil transportation, with no direct removal, and that the rate of transportation varies with the sine of the slope angle. The evolution commences with a rounding of the initial break of slope at this stage; the lower part of the rectilinear slope undergoes no retreat, since the rate of transportation across it is constant. The convexity extends progressively to the base of the slope. Following this, the angle of the lowest part of the slope decreases, but no concavity develops. Throughout the evolution the convexity extends back into the level ground lying above the slope. The principal features in the evolution of this model are therefore first, that the evolution is entirely by slope decline; secondly, no concavity is developed, even in the later stages of evolution; and thirdly, a long and smoothly-curved convexity is formed, the curvature of which becomes progressively less as evolution advances. During the later stages the curvature on this convexity is very low over the upper part; it increases gradually to the position of the initial break of slope, and decreases again below this.

Table III. *Assumptions on which the slope models are based. The values given are for unit time. No direct removal occurs except where stated.*

Model	Property	Assumptions	Values used in computation
1	Initial slope Conditions at slope foot Rate of transportation	Rectilinear, 35°, with level ground above No erosion, unimpeded basal removal Varies with sine of slope angle	Height 700 $S = 1000 \sin \alpha$
2	Initial slope Conditions at slope foot Rate of transportation	As model 1 As model 1 Varies with square of slope angle	As model 1 $S = \frac{\alpha^2}{100}$
3	Initial slope Conditions at slope foot Rate of transportation	As model 1 As model 1 Varies with sine of slope angle and with distance from crest	As model 1 $S = 1000 \sin \alpha \cdot \frac{D}{100}$
4	Initial slope Conditions at slope foot Rate of transportation	As model 1 As model 1 Varies with sine of slope angle and with distance from crest; downslope increase more rapid on the concavity	As model 1 $S = 1000 \sin \alpha \cdot \frac{D}{100}$ with distances on concavity doubled
5	Initial slope Conditions at slope foot Rate of transportation Rate of direct removal	As model 1 As model 1 Nil Uniform on all slopes above 0°	As model 1 $R = 10$
6	Initial slope Conditions at slope foot Rate of transportation Rate of direct removal	As model 1 As model 1 Nil Varies with sine of slope angle	As model 1 $R = 10 \sin \alpha$
7	Initial slope Conditions at slope foot Rate of transportation	Rectilinear, 35°, with level ground above and below No erosion, no basal removal Varies with sine of slope angle	Height 700 $S = 1000 \sin \alpha$
8	Initial slope Conditions at slope foot Rate of transportation	As model 7 As model 7 Varies with sine of slope angle and with distance from crest	As model 7 $S = 1000 \sin \alpha \cdot \frac{D}{100}$
9	Initial slope Conditions at slope foot Rate of transportation Rate of direct removal	As model 1 As model 1 Varies with sine of slope angle Varies with sine of slope angle	As model 1 $S = 1000 \sin \alpha$ $R = 2 \sin \alpha$

Table III (continued)

Model	Property	Assumptions	Values used in computation
10	Initial slope Conditions at slope foot Rate of transportation Rate of weathering	As model 1 As model 1 Varies with sine of slope angle and with soil depth Varies with sine of slope angle and inversely with soil depth	As model 1 $S = 2000 \sin \alpha \cdot H$ $W = \frac{2000 \sin \alpha}{H}$
11	Initial slope Conditions at slope foot Rate of transportation Rate of weathering	As model 1 As model 1 Varies with sine of slope angle and with soil depth Varies inversely with soil thickness; nil on 0°	As model 1 $S = 2000 \sin \alpha \cdot H$ $W = \frac{10000}{H}$
12	Initial slope Conditions at slope foot Rate of transportation	Level ground Vertical erosion at uniform rate Varies with sine of slope angle	 $E = 2$ $S = 1000 \sin \alpha$
13	Initial slope Conditions at slope foot Rate of transportation	Level ground Vertical erosion at a uniform, rapid, rate Varies with sine of slope angle and with distance from crest	 $E = 10$ $S = 1000 \sin \alpha \cdot \frac{D}{100}$
14	Initial slope Conditions at slope foot Rate of transportation	Level ground Vertical erosion at a uniform, slow, rate Varies with sine of slope angle and with distance from crest	 $E = 2$ $S = 1000 \sin \alpha \cdot \frac{D}{100}$
15	Initial slope Conditions at slope foot Rate of transportation	Level ground 3 periods of vertical erosion; 3 periods of no erosion, unimpeded basal removal Varies with sine of slope angle and with distance from crest	 $T = 0–14$, $E = 20$ $T = 15–85$, $E = 0$ $T = 86–99$, $E = 40$ $T = 100–163$, $E = 0$ $T = 164–177$, $E = 40$ $T = 177–185$, $E = 0$ $S = 1000 \sin \alpha \cdot \frac{D}{100}$
16	Initial slope Conditions at slope foot Rate of transportation	Level ground 3 periods of vertical erosion; 3 periods of no erosion, unimpeded basal removal Varies with sine of slope angle	 $T = 0–140$, $E = 2$ $T = 141–1290$, $E = 0$ $T = 1291–1430$, $E = 2$ $T = 1431–2010$, $E = 0$ $T = 2011–2150$, $E = 2$ $T = 2151–2290$, $E = 0$ $S = 1000 \sin \alpha$

α = angle of section upslope of point
D = horizontal distance from crest
S = volume of soil transported
R = depth of ground lost by direct removal
W = volume of soil weathered
H = soil depth
E = amount of vertical erosion
T = total time elapsed

Table IV.

Profile forms developed by the slope models. The form at one selected time is shown for each model. The upper row of figures gives the angles of successive sections of the slope, in degrees. The crest of the slope is to the left and the base to the right. In models 1—11 the vertical line marks the position of the break between 0° and 35° on the initial slope. The lower row of figures gives the difference in angle between adjacent sections of the profile; negative values represent concavities.

Model			
1	Angle	0 1 1 2 3 5 7 10 13 16	20 23 26 28 30
	Curvature	1 0 1 1 2 2 3 3 3	4 3 3 2 2
2	Angle	0 2 9 15 20	24 27 29 30 31
	Curvature	2 7 6 5	4 3 2 1 1
3	Angle	0 1 6 13 18 23	26 27 27 25 22
	Curvature	1 5 7 5 5	3 1 0 —2 —3
4	Angle	0 1 2 6 9 12 15 17 19 20	21 21 19 16 15
	Curvature	1 1 4 3 3 3 2 2 1	1 0 —2 —3 —1
5	Angle	0 0 0 35 35 35	35 35 0 0 0
	Curvature	0 0 35 0 0	0 0 —35 0 0
6	Angle	0 2 7 13 20 26 31 35	35 35 35 35 33 28 16 6 0 0
	Curvature	2 5 6 7 6 5 4	0 0 0 0 —2 —5 —12 —10 —6 0
7	Angle	0 1 2 4 6 9 13 17 20 22	22 20 17 13 9 6 4 2 1 0
	Curvature	1 1 2 2 3 4 4 3 2	0 —2 —3 —4 —4 —3 —2 —2 —1 —1
8	Angle	0 7 20	29 33 29 22 16 11 6 3 1 0
	Curvature	7 13	9 4 —4 —7 —6 —5 —5 —3 —2 —1
9	Angle	0 1 3 5 8 12 16 20 24	26 26 22 17 12
	Curvature	1 2 2 3 4 4 4 4	2 0 —4 —5 —5
10	Angle	0 2 4 7 13 20 24	27 30 29 25 9
	Curvature	2 2 3 6 7 4	3 3 —1 —4 —16
11	Angle	0 3 6 9 14 21	27 29 31 28 3
	Curvature	3 3 3 5 7	6 2 2 —3 —25
12	Angle	0 1 2 4 6 9 13 18 25 35 35	
	Curvature	1 1 2 2 3 4 5 7 10 0	
13	Angle	0 3 6 9 12 15 18 21 23 26 29 32 35 35	
	Curvature	3 3 3 3 3 3 3 2 3 3 3 3 0	
14	Angle	0 2 4 5 6 7 7 8 9	
	Curvature	2 2 1 1 1 0 1 1	
15	Angle	0 1 3 6 8 10 13 16 19 23 28 31 30	
	Curvature	1 2 3 2 2 3 3 3 4 5 3 —1	
16	Angle	0 1 1 1 2 3 4 5 7 9 11 13 16 19 23 29 31 32	
	Curvature	1 0 0 1 1 1 1 2 2 2 2 3 3 4 6 2 1	

Model 2 shows the evolution under similar conditions, but assumes that the rate of transportation varies with the square of the slope angle. This precise relationship was selected arbitrarily, but is taken as an example of conditions in which the rate of transportation is very much faster on steep slopes than on gentle ones. The evolution is very similar to that of model 1, although the final slope form differs in having its maximum curvature close to the crest. The same type of evolution is again found to occur when other numerical relationships are assumed. It is therefore concluded that under any conditions in which the rate of downslope transportation varies only with the slope angle, and assuming that

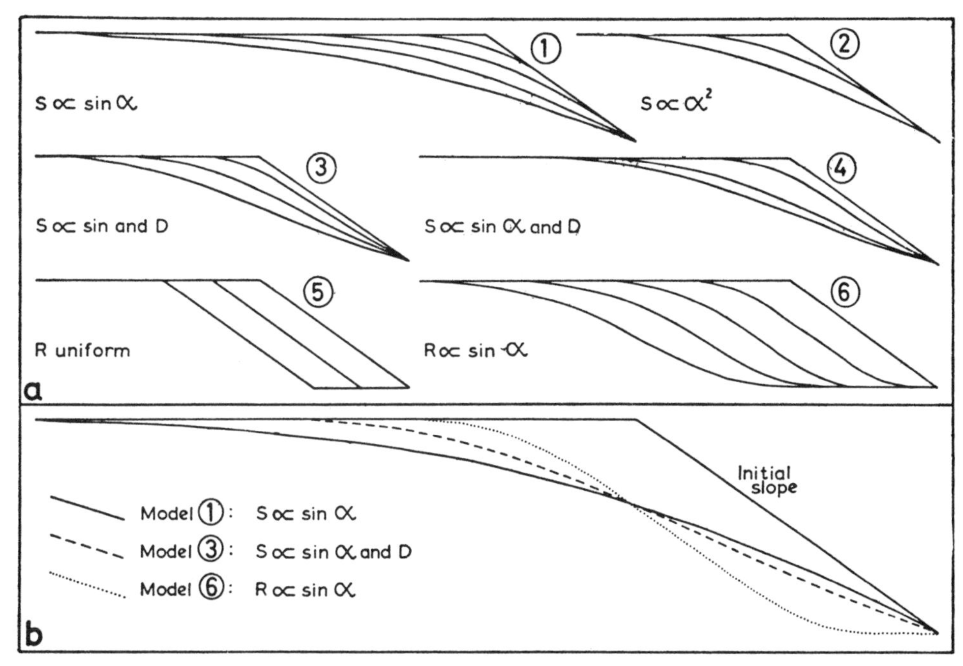

Fig. 2

a) Models 1—6. b) Comparison of profile forms developed in models 1, 3, and 6. The assumptions on which each model is based are given in table III. S = rate of transportation, D = distance from slope crest, R = rate of direct removal, α = slope angle.

no direct removal takes place, an initial rectilinear slope evolves by slope decline, with the development of a long and smoothly-curved convexity but no concavity.

The second type of action of denudational processes is considered in models 3 and 4. In model 3 the rate of transportation is assumed to vary with the sine of the slope angle and also to increase linearly with distance from the crest. The early stages of evolution consist of parallel retreat of the rectilinear slope, combined with the development of a convexity and a concavity. These curved parts of the profile progressively extend in length until they unite. After this stage the angle of the steepest section progressively decreases, although there is still a substantial element of parallel retreat in the evolution. The convexity is relatively shorter than in model 1, and its point of maximum curvature migrates upslope, keeping close to the crest. A concavity remains present on the lower part of the slope, but the angle of the lowest section never becomes more than 5° gentler than that of the steepest part.

Model 4 is based on the same assumptions as model 3, except that the rate of transportation is assumed to increase downslope more rapidly over the concave sector of the profile. This was suggested by the field observation that soil texture is frequently much finer on a concavity, with consequent effects on both soil creep and slope wash. The evolution is very similar to that of model 3, the concavity becoming slightly more strongly developed. It is concluded that where the rate of transportation varies with both the sine of the slope angle and distance from the crest, the evolution of an initial rectilinear slope shows the following features: first, the early stages of evolution are by parallel retreat, but following the union of the convexity and concavity a progressive decrease in the angle of the steepest part of the slope takes place; and secondly, a concavity develops, but the lowest part of this never acquires a very much smaller angle than that of the maximum slope.

Models 5 and 6 are based on the third type of action of denudational processes, in which only direct removal takes place. In model 5 the rate of removal is assumed to be uniform on all angles above 0°, and a simple case of parallel retreat occurs. In model 6 the rate of removal varies with the sine of the slope angle. Parallel retreat is again the dominant feature of the evolution, although a convexity and concavity also develop. There is thus a causal association between processes which act by direct removal and evolution by parallel retreat. This is in agreement with the deductive system of W. Penck (1924), who implicitly assumed a manner of action of processes which was in effect one of direct removal.

A comparison of models 1, 3, and 6, representing the three types of action of denudational processes, is shown in fig. 2b. The profile of each model is shown for stages having comparable degress of development. The two models which have evolved by downslope transportation have both declined considerably in angle; they differ from each other in the greater length of the convexity on model 1 and the presence of a concavity on model 2. The form resulting from direct removal

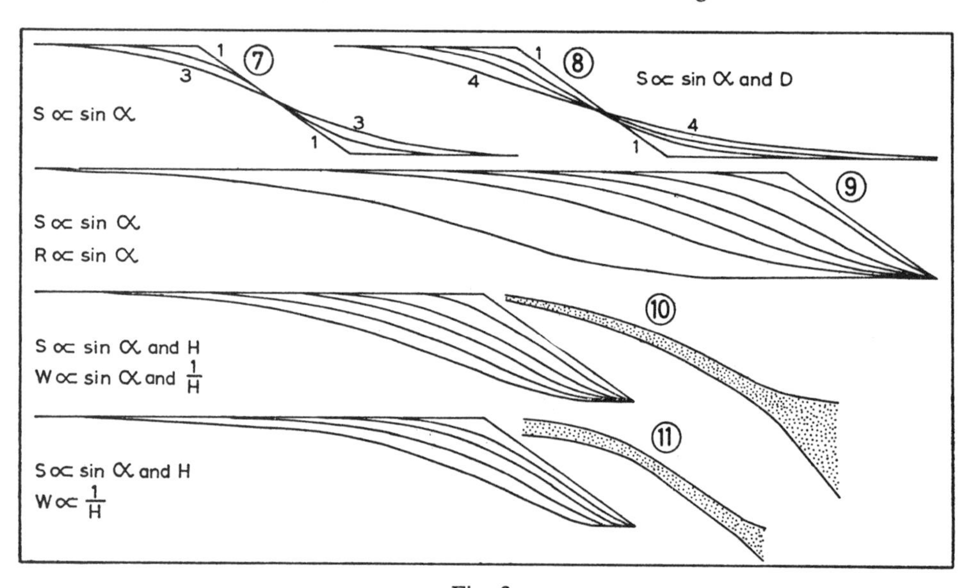

Fig. 3

Models 7—11. W = rate of weathering, H = soil thickness. Other symbols as in fig. 2.

differs greatly from the other two profiles; it is only slightly gentler in angle than the initial slope, and the basal concavity is very much more strongly developed than in model 3.

The development of accumulation features as a consequence of downslope soil transportation is shown in models 7 and 8 (fig. 3). An initial 35° rectilinear slope is again assumed, but with level ground below; no removal of weathered material from the base of the slope takes place. This is representative of field conditions in which a flood-plain or river terrace occurs at the slope foot. In model 7 the rate of transportation varies with the sine of the slope angle. The resulting profile evolution is symmetrical, the ground loss from the convexity being matched by accumulation on the concavity. Thus in the case of no basal removal a concavity can be developed by transportation, the rate of which varies with the slope angle, but this could be distinguished in the field as an accumulation landform. The main feature of model 8, in which the rate of transportation increases with distance from the crest, is a long, gently-sloping extension of the lower part of the concavity.

Model 9 is an example of slope evolution when both downslope transportation and direct removal occur. The same initial slope as in models 1—6 is assumed, with unimpeded basal removal. The rates of both transportation and removal are assumed to vary with the sine of the slope angle. The evolution is intermediate between that of models 1 and 6, showing elements of both parallel retreat and slope decline. The occurrence of parallel retreat on the steepest part of the slope results in the formation of a concavity at the foot. A long convexity is also developed. The relative importance of parallel retreat and slope decline depends on the quantitative relationship between rates of transportation and removal. A constant feature, however, is that for any given set of assumptions, ground loss by direct removal becomes of relatively greater importance in the later stages of evolution.

Conditions in which the rates of weathering and of downslope transportation both vary with the thickness of the regolith cover are investigated in models 10 and 11. The assumptions are based partly on the concept of the balance of denudation, proposed by Jahn (1954). In these models the rate of transportation becomes greater with increase in regolith thickness, a condition which is probably applicable to the process of soil creep: in addition, this rate varies with the slope angle. The rate of weathering is assumed to vary inversely with soil thickness: in model 10 it varies also with the slope angle. Both models show a similar evolution. This commences with a ground loss from the whole of the rectilinear slope, but the amount of loss decreases slightly downslope. There is consequently a proportion of parallel retreat combined with a decline in angle; the evolution differs from that of models 3, 4, and 6 in that decline in angle commences immediately, before the joining of the convexity and concavity. A short concavity with a high curvature develops, the lowest part of which reaches 0°. The convexity progressively extends in length and decreases in curvature, as in model 1.

The relative thickness of the regolith on different parts of the slope for selected stages of models 10 and 11 is also shown in fig. 3; the technique of computation does not, without additional assumptions, yield values for absolute thickness. In model 10 there is a continuous downslope increase over the whole of the profile; this increase is gradual on the convexity and maximum slope but rapid on the concavity. Model 11 differs in having a slight downslope decrease in regolith thickness across the convexity; below this the variation is similar to that of model 10, with a slight downslope increase in thickness across the maximum slope and a rapid increase on the concavity.

[*Editor's Note:* Material has been omitted at this point.]

References

Ahnert, F.: Zur Frage der rückschreitenden Denudation und des dynamischen Gleichgewichts bei morphologischen Vorgängen. Erdkunde 8, 61—64. 1954.

Bakker, J. P. and Le Heux, J. W. N.: Projective-geometric treatment of O. Lehmann's theory of the transformation of steep mountain slopes. Proc. Acad. Sci. Amst. 49, 533—547. 1946.

Baulig, H.: Le profil d'équilibre des versants. Ann. Géogr. 49, 81—97. 1940. Reprinted in: Essais de Géomorphologie, 125—147. Paris 1950.

Birot, P.: Évolution des versants. Essai sur quelques problèmes de morphologie générale. Lisbon, 17—77. 1949.

Fair, T. J.: Slope form and development in the interior of Natal. Trans. geol. Soc. S. Afr. 50, 105—120. 1947.

—: Slope form and development in the coastal hinterland of Natal. Trans. geol. Soc. S. Afr. 51, 37—53. 1948.

Fisher, O.: On the disintegration of a chalk cliff. Geol. Mag. 3, 354—356. 1866.

Gerber, E.: Zur Morphologie wachsender Wände. Z. Geomorph. 8, 213—223. 1934.

Gilbert, G. K.: The convexity of hilltops. J. Geol. 17, 344—350. 1909.

Götzinger, G.: Beiträge zur Entstehung der Bergrückenformen. Geogr. Abh. Bd. 9, Heft 1, 174 pp. 1907.

Jahn, A.: Denudacyjny bilans stoku. Czas. geogr. 25, 38—64. 1954.

Lane, A. C.: Can U-shaped valleys be produced by removal of talus? Bull. geol. Soc. Amer. 26, 75. 1915.

Lehmann, O.: Morphologische Theorie der Verwitterung von Steinschlagwänden. Vjschr. naturf. Ges. Zürich 78, 83—126. 1933.

—: Über die morphologischen Folgen der Wandwitterung. Z. Geomorph. 8, 93—99. 1934.

Louis, H.: Probleme der Rumpfflächen und Rumpftreppen. Verh. u. wiss. Abh. d. 25. Dtsch. Geographentages zu Bad Nauheim 1934, 118—137. 1935.

Morawetz, S.: Eine Art von Abtragungsvorgang. Petermanns geogr. Mitt. 78, 231—233. 1932.

Penck, W.: Morphological analysis of land forms [1924]. Translated by H. Czech and K. C. Boswell, 429 pp. London 1953.

Savigear, R. A. G.: Some observations on slope development in South Wales. Trans. Inst. Brit. Geogr. 18, 31—51. 1952.

—: Technique and terminology in the investigation of slope forms. Union géogr. int., premier rap. Comm. ét. versants, 66—75. Amsterdam 1956.

Scheidegger, A. E.: Theoretical geomorphology. 333 pp. Berlin 1961.

Wurm, A.: Hangentwicklung, Einebnung, Piedmonttreppen. Z. Geomorph. 9, 57—87. 1935—36.

Young, A.: Soil movement by denudational processes on slopes. Nature, 188, 120—122. London 1960.

40

Reprinted from *Jour. Glaciol.*, 2(12), 82–93 (1952), by permission of the International Glaciological Society

THE MECHANICS OF GLACIER FLOW

By J. F. Nye
(Cavendish Laboratory, Cambridge)

Abstract. The flow of valley glaciers is examined in the light of recent laboratory experiments on the behaviour of ice under load. Simple expressions are given for the velocity distributions in some cases of laminar flow, and the modification of a pure laminar flow theory necessary to explain the formation of transverse crevasses and thrust planes is considered. The paper ends with some remarks about the formation of crevasse patterns on the surfaces of glaciers. The statical equilibrium of a circular ice cap is discussed in an appendix.

1. The Mechanical Properties of Ice

Present knowledge of the mechanical properties of ice suggests a re-examination of the theory of the flow of long valley glaciers. It has sometimes been assumed that ice under stress behaves like a very viscous Newtonian liquid: in other words, that for ice, as for a liquid, there is a proportional relationship between rate of strain (velocity gradient) $\dot{\gamma}$ and shear stress τ, as shown by curve B in Fig. 1 (p. 83), although the coefficient of viscosity for ice is much greater than for a normal liquid.* Unfortunately, this assumption, although mathematically simple, does not represent the real behaviour of ice very well. A constant viscosity is not observed with other polycrystalline materials such as metals, and it would be surprising if ice were an exception. In fact the matter seems now to have been put beyond doubt by careful laboratory experiments carried out by Mr. J. W. Glen.[11] He finds that, just as with a metal, applying a sustained constant stress (he actually applies a uniaxial compressive stress) to a specimen of ice causes it to deform permanently, and that after a few hours (the "transient" period) the rate of deformation settles down to a steady value. If the experiment is repeated with different values of the stress, the relationship between shear stress τ and the rate of shear strain $\dot{\gamma}$ in the specimen is given by a curve of the general shape shown at A in Fig. 1. At low shear stresses the rate of strain is small; for higher shear stresses, however, the rate increases very rapidly, so that a small increase of stress produces a large increase in strain-rate. This is the type of curve, then, on which a theory of glacier motion should be based.

In a glacier the state of stress is not uniaxial, as in Glen's experiments, but triaxial, and in particular there is a hydrostatic pressure acting deep in the ice. The pressure reaches about 30 atmospheres in the Mer de Glace and 300 atmospheres or more in the Greenland ice cap. It has been suggested by Streiff-Becker, by Haefeli and by Demorest that this pressure may make the ice more deformable at depth, that is, reduce the shear stress needed to produce a given rate of strain. There seems to be no direct experimental evidence on the behaviour under shear stress of ice or any other polycrystalline substance already subjected to a high pressure and very near the melting point, and I think one should keep an open mind on what might occur. It may be remarked, however, that a pressure effect is not observed in metals far from the melting point, and that with liquids the viscosity is substantially independent of pressure. The present discussion is founded on the assumption that any pressure effect is negligible in ice. Should the existence of such an effect be later proved by experiment the necessary modifications of the expressions in Section 2 would be simple; the changes that would occur in the results of Section 3 are not so obvious. (See also *Journal of Glaciology*, Vol. 2, No. 11, 1952, p. 52–53.)

* A paper containing several very apposite and simple calculations on glacier mechanics, not only with this assumption but also allowing for a change of viscosity with depth, was given by J. W. Evans[1] in 1913.

A possible effect of hydrostatic pressure on the shear stress necessary to produce a given rate of deformation should be distinguished from the diminishing slope of Glen's curve A in Fig. 1 of this paper, which might be described as a lowering of viscosity with increasing shear stress.

For the purpose of calculation it is convenient to express curve A analytically. Glen finds that for polycrystalline ice with random crystallographic orientation a power law gives a good fit over the range of stresses $\tau=0{\cdot}8$ to $5{\cdot}5$ bars *:

$$\dot{\gamma}=\left(\frac{\tau}{B}\right)^n \quad \ldots \ldots \ldots \ldots \quad (1)$$

where B and n are constants. (This law was suggested by Perutz.[2]) For ice at $-1{\cdot}5^\circ$ C., if $\dot{\gamma}$ is expressed as shear strain per year and τ is in bars, $B=1{\cdot}62$ and $n=4{\cdot}1$, but the values of the constants depend rather sensitively on the amount of bubbliness in the ice and on the temperature. In this formula both τ and $\dot{\gamma}$ are always to be taken as positive.

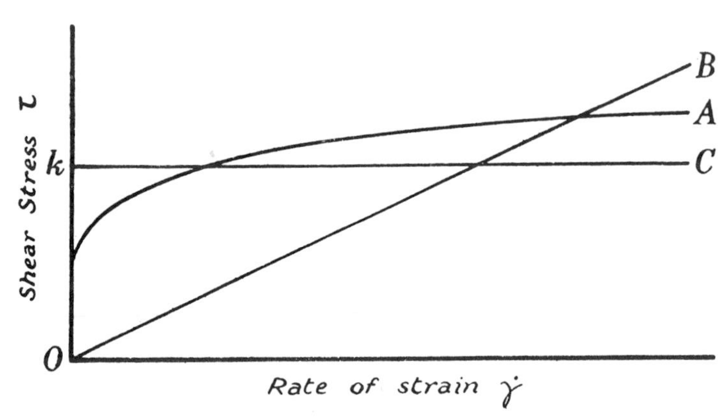

Fig. 1. Relations between the rate of strain $\dot{\gamma}$ and the applied shear stress τ: A, ice (after Glen); B, liquid of constant viscosity; C, simplified flow law used in section 4

2. Laminar Flow

A liquid of constant viscosity obeys the law (1) with the value of $n=1$. With the more general flow law we are in a position to attempt a recalculation of the distribution of velocities within a valley glacier on the same lines as Somigliana's calculation[3] for $n=1$.

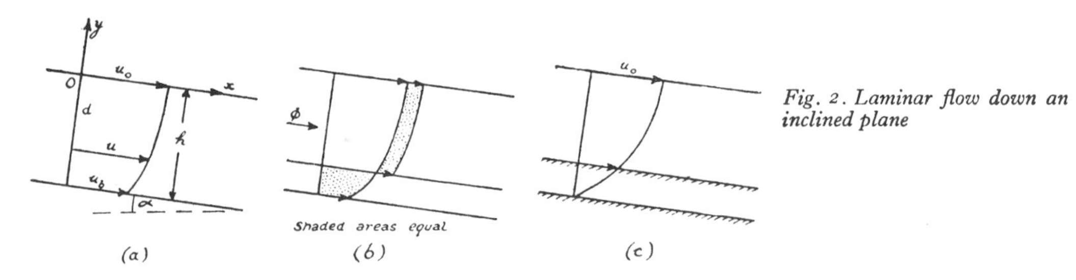

Fig. 2. Laminar flow down an inclined plane

The simplest case to start with is that of flow down a uniform plane slope (Fig. 2*a*, above). Axes are taken as shown, with the origin on the surface and Ox down the line of greatest slope. Oz is horizontal and perpendicular to the plane of the diagram. It is assumed for the moment that the lines of flow are everywhere parallel to the bed and that conditions are the same on all sections perpendicular to Ox. The shear stress on a layer at a depth d, which is measured perpendicular to the surface, is

$$\tau_{xy}=\rho g d \sin \alpha$$

* 1 bar$=10^6$ dynes/cm.$^2\simeq 1$ Kg./cm.$^2\simeq 1$ atmosphere.

where ρ is the density, assumed constant, g is the acceleration due to gravity and α is the angle of the slope. $\dot{\gamma}$ in (1) is here du/dy, where u is the velocity. It then follows, by integration, that the difference between the velocity u_0 of the top layer and the velocity of the layer at depth d is given by

$$u_0 - u = \frac{K}{n+1} \sin^n \alpha \, . \, d^{n+1} \qquad (2)$$

where $K = \left(\frac{\rho g}{B}\right)^n$.

The relative velocity between the top and bottom layers is thus

$$u_0 - u_b = \frac{K}{n+1} \sin^n \alpha \, . \, h^{n+1} \qquad (3)$$

where u_b and τ_b are respectively the velocity and the shear stress on the bed, and h is the total depth measured perpendicular to the bed.

ϕ, the volume passing through any cross-section in unit time, for unit thickness in the z direction, is given by a further integration:

$$\phi = u_b h + \frac{K}{n+2} \sin^n \alpha \, . \, h^{n+2} \qquad (4)$$

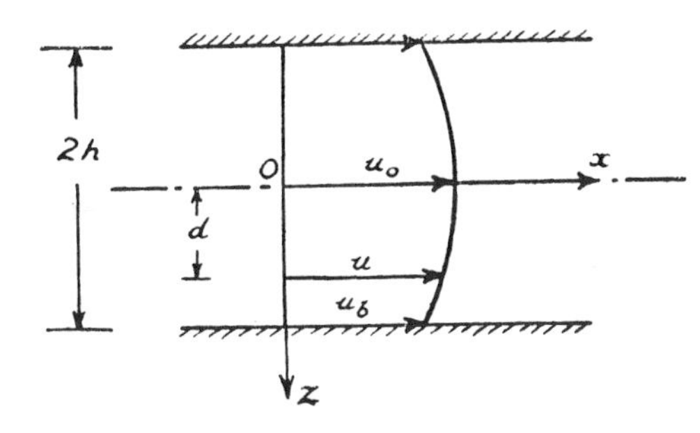

Fig. 3. Laminar flow in an infinitely deep channel of finite width. Plan view

One sees from equation (2) that if K, n, α and u_0 are given one can calculate the velocity at any required depth. For the physical problem, on the other hand, we may regard ϕ, K, n and α as given. Equation (4) does not enable h to be found, though, because u_b is not yet determined. One could have the same ϕ for different values of h, as indicated in Fig. 2*b* (p. 83).

The physical properties of ice used in the calculation so far do not allow a prediction of how fast the glacier slips on its bed. All we can calculate are differential velocities within the ice. Therefore, on these assumptions, one cannot predict the depth of a glacier from a knowledge of its surface velocity and slope; the two cases indicated in Fig. 2*c* (p. 83), for instance, could not be distinguished. On the other hand it would evidently be possible to find an upper limit for the depth, and if ϕ were known as well as the surface velocity and slope, the depth could be deduced.

3. Effect of the Valley Sides

We now have to ask what effect the sides of the valley will have on these results. One may first notice that exactly the same equations (2), (3) and (4) apply to an infinitely deep, narrow glacier (Fig. 3, above), if the symbols are given the meanings:

u_0 = velocity on the central plane,
u = velocity at a distance d from the central plane,
u_b = velocity at the edges,
α = slope of the surface, as before,
h = the half width.

K and n have the same meanings as before.

A case intermediate between the very wide and the very narrow valley would be a bed formed from one half of a circular cylinder (Fig. 4, below). This is an easy problem to treat because the surfaces of maximum shear are all half-cylinders parallel to the bed. Simple statics shows that the variation of shear stress τ with depth is still linear but that the rate of increase is just half as rapid as with a very wide valley of the same slope. Thus

$$\tau = \tfrac{1}{2}\rho g r \sin \alpha$$

where r is the distance from the x axis of the point considered. Equations (2), (3) and (4) take the forms

$$u_0 - u = \left(\frac{1}{2}\right)^n \frac{K}{n+1} \sin^n \alpha \,.\, r^{n+1} \quad \text{(2)}'$$

$$u_0 - u_b = \left(\frac{1}{2}\right)^n \frac{K}{n+1} \sin^n \alpha \,.\, R^{n+1} \quad \text{(3)}'$$

$$\text{Total rate of discharge} = \frac{1}{2}\pi R^2 u_b + \frac{\pi K}{n+3}\left(\frac{1}{2}\right)^{n+1} \sin^n \alpha \,.\, R^{n+3} \quad \text{(4)}'$$

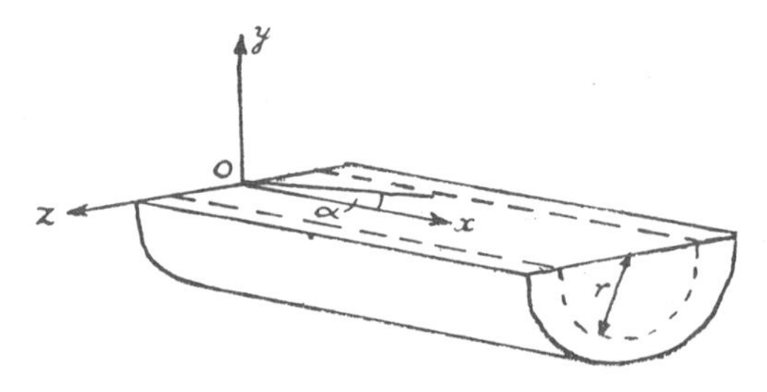

Fig. 4. Flow in a channel formed from a half-cylinder

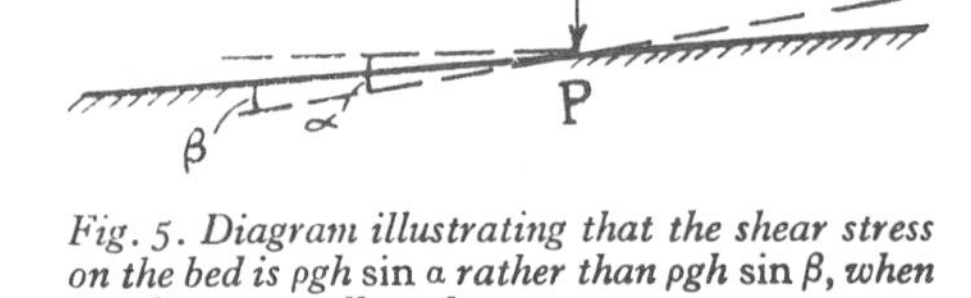

Fig. 5. Diagram illustrating that the shear stress on the bed is $\rho gh \sin \alpha$ rather than $\rho gh \sin \beta$, when $(\alpha - \beta)$ is a small angle

R is here the maximum depth, the radius of the cylinder. The same remarks about the possibility of finding, from a knowledge of u_0, an upper limit to the depth, but not an exact figure, apply to this case equally.

Somigliana, treating the case with $n=1$, was able to give an exact analytical solution for a bed whose cross-section formed a semi-ellipse, and he also made calculations for beds of more complex shape. Unfortunately, with the more general flow law (1) more complex cross-sections than a semi-circle do not readily lend themselves to exact analysis. We therefore have to resort to approximate methods.

In a glacier of arbitrary but constant cross-section, flowing uniformly, one may find the average value of the shear stress, τ_{av}, at the bed by simple resolution of forces. If the area of cross-section perpendicular to the bed is A and the perimeter of this cross-section is p, then, for unit length of valley, a force $\rho g A \sin \alpha$ due to the weight is balanced by a force $\tau_{av} p$ due to the resistance of the bed. Therefore

$$\tau_{av} = \rho g \frac{A}{p} \sin \alpha \quad \text{(5)}$$

(A/p is analagous to the "hydraulic radius" of a river valley or channel, but it should be noted that A and p are measured here not in a vertical plane but on a section perpendicular to the bed.)

The values of A and p are not known for many existing glaciers. On the other hand, there are a number of glaciers whose depths are now known. Knowing the depth and width of a glacier one may make a good guess at the value of A/p. The procedure I have used (see also Koechlin[4]) is to put a parabola through the three known points, the lowest point on the bed and the two marginal points on the surface, to assume that the upper boundary of the cross-section is a horizontal line and hence to calculate A/p. One also needs to know α. But in real glaciers the slope of the bed is not always the same as the slope of the surface. So which slope should be taken in the formula? The following argument makes it plausible that the surface slope is the right one to use. For fuller arguments leading to the same conclusion see reference 5.

Consider a wedge-shaped block of ice whose surface has a slope α resting on a slope of inclination β (Fig. 5, p. 85). Provided $(\alpha-\beta)$ is small the shear stress acting on the bed at the point P may be approximately calculated. By the argument used for the parallel-sided slab, the shear stress at P on a plane drawn parallel to the surface is $\rho gh \sin \alpha$ where h is the depth of P. Since the bed is near a plane of maximum shear stress, the shear stress on the bed itself will only differ from $\rho gh \sin \alpha$ by a small amount, particularly if $(\alpha-\beta)$ is small, since $(\alpha-\beta)$ is the angle through which one has to rotate the axes of reference of the stress tensor, α and not β is therefore the angle to be used.

The calculation of τ_{av} using the data for sixteen Alpine glaciers collected by Mercanton [6] gives values of τ_{av} ranging from 0·49 bars, for the Unteraar Glacier at 2100 m. in 1945–47 (the last in the list), to 1·51 bars, for the Grenz Glacier at 2700 m. in 1948–49 (the fifth in the list). Considering the comparatively wide range of widths and depths involved and, particularly, the wide range of slopes, the spread of values of τ_{av} is not very large—a result which might have been anticipated from the form of the curve A in Fig. 1.

One way of proceeding further is to assume that the linear variation of τ_{xy} with depth that was found for a circular cylindrical bed and for a plane bed also holds for these more complicated cross-sections, and that the rate of increase of τ_{xy} with depth on the central vertical plane $z=0$ is fixed by putting τ_b, the actual shear stress on this part of the bed equal to τ_{av}, as is also rigorously true for the cylinder and the plane. This assumption gives a formula analogous to (3) for the relative velocity of top and bottom

$$u_0-u_b=\frac{1}{n+1}\left(\frac{\tau_b}{B}\right)^n h$$

In theory, therefore, knowing the value of u_0 one could predict the value of u_b. The calculation is not very reliable because the flow may not be laminar, as discussed in the next section, because the assumption of linearity for τ_{xy} may be at fault, because τ_b may not be equal to τ_{av}, and because the values of B and n are not yet precisely known for the temperatures and types of ice in the glaciers considered. Nevertheless, in spite of this formidable, and perhaps rather pessimistic, list of pitfalls, it is encouraging that when the calculation is made for the sixteen glaciers listed by Mercanton, with the values of B and n found by Glen for $-1{\cdot}5°$ C., the relative speeds of top and bottom always come out less than the observed surface velocities and give reasonable figures for the bottom velocities, ranging from 4 m./yr. for the Arolla Glacier at 2210 m. in 1908–9 (the second in Mercanton's list) to 79 m./yr. for the Rhône Glacier at 2520 m. in 1945–47 (the ninth in the list). If the calculation is done the other way round and used to predict an upper limit for the depth from the measured surface velocities, the calculated depth is always greater than the measured depth.

From this discussion one point stands out. No satisfactory answer, without *ad hoc* assumptions, seems to have been given to the question of what ultimately determines u_b. But, until this question is answered, mathematical analysis may be able to explain the observed relative velocities within glaciers but it will not be able to predict at all the absolute velocities of movement. If u_b was found in practice to be much less than u_0 the matter would be less serious. In fact this is by no means the case, and indeed it seems that sometimes the major contribution to the surface velocity comes from u_b and only a small part from differential movement within the ice.

4. Compressive and Extending Flow

In Sections 2 and 3 attention was restricted to pure laminar flow, in which all points move parallel to the bed. In such a state of flow the surfaces along which shearing takes place, in a differential sense, are in all places parallel to the bed. On the central vertical plane, assumed here to be a plane of symmetry, the stress components acting are τ_{xy}, σ_x, σ_y and σ_z. (Throughout this paper tensile stresses are counted as positive.) τ_{xy} increases linearly with depth and $\sigma_x=\sigma_y=\sigma_z=\rho gy \cos\alpha-A$, where A is the atmospheric pressure. The three normal pressures must be equal as they would otherwise cause a longitudinal or transverse extension or compression, which is contrary to the original assumption. The stress in the surface at $z=0$ is therefore a pure hydrostatic pressure A. It follows that a pure laminar flow theory cannot include any explanation of transverse crevasses or shear faults on the central axis of a glacier.

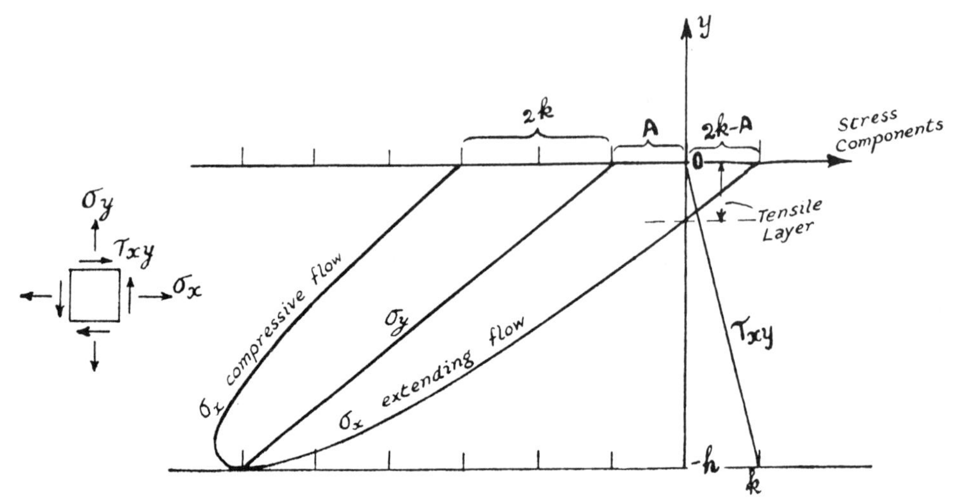

Fig. 6. Stress distribution in compressive or extending flow of a parallel-sided slab on a slope of angle α. *In the figure* α *has been taken as* $\cot^{-1} 5=11\cdot3°$. *This diagram differs slightly from the one given in reference 7 in that the effect of atmospheric pressure has been included by moving the curves for* σ_x *and* σ_y *an amount* A ($\simeq k$) *to the left*

To explain these phenomena we need to study a more complex type of flow in which an excess longitudinal stress is allowed. The mathematics of this are rather more complicated, and in the exact analysis a difficulty is met similar to that mentioned for laminar flow, namely that of postulating a suitable, and physically plausible, law to give the velocity or the shear stress on the bed. The difficulty is avoided, and the main features of the analysis are retained, if one assumes a simpler flow law. The one chosen is shown by curve C in Fig. 1. It is the special case of law (1) obtained by putting n infinite. This is equivalent to assuming that the rate of strain is very small up to a certain shear stress, k ($\simeq 1$ bar), and that shear stresses greater than this do not occur. (For n infinite, $B=k$.) As the flow theory based on this simplified law has already been published elsewhere [7] the details will not be given here but only some of the results. It was assumed that the effect of the drag of the valley sides was negligible, as would be the case in a very wide valley. It seems likely, though, that the main results apply on the vertical axial plane of any valley glacier.

The analysis shows that there are two possible sorts of flow. One sort gives a longitudinal stress σ_x which is compressive throughout the depth of the glacier (Fig. 6, above) and always more compressive than σ_y; the other gives a longitudinal stress which, although compressive at depth, is tensile in an upper surface layer, and is always more tensile (that is, algebraically greater) than σ_y. The result is that in the first case the forward velocity of the glacier decreases as one goes down glacier, because the ice is being compressed, and in the second case the velocity increases because the ice is being extended. One could call the first case "compressive flow" and the second "extending flow." (In the paper referred to above I used the terms "passive" and "active" flow, borrowed from soil mechanics, to describe these states. I think the new terms are preferable as being more graphic and less likely to lead to confusion.)

An immediate result is that one could expect transverse crevasses to form during extending flow but not during compressive flow. When the effect of atmospheric pressure is taken into account (which is strictly necessary because it is of the same order of magnitude as k) the theoretical thickness of the zone of tensile stress is approximately $(2k-A)/\rho g \simeq k/\rho g \simeq 11$ m., but of course this figure is only to be thought of as giving a rough approximation. Crevasses might be expected to open up to about this depth. Their presence would then modify the stress distribution and they might propagate to greater depths. On the wall of a crevasse at a depth $2k/\rho g \simeq 23$ m., however, the pressure from above $(A+2k)$ would exceed the lateral pressure, A, by $2k$, which is the yield stress in compression, and so below this the crevasse would close up comparatively rapidly.

Another distinction between the two types of flow can be made by considering the slip-line fields (Fig. 7, below). A slip-line field is represented by two families of curves drawn so that their directions at any point give the two perpendicular directions of maximum shear stress (that is, the two directions in which the tendency to shear is greatest). In both the present cases they are

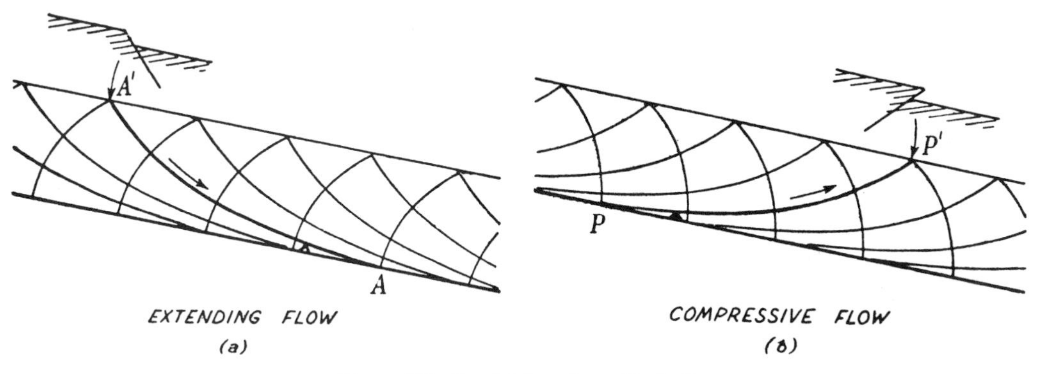

Fig. 7. Slip-line fields and possible faults

parallel and perpendicular to the bed at the bottom but they turn so as to emerge at 45 degrees to the surface. The curves are parts of cycloids, and the field in compressive flow is the mirror image of the field in extending flow. The importance of the slip-lines is that they show the directions in which the ice has the greatest tendency to fracture by shear. If the ice had no structure of its own one would expect that if shear fractures—faults—occurred, they would run along the directions of the slip-lines. Actually, of course, the ice is not equally strong in all directions and the laminar structure which every glacier possesses probably provides surfaces of weakness. One could say, therefore, that the closer any surfaces of weakness are to the slip-line directions, the more liable they are to give shear faulting.

In practice, by chance, the banding in glaciers often tends to run roughly the same way as one set of slip-lines (PP′ in Fig. 7) and this is probably why thrust planes are formed in glaciers with just about this shape. Shear displacement exactly along a slip-line would give the step shown at P′ in the figure. Faulting along the orthogonal set of slip-lines is not seen so often, nor is the corresponding faulting in the extending solution often observed. This may be because, first, it could be obscured on the surface by crevassing and, secondly, because the necessary slip-lines are not parallel to any structural surfaces of weakness. There may be a few examples of it, however, such as the one shown in Fig. 6 of reference 7.

In the calculations with the simplified flow law, unlike those for laminar flow, which were made with the more realistic flow law, it is possible to allow for the effect of accumulation and ablation, and also for changes in the slope of the glacier bed. In fact, which type of flow occurs, compressive or extending, depends on just these two factors. If ϕ is the rate of discharge, as defined in Section 2, $d\phi/dx$ is the rate of addition of ice to the upper surface of the glacier. Positive $d\phi/dx$ represents accumulation; negative $d\phi/dx$ represents ablation. The second factor is measured by R, the radius of curvature of the bed. If R is positive the bed is convex, and if R is negative the bed is concave. The criteria for the two types of flow may then be expressed as:

$$\text{Extending flow if} \quad \left(\frac{d\phi}{dx}+\frac{\phi}{R}\cot\alpha\right) \text{ is positive.}$$

$$\text{Compressive flow if} \quad \left(\frac{d\phi}{dx}+\frac{\phi}{R}\cot\alpha\right) \text{ is negative.}$$

Thus, if there were no changes in the slope of the bed, R would be infinite everywhere, and so where $d\phi/dx$ was positive (accumulation area) extending flow would occur, and where $d\phi/dx$ was negative (ablation area) flow would be compressive. At the other extreme, if there were neither accumulation nor ablation, $d\phi/dx$ would be zero and it would then be the sign of R that decided the type of flow. A convex bed (R positive) would give extending flow, while a concave bed (R negative) would give compressive flow. In general, both factors are present; Fig. 8 (p. 90) shows the result expected in an idealized glacier valley. (The upper diagram in the figure shows the velocity distribution for the simplified flow law; for details reference should be made to the original paper.)

The following table summarizes the matter:

Compressive flow	*Extending flow*
Effects	
Upper layer in compression	Upper layer in tension
No crevasses	Transverse crevasses
Thrust planes	(Other shear faults ?)
Conditions	
$\frac{d\phi}{dx}$ negative. Ablation area	$\frac{d\phi}{dx}$ positive. Accumulation area
R negative. Concave bed	R positive. Convex bed
$\left(\frac{d\phi}{dx}+\frac{\phi}{R}\cot\alpha\right)$ negative	$\left(\frac{d\phi}{dx}+\frac{\phi}{R}\cot\alpha\right)$ positive

As mentioned at the beginning of this section, the theory just described, giving the two types of flow, is most easily developed for the simplified flow law with τ constant. But it can be shown that, by using the observed curve A, and making a suitable generalization to take account of the three-dimensional state of stress and deformation, the principal qualitative features of the simpler theory remain: in particular, the upper tensile layer and the general shape of the slip-lines.

I believe, therefore, that we ought to think of the flow of a glacier in a gently undulating, parallel-sided valley as a laminar flow of the simple type discussed in Section 2, to which is added a longitudinal extension or compression according to the curvature of the bed and according to whether there is accumulation or ablation at the surface. It is this last component of the motion that is related to the formation of thrust planes and transverse crevasses.

5. The Theory of Crevasse Patterns

The general explanation of the formation of crevasse patterns on glaciers was given in a fine paper by Hopkins [8] in 1862, which must have been one of the first applications of the general analysis of stress to the distribution of stress in a continuously deforming body. The principles used in this section are essentially the same as those employed by Hopkins, except for the introduction of a result from the modern theory of plasticity. We neglect here the effect of atmospheric pressure, although strictly this is not permissible because its magnitude is comparable with that of the shear stress; the general effect of the pressure would be to reduce the area of crevasse fields.

Let us first assume (Fig. 9, p. 91) that, if σ_x is compressive on the surface of the glacier at the centre, it is also compressive, with much the same value, at other points of the surface right up to the margins; and, similarly, that there will also be regions where σ_x at the surface is tensile with much the same value right across the glacier.

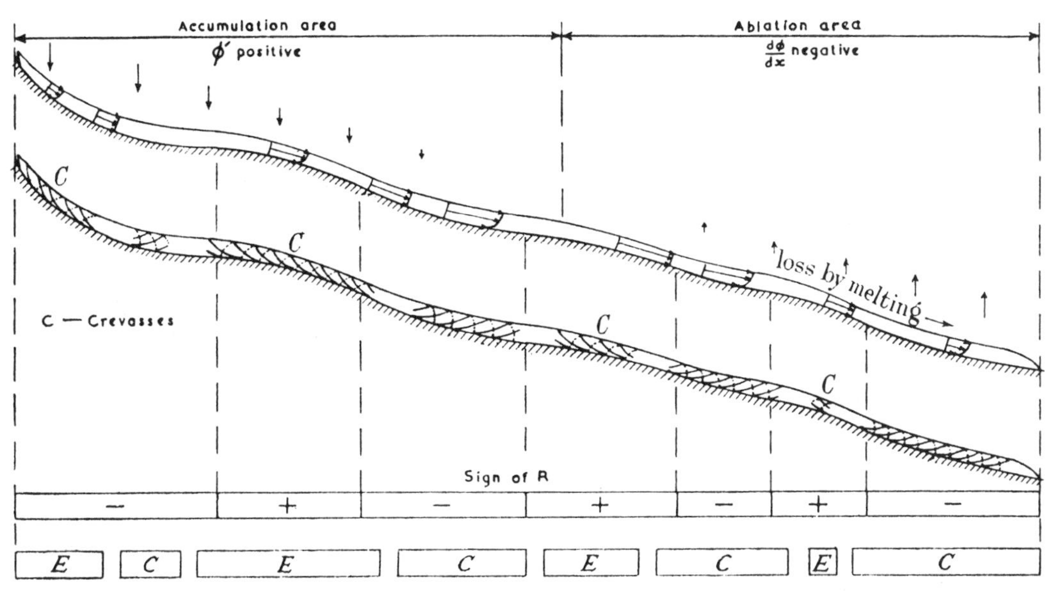

Fig. 8. Longitudinal section of an ideal glacier valley. The sign of R and the type of flow are given at the bottom of the figure. The lower diagram shows the slip-line field; the upper diagram shows the velocity distribution calculated with the simplified flow law C in figure 1

If σ_x is compressive the ice will tend to expand sideways in the z direction. There are two extreme cases to consider. If the sides of the valley are sufficiently steep to prevent lateral expansion a transverse compressive stress σ_z will be set up, and the theory of plasticity [9] shows that $\sigma_z = \frac{1}{2}\sigma_x$ (both negative). Alternatively, the valley sides may be less steep and $|\sigma_z|$ would be less than this, or even zero. The same would be true if σ_x were tensile: with steep valley sides transverse contraction by general downward movement might not be possible and a transverse tensile stress $\sigma_z = \frac{1}{2}\sigma_x$ would be set up at the surface; or, alternatively, σ_z might be less than this and even zero. In all cases $0 \leqslant |\sigma_z| \leqslant \frac{1}{2}|\sigma_x|$.

The only shear component of stress on the surface is τ_{zx}. This is zero at the middle and increases, in absolute magnitude, towards the margins. The result obtained (for instance by using the Mohr circle construction [10]) when τ_{zx} is added to the other components, σ_x and σ_z, is shown schematically in Fig. 9*a*, *b* and *c*, p. 91. The lines show the direction of possible crevasses. They are drawn at all points where a tensile stress can exist. When $\sigma_x = 0$ (Fig. 9*b*) the principal axes of stress are everywhere at 45 degrees to the edge. One principal stress is tensile and the other is compressive, both being of magnitude $|\tau_{zx}|$; the tension, of course, decreases to zero at the centre.

The effect of a longitudinal compressive stress (Fig. 9*a*) is to swing the direction of maximum tensile stress more transverse to the line of flow, so that any crevasses would make angles of less than 45 degrees with the margin. The line of the crevasses should curve in the sense shown in the diagram, assuming σ_x constant across the surface. The existence of a tensile stress in this case, however, depends upon $|\tau_{zx}|$ reaching a certain value, and so the tendency to crevassing dies away towards the centre. If transverse expansion is prevented the necessary value of $|\tau_{zx}|$ is $|\sigma_x/\sqrt{2}|$.

A longitudinal tensile stress (Fig. 9*c*) will swing the direction of maximum tension towards the line of flow. The crevasses would therefore meet the edges at angles greater than 45 degrees, and would curve across the glacier in the way shown. If, in this case, a transverse tensile stress (σ_z) were developed, there would be a central strip of the glacier surface where both principal stresses were tensile. This can only happen at places where τ_{zx} is not too large, and so the strip may not extend to the margins. If transverse contraction were prevented the condition for both principal stresses to be tensile would be $|\tau_{zx}|<|\sigma_x/\sqrt{2}|$. In this central strip one might expect to see crevasses not only transverse to the direction of flow but in many other directions as well; in particular, crevassing would be possible in the longitudinal direction, as indicated by the broken lines.

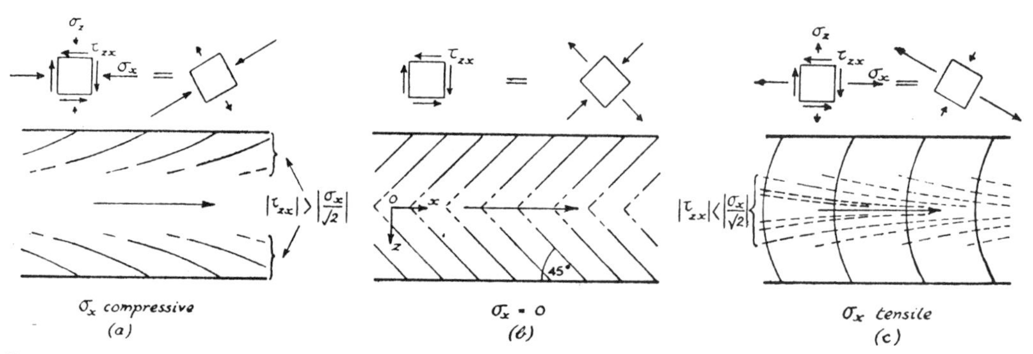

Fig. 9. The unbroken lines show the theoretical positions and directions of crevasses in three possible cases. The diagrams at the top indicate the stresses acting near the margin shown uppermost in the figure

A longitudinal compressive stress could be caused by ablation or a concave bed as discussed in the last section, or by a narrowing of the glacier valley, or on the inside of the bend when a glacier changes direction. In a similar way a longitudinal tension could be set up by accumulation, a convex bed, a widening of the valley or on the outside of a bend in the valley.

An examination of many photographs, including some hundred aerial photographs taken in Alaska by Mr. Maynard Miller, who was kind enough to lend them to me, shows that although crevasse fields are often highly complicated in their details the main features of many of them can be explained on the above lines.

6. Acknowledgements

I should like to thank Mr. J. W. Glen for allowing me to use the unpublished results of his laboratory experiments on ice and I am also grateful to him and to Mr. W. V. Lewis for kindly reading and commenting on the first draft of this paper.

MS. received 3 *January* 1952

APPENDIX

PROFILE OF AN IDEAL CIRCULAR ICE CAP

Orowan [12] and Hill [7] have considered the quasi-static equilibrium of an ice cap resting on a horizontal base. They used curve C of Fig. 1 (p. 83) as an approximation to the plastic behaviour of ice and treated the case where the ice cap was very long in one horizontal direction, so that plastic spreading took place entirely in the other horizontal direction at right angles. In such an ice cap each half of the theoretical surface profile in the direction of flow is approximately part of a parabola. The equation is

$$h=\sqrt{2h_0x} \qquad \text{(A.1)}$$

where h is the height at a distance x from the nearer of the two edges; $h_0=k/\rho g$. If $k=1$ bar, $h_0=11{\cdot}3$ m. and so $h=\sqrt{23x}$, where distances are measured in metres. The equation does not hold very near the centre or the edges of the cap.

It is interesting to consider what would happen if the two horizontal axes of the cap were comparable in length, and for this purpose we may take an ice cap that is circular in plan.

Consider the small prism-shaped element of such an ice cap shown shaded in plan and elevation in Fig. 10 (below). The prism, which is of height h, is taken to be at a distance r from the centre O

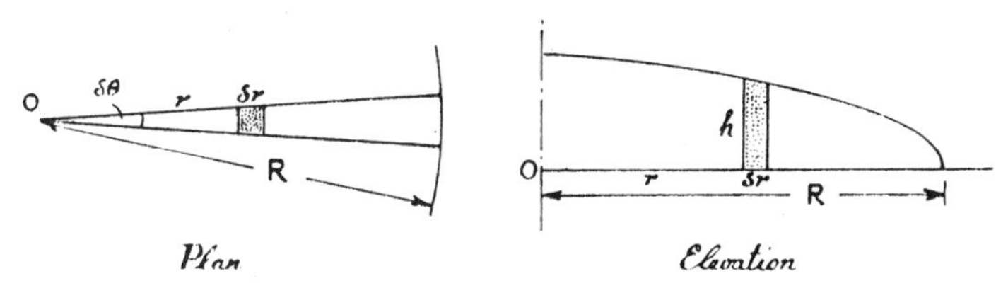

Fig. 10. Ideal circular ice cap on a horizontal bed

of the ice cap. If the ice is spreading out radially there is an inward component of force exerted by the floor on the base of the prism. If we assume a perfectly rough bed which exerts a constant shear stress k, the force is

$$kr\,\delta\theta\,\delta r$$

If $h \gg h_0$, as will be true except very near the edges of the cap, the normal pressure on the vertical faces of the prism increases from approximately 0 at the surface to approximately ρgh at the base. The average pressure is thus $\frac{1}{2}\rho gh$. This results in a component of force acting radially outwards of

$$\tfrac{1}{2}\rho gh^2\,\delta r\,\delta\theta - \delta(\tfrac{1}{2}\rho gh^2 r\,\delta\theta) = -\rho ghr\,\delta h\,\delta\theta$$

The average pressure on the faces parallel and perpendicular to the radial direction may differ by an amount of order $2k$, which would give an outward force of order $2kh\,\delta r\,\delta\theta$. However, this is small compared with the term $kr\,\delta\theta\,\delta r$, provided $h \ll r$, and so we neglect it.

Equating the radial forces to zero and proceeding to the limit we find that, approximately,

$$\frac{dh}{dr} = -\frac{k}{\rho gh} = -\frac{h_0}{h} \qquad \text{(A.2)}$$

where $h_0 \ll h \ll r$. This integrates to the parabola,

$$h = \sqrt{2h_0(R-r)} \qquad \text{(A.3)}$$

Where R is the radius of the cap, as the equation of the profile. The complete profile is thus part of the surface formed by rotating the parabola of Fig. 10 (p. 92) about a vertical axis through O. By comparing equation (A.3) with equation (A.1) we see that the profile taken through the centre of a circular ice cap of diameter $2R$ is identical with the transverse profile of a very long ice cap of width $2R$.

It is perhaps surprising, at first sight, that the dimension transverse to the line of flow does not affect the profile, but the reason for this is clear when one considers a more general case.[13] It can be shown that the result just proved for a circular cap is merely a special case of a general theorem applicable to ice caps on uneven bases of irregular outline. The general theorem is that (1) seen in plan view, flow takes place in the direction where the downward slope of the *surface* is greatest; (2) the surface gradient α is connected approximately with the ice thickness h at each point by the equation

$$\alpha = \frac{h_0}{h} \qquad \text{(A.4)}$$

For the circular ice cap on a horizontal bed, $\frac{dh}{dr} = -\alpha$, and equation (A.4) is identical with (A.2). For an ice cap of irregular shape resting on a horizontal bed the equation of the profile is evidently

$$h = \sqrt{2h_0 s} \qquad (A.5)$$

where s is the distance from the edge *taken along a line of flow*. Equation (A.5) embraces both (A.1) and (A.3). The derivation of equation (A.4) and a discussion of its possible application to the Pleistocene ice-sheets and to Greenland are given in reference 13.

MS. received 28 *April* 1952

REFERENCES

1. Evans, J. W. The wearing down of the rocks. Presidential address to the Geologists' Association. *Proc. Geol. Ass.*, Vol. 24, 1913, p. 241–300.
2. Perutz, M. F. Glaciology—the flow of glaciers. *The Observatory*, Vol. 70, 1950, p. 64–65.
3. Somigliana, C. Sulla profondità dei ghiacciai. *Atti della Reale Academia Nationale dei Lincei, Rendiconti, Classe di Scienze fisiche, matematiche e naturali*. Vol. 30, Serie 5, 1921, 1° semestre, p. 291–96, 323–27, 360–64; 2° semestre, p. 3–7.
4. Koechlin, R. *Les glaciers et leur mécanisme*, Lausanne: Rouge et Cie., 1944, p. 103.
5. Nye, J. F. A comparison between the theoretical and the measured long profile of the Unteraar Glacier. *Journ. Glac.*, Vol. 2, No. 12, 1952, p. 103-07.
6. Mercanton, P. L. Examen de quelques formules pour la prédétermination de l'épaisseur du glacier, à l'occasion de sondages récents. *Geofisica pura e applicata*, Vol. 16, 1950, p. 170–74.
7. Nye, J. F. The flow of glaciers and ice-sheets as a problem in plasticity. *Proc. Roy. Soc.*, A, Vol. 207, 1951, p. 554–72.
8. Hopkins, William. On the Theory of the Motion of Glaciers. *Phil. Trans.*, Vol. 152, 1862, Part II, p. 677–745.
9. Hill, R. *The mathematical theory of plasticity*, Oxford: Clarendon Press, 1950, p. 129.
10. Nádai, A. *Plasticity*, New York: McGraw-Hill, 1931, p. 44–46.
11. Glen, J. W. Experiments on the deformation of ice. *Journ. Glac.*, Vol. 2, No. 12, 1952, p. 111–14.
12. Orowan, E. Discussion. *Journ. Glac.*, Vol. 1, 1949, No.5, p. 231-36.
13. Nye, J. F. A method of calculating the thicknesses of the ice-sheets, *Nature*, Vol. 169, 1952, p. 529.

41

Reprinted from *Spatial Analysis in Geomorphology,* R. J. Chorely, ed., Methuen, London, 1973, pp. 372-375, 388-389

Digital simulation of drainage basin development

B. SPRUNT

Department of Geography, Portsmouth Polytechnic

[*Editor's Note:* In the original, material precedes this excerpt.]

The nature of simulation studies

The term simulation has been applied with a variety of shades of meaning in geomorphological research, as in other fields of research. A very general view of the nature of simulation regards simulation as 'essentially a working analogy' (Chorafas 1965). This is too general for many applications so it is convenient here to restrict attention to what is meant by the nature of simulation as it arises when digital computers are used in the study of systems.

Parts of a system which interact and behave collectively like the system are called components of the system. Components may sometimes be systems themselves, i.e. subsystems. At a particular instant in time, the system is in a specific state which is determined by the instantaneous states of its components. A record of the state of a system is a state description and applies to an instant in time so that a succession of state descriptions in chronological order may be regarded as a state history of the system.

Essential to simulation studies is the concept of a model of a system. Models in geomorphology have been discussed at length elsewhere (Chorley 1967), so that only a few comments will be made here. There may be several possible models of a system but no single model can represent all aspects of a real system. Even when a suitable model of a system has been constructed, that model does not represent the activity or behaviour of the system. It is a state history which provides the necessary representation of behaviour of the system.

Using the terminology introduced above, Evans, Wallace and Sutherland (1967) have produced a definition of simulation which states: 'Given a system and a model of that system, simulation is the use of the model to produce chronologically a state history of the model, which is regarded as a state history of the modeled system.'

From the state history produced by the simulation study it is possible to make an assessment of the adequacy of the model used. If the state history represents the behaviour of the real system in a way that is sufficiently accurate for the purposes of the study then it may be assumed that the model is an adequate one. Without a statement of the purpose of the study and a subsequent testing of the adequacy of the model, a simulation study has little value except as an educational exercise.

Investigation of the correspondence between the components of the model and the components of the system is very important when interpreting the results of a simulation study. It is possible to regard the model as a black box which produces an adequate state history, but the behaviour of the components of the model may not correspond to the behaviour of the components of the real system. In terms of drainage basin simulation, the study may produce a succession of landform surfaces which look real and are an adequate representation of reality, but the network and flow characteristics which produced them may not be of the same form as real stream networks.

Simulation by digital computer has the great advantage that it is completely repeatable and free from the physical limitations of the system being studied. Quantitative data for collection and processing are readily available at all stages of the simulation study making it possible to monitor the behaviour of the system in graphical form via on-line displays.

Inherent in the nature of simulation is the fact that a state history represents a sequence of discrete changes of state within the system being simulated and that two consecutive state descriptions are separated by a finite time step. The consequences of these two facts are discussed in detail by Evans *et al.* (1967, 118–25), and are summarised in the following paragraph.

The choice of the size of the time step in relation to the separation time of events which change the state of the system may be critical in determining the form of the state history produced. Ideally, a 'next event' formulation is required in which the time step is chosen so that events are processed as they occur. Unfortunately this does not appear to be practicable in drainage basin simulation so that a 'fixed time step' formulation has to be used. An important consequence of the fixed time step formulation is that events occurring within the time step are batch processed so that the choice of the time step may radically alter the interrelationships of event occurrences. One further consequence of using the fixed time step formulation is that almost any time step chosen will enable the model to go through the process of simulating.

These comments emphasise the need for a precise statement of the assumptions made in a simulation study, especially if it is the purpose of the study to test a particular hypothesis. Without this information, it is impossible to make a critical assessment of the adequacy of another person's simulation study.

The drainage basin as an open system

A general systems approach to geomorphological problems (Chorley 1962) provides a suitable conceptual framework for building models of drainage basins. The basic concept of the drainage basin as a system is central to all morphometric studies, both real and simulated. Strahler (1964, 4–40) writes:

> Of fundamental importance is the concept of the drainage basin as an open system tending to achieve a steady state of operation. . . . An open system imports and exports matter and energy through system boundaries and must transform energy uniformly to maintain operation. In a drainage basin the land surface within the limits of the basin perimeter constitutes a system boundary through which precipitation is imported. Mineral matter supplied from within the system and excess precipitation leave the system through the basin mouth. In a graded drainage basin the steady state manifests itself in the development of certain topographic characteristics which achieve a time-independent state.

These are the ideas which a successful simulation study must try to illuminate. Early attempts will inevitably make simulation runs simply to see what happens, but ultimately it is hoped that they will be more concerned with energy distributions as well as with the dynamic carving of surfaces.

Chorley (1967, 78) highlights another important feature of geomorphic systems:

> Geomorphic systems can all be considered part of 'supersystems' (e.g. whole landform assemblages) and as being composed of 'subsystems' (e.g. slope or channel segments). . . . In geomorphology subsystems are commonly combined by 'cascading' the output of one subsystem into another to form its input.

This cascading effect is a central feature of the writer's simulation model described in a later section of this chapter.

The distinctive feature of a drainage basin system which makes it a particularly *geographical* system is the characteristic way in which the elements of its three-dimensional surface are spatially interrelated by an implied transportation network resting on the surface. One of the most difficult problems of drainage basin simulation is to simulate the delicate interdependence between the surface and the network. Any adequate simulation of drainage basin development must solve this problem but no published work has yet achieved a satisfactory solution.

Development of simulation models

During the 1960's, simulation studies relevant to drainage basin development followed two main lines of investigation. On the one hand, network models were developed in order to simulate the spatial variation of network characteristics in the landscape, and were, in the majority of cases, unrelated to any underlying surface form. On the other hand, models of slope development in two dimensions provided a more fundamental approach by concentrating on the simulation of erosion processes from which a characteristic form might be produced. Network models have tended to be stochastic in nature, while process models have been predominantly deterministic.

An account of simulation studies which use network models has already been published (Haggett and Chorley 1969). Some aspects of these studies are repeated here because they serve to illuminate the nature and purpose of simulation when applied to three-dimensional models of drainage basins.

Simulations of slope development are of an entirely different nature, using numerical methods of solution to differential equations in order to produce sequences of slope profiles. The profiles may be regarded as state histories of points on a stream or hillslope which indicate stages of development with time.

[*Editor's Note:* Material has been omitted at this point.]

References

CHORAFAS, D. N. (1965) *Systems and Simulation* (Academic Press, New York).

CHORLEY, R. J. (1962) Geomorphology and general systems theory; *United States Geological Survey Professional Paper* 500-B, 10p.

CHORLEY, R. J. (1967) Models in geomorphology; In Chorley, R. J. and Haggett, P (Eds.), *Models in Geography* (Methuen, London), 59–96.

EVANS, G. W., WALLACE, G. F. and SUTHERLAND, G. L. (1967) *Simulation Using Digital Computers* (Prentice-Hall, New Jersey).

HAGGETT, P. and CHORLEY, R. J. (1969) *Network Analysis in Geography* (Arnold, London), 348p.

STRAHLER, A. N. (1964) Quantitative geomorphology of drainage basins and channel networks; In Chow, V. T. (Ed.), *Handbook of Applied Hydrology* (McGraw-Hill, New York).

42

Reprinted from Symp. River Morphol. (General Assembly of Bern, 1967), *Rept. Publ. Staff Members,* Centre S.E. Asian Stud., The University of Hull, 2nd Ser., No. 2, 17-29 (1968)

NATURAL AND MAN-MADE EROSION IN THE HUMID TROPICS OF AUSTRALIA, MALAYSIA AND SINGAPORE

Ian DOUGLAS
Department of Geography
University of Hull, U.K.

ABSTRACT

Under natural rain forest conditions in the humid tropics there is a striking contrast between the great depth of weathered rock material and soil, often exceeding 30 metres, and the very small concentrations of suspended and dissolved matter carried in stream waters. The dense forest vegetation which exists in a delicate ecological balance with the soil, exerts a protective effect against mechanical processes such as raindrop splash and slope wash, but favours the chemical attack on minerals. Variations in the nature of the vegetation cover affect the rate of erosion under natural conditions. In the highlands where, at an altitude of around 2,000 metres, the vegetation has but a single tree storey about 7 metres high, the erosive effects of intense rain are more marked. Where rainforest is replaced by less dense formations in areas of nutrient deficiency soil erosion is greater.

Once the vegetation is removed the rate of erosion is greatly increased. Comparisons between rates of erosion determined on headwater streams, where interference is minimal, and at stations further downstream, where considerable human activity affects river behaviour, are used to illustrate the effects of forestry, grazing, tree crops, cultivation, mining and urban development. A great volume of erosion occurs during the construction stage when large areas of soil are left exposed to the effects of splash and surface runoff. These conditions provide very rapid rates of runoff to streams, producing much higher flood peaks than occur natural conditions with consequent damage to stream banks and riverine structures.

More permanent are the effects of changes in land use on river regimes and on the course of erosion. With increasing human occupance of the catchment areas of the humid tropics, erosion becomes concentrated into short duration flood events which may cause disruption to river users and flood plain dwellers in the lower parts of the catchment areas.

The deep weathering profiles developed under tropical rain forest demonstrate that under hot, humid conditions abundant rock material is transformed into a state suitable for erosion by running water and trasportation by streams. Yet, despite the intense rainstorms which occur in the humid tropics, the weathered material is not eroded as rapidly as it becomes available. Thus the characteristic deep weathering profiles, sometimes more than 30 metres thick, develop. This rotted rock material is protected by the dense rain forest vegetation, with which it exists in a delicate ecological balance. The foliage breaks the force of falling raindrops and roots give stability to the soil. At same time the CO_2 given off by the vegetation and other acids in the soil combine with the rainwater in promoting the chemical attack on the rock minerals. Some erosion takes place under the forest, by suface wash, particularly as a result of water streaming down the tree trunks during heavy rain, by the splash effects of drips from leaves, by the movement of water and colloids laterally through the upper layers of the soil, and by soil creep. Such erosion is however not as rapid as the preparation of rock material by chemical weathering.

The net results of this type of weathering and denudation is that the streams carry far less sediment than the soils in their catchment areas would appear to warrant. At all times the concentrations of suspended sediment in rain forest streams under natural conditions are low, but the frequency of high stream flows, brought about by short duration but intense rainstorms, is such that a considerable amount of material is evacuated per year.

The importance of the vegetation cover in the erosion of humid tropical areas is demonstrated when the structure of the vegetation is changed by altitude or by nutrient or moisture deficiency in soils. On the Cameron Highlands and other elevated areas of West Malaysia, the forest, instead of having a complex canopy reaching up to over 30 metres above the soil, has a single tree storey about 7 metres which is much more easily penetrated by falling raindrops than the dense lowland canopy. Erosion under this montane forest is thus more effective than in the lowlands, and the depth of the weathering profile less than 5 metres, even on the plateau surfaces of the highland area. River beds contain much coarser sediments than do similar sized, equally steep streams in catchments with lowland rain forest vegetation.

The effects of nutrient deficiency on vegetation distribution are particularly marked in north-east Queensland, where the area which can truly be defined as humid tropical is limited to a small mountainous area within 50 km of the east coast. Here outcrop three major groups of rocks, late Tertiary basalts, Mesozoic granites and metamorphosed Palaeozoic sediments. The greater nutrient content of the krasnozemic soils developed on the basalts has enabled rain forest to exist in much drier areas than it would be able to do so on granite or metamorphic rock-derived soils. There thus occur several localities where there is an abrupt transition from tropical rain forest to open sclerophyll woodland. A short belt of wet sclerophyll forest exists at the margin of the rain forest, and has such a dense shrub and ground flora, that this wet sclerophyll forest is as effective in preventing erosion as the rain forest itself. The open woodland however has a very different structure, the trees are so widely spaced that their canopies are not continuous, soils are extremely thin and have a discontinuous grass cover. Thus intense rain beats on the soil surface with little loss of momentum. Very rapidly all the streams in such areas begin to be discoloured by suspended sediment washed from the soil surface. The very lack of nutrients which causes the relatively sparse vegetation is thus perpetuated by the erosion of the upper, organic layers of the soil during wet season rain storms.

Thus under natural conditions a very important cycling of nutrients exists between plants and soils, the losses by erosion are but a very small fraction of the total nutrient mobility under tropical rain forest (Nye, 1961). The stability of this tropical rain forest ecosystem is essential for the maintenance of the fertility of the soils, the water yields of the catchment areas and the prevention of sedimentation in the lower reaches of the rivers. Once any element in this delicately balanced ecological system is disturbed, a whole series of complex consequences follow.

MAN'S EFFECTS ON CATCHMENT AREAS IN THE HUMID TROPICS

Once the vegetation of the tropical rain forest is disturbed, soil deterioration and erosion set in (Nye and Greenland, 1960). The contrasts between catchments little affected by man and those with considerable areas of cultivation in the humid tropics have been discussed by Shallow (1956) in the Cameron Highlands of West Malaysia and by Van Dijk and Vogelzang (1948) in west Java, Indonesia. Shallow showed that

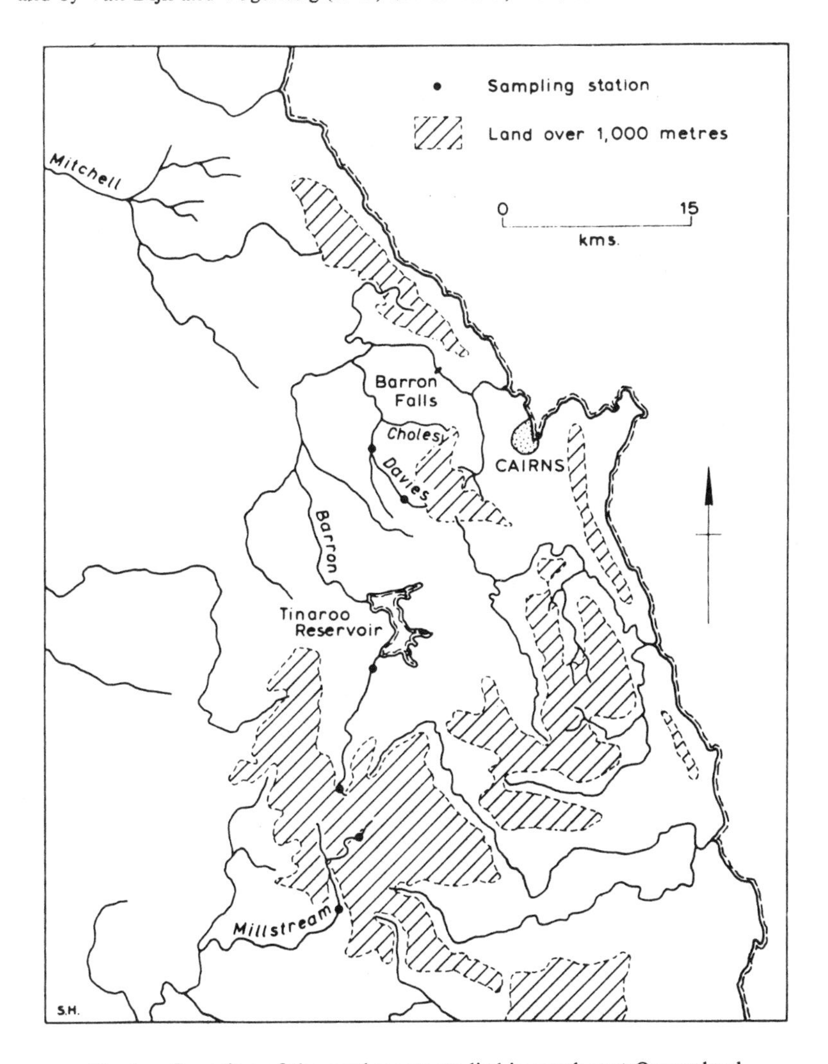

Fig. 1 — Location of the catchments studied in north-east Queensland.

whreas a catchment with 94% of its area under natural vegetation had a sediment yield of 21.1 $m^3/km^2/yr$, a neighbouring catchment with only 64% of the area covered by natural forest had a yield of 103.1 $m^3/km^2/yr$. Van Dijk and Vogelzang compared the sediment yields of the Tjiloetoeng catchment in west Java in the water years 1911–1912 and 1934–35. The total annual rainfall in the two periods was 1797 and 1941 mm respectively, 86% of the total falling in the wet, monsoon season in 1911–1912 and 94% in the later period. The suspended sediment yields in the two periods were 900 $m^3/km^2/yr$ and 1,900 $m^3/km^2/yr$ respectively. The increased rate of soil erosion in the latter period is not due to the slight increase in rainfall but in the authors' own words, "In the Tjiloetoeng basin the gradually increasing deforestation, reckless cultural methods and pasturing after 1917 caused a doubled soil erosion".

These comparative observations have been extended by investigations in northeast Queensland and in Selangor, Malaysia. Four catchments were examined, two stations on each catchment being selected for observations of streamflow and suspended sediment concentrations. The upper station on each river was chosen so that the catchment above it had conditions as near natural as possible, while the lower station was chosen so that the catchment above it would be as homogeneous geologically as possible. The actual sites of the lower stream gauging stations were permanent recording stations operated by local stream gauging authorities.

Three of the catchments were in northeast Queensland (fig. 1) where there is a markedly seasonal distribution of rainfall. River regimes are greatly affected by tropical cyclones, which bring large amounts of rainfall in very short periods, and by heavy thunderstorm rain during the wet season, which extends from December to April. The hydrographs of these rivers show rapid rises and falls following major rainstorms. The rise from base flow to flood peak and back to baseflow may take as little as 10 to 12 hours. The rise to flood peak being particularly abrupt. Most of the work of sediment transport is done by these sudden flood flows, and therefore a particular effort was made to obtain frequent observations during flood flows.

The fourth catchment, the Sungei Gombak, is in Selangor, West Malaysia (fig. 2), and experiences two seasonal rainfall maxima associated with the south-west and north-east monsoons. Hydrographs show the same rapid rise and fall following heavy rain as the Queensland rivers.

The characteristics of the catchments studied are summarised in figure 3 and the results of the observations in figure 4. Despite the steepness of the upper catchment areas, and their higher rainfalls, all the lower catchments had much greater mean annual sediment loads than the headwater streams. It would be expected that under natural conditions higher sediment loads would be recorded in the relatively steep headwater sections. In the larger catchments there is more opportunity for sediment to be deposited along the stream channel. The slopes of the regressions of suspended sediment load on discharge were calculated from individual determinations of instantaneous discharge and suspended sediment concentration. Each pair of observations was tested by analysis of variance to see whether the differences between the regression coefficients and the distance between the regression lines for the upper and lower stations were significant. These statistical calculations are summarised in figure 5.

Each individual correlation of suspended sediment load with discharge is highly significant. This is partly to be expected as the suspended sediment load is calculated as the product of the suspended sediment concentration multiplied by the discharge. It is apparent from the graphs in figure 4 that there is a wide variation in the suspended sediment load carried at any given discharge. On all these relatively small catchments the suspended sediment load carried at any given discharge was much higher on the rising than on the falling stage of a flood (fig. 6). Nevertheless these regressions based on observations through several flood events are accurate enough for a preliminary estimate of the annual sediment load in the period during which observations were made.

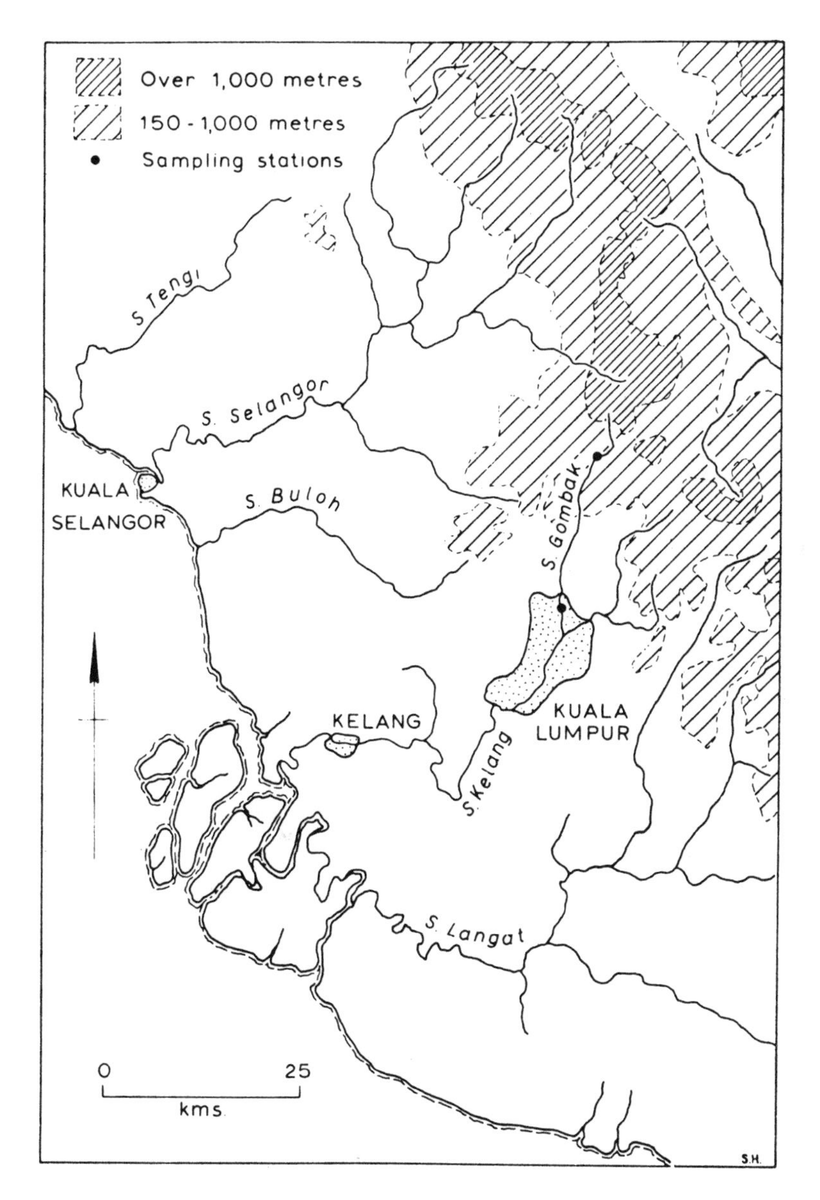

Fig. 2 — Location of the Gombak in Selangor, Malaysia.

Catchment	Barron		Davies		Millstream		Gombak	
	Upper	Lower	Upper	Lower	Upper	Lower	Upper	Lower
Area km^2	11.91	225.33	14.17	105.67	19.81	91.95	26.50	140.00
Mean Annual Rainfall mm	1778	1562	1397	1270	2032	1900	2455	2360
Catchment Relief m	274	587	718	924	274	412	1235	1497
Annual Total Water Discharge m^3/km^2		656903				621165		856802
Estimated Suspended Sediment Load Calendar Years 1963-1964 $m^3/km^2/yr$	5.65	13.60	2.02	3.75	6.15	12.25		67.26

Lithology of catchment areas

Barron	Upper	100% basalt		
	Lower	54% basalt	42% granite	
Davies	Upper	100% granite		
	Lower	82% granite	18% metamorphic	
Millstream	Upper	86% rhyolite	14% basalt	
	Lower	63% rhyolite	37% basalt	
Gombak	Upper	55% metamorphic	45% granite	
	Lower	52% granite	35% alluvium	13% metamorphic

Fig. 3 — Characteristics of the catchments studied.

Only in the case of the Millstream is there a really significant difference between the regression coefficients for the upper and lower stations on the same river. Nevertheless, in all cases the difference between the regression lines for upper and lower stations would be insignificant in less than 0.1 per cent of trials. In nearly all cases the sediment load at the upper station at any given discharge is less than that at the lower station.

CAUSES OF CONTRASTS WITHIN CATCHMENTS

The causes of these differences in sediment loads carried past the upper and lower stations are all related to the activities of man, but in various ways, best explained by taking each catchment in turn.

The Barron River Catchment

The head of the Barron River is high in the basalt country of the Atherton Tableland, where the original rain forest vegetation is still intact, except for a few clearings

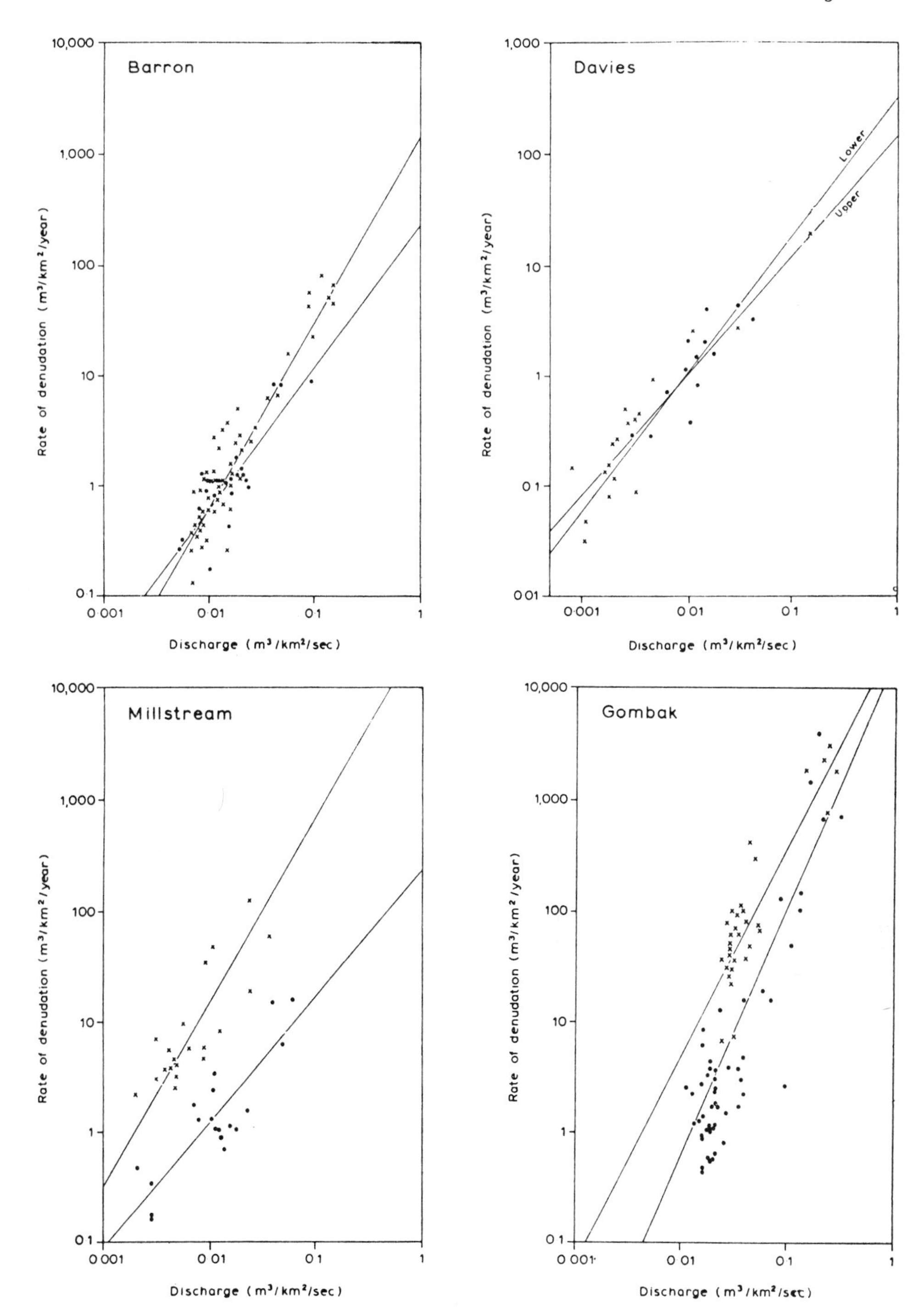

Fig. 4 — Results of discharge and suspended sediment observations.—Observations at the upper stations are represented by closed circles, those at the lower stations by crosses. The regression equations are given on figure 5.

Regression equations of suspended sediment load on discharge

BARRON	Upper	$\log 10y = -1.016 + 1.283 \log 3600x$
	Lower	$\log 10y = -1.558 + 1.677 \log 3600x$
DAVIES	Upper	$\log 10y = -0.718 + 1.085 \log 3600x$
	Lower	$\log 100y = 0.056 + 1.276 \log 3600x$
MILLSTREAM	Upper	$\log 100y = 0.358 + 1.113 \log 3600x$
	Lower	$\log 10y = -1.513 + 1.687 \log 3600x$
GOMBAK	Upper	$\log y = 4.241 + 2.339 \log x$
	Lower	$\log y = 4.490 + 1.885 \log x$

Correlation coefficients and their significance

			Level of Significance
BARRON	Upper	0.860	0.001
	Lower	0.857	0.001
DAVIES	Upper	0.843	0.001
	Lower	0.893	0.001
MILLSTREAM	Upper	0.891	0.001
	Lower	0.878	0.001
GOMBAK	Upper	0.846	0.001
	Lower	0.880	0.001

Levels of significance of differences between upper and lower stations

	Difference between regression coefficients	Distance between regression lines
BARRON	0.20	0.001
DAVIES	N.S.	0.001
MILLSTREAM	0.01	0.001
GOMBAK	0.20	0.001

N.S. denotes not significant.

Levels of significance are determined from Fisher R. A. and Yates F. (1963) *Statistical tables for biological, agricultural and medical research*, Edinburgh, Oliver and Boyd, 6th Edition, 146 p.

Fig. 5 — Statistical results of the relationship between suspended sediment load and discharge.

for cattle pasture. A few kilometres below the source the river enters the granite outcrop of the eastern slope of the Herberton Range. The rain forest only survives on the flatter parts of the granite outcrop, and on the steeper, better drained slopes, gives way to open sclerophyll woodland, which occupies about 25% of the catchment above the lower station. The remaining 75% was originally under the same rain forest as that around the headwaters, but a large proportion of the forest has been cut down in the last 60 years, and dairy farming is practised on the steeper slopes, with maize, potato and peanut cultivation on the flatter slopes. This change in land use has introduced several forms of accelerated erosion. On the very steep slopes where the forest has given way to pasture grasses, soil creep occurs, and at the base of the slope streams undercut their banks. The cattle, walking up and down slope between watering points and grazing areas, accelerate this soil creep. The grass cover permits more rapid runoff to streams

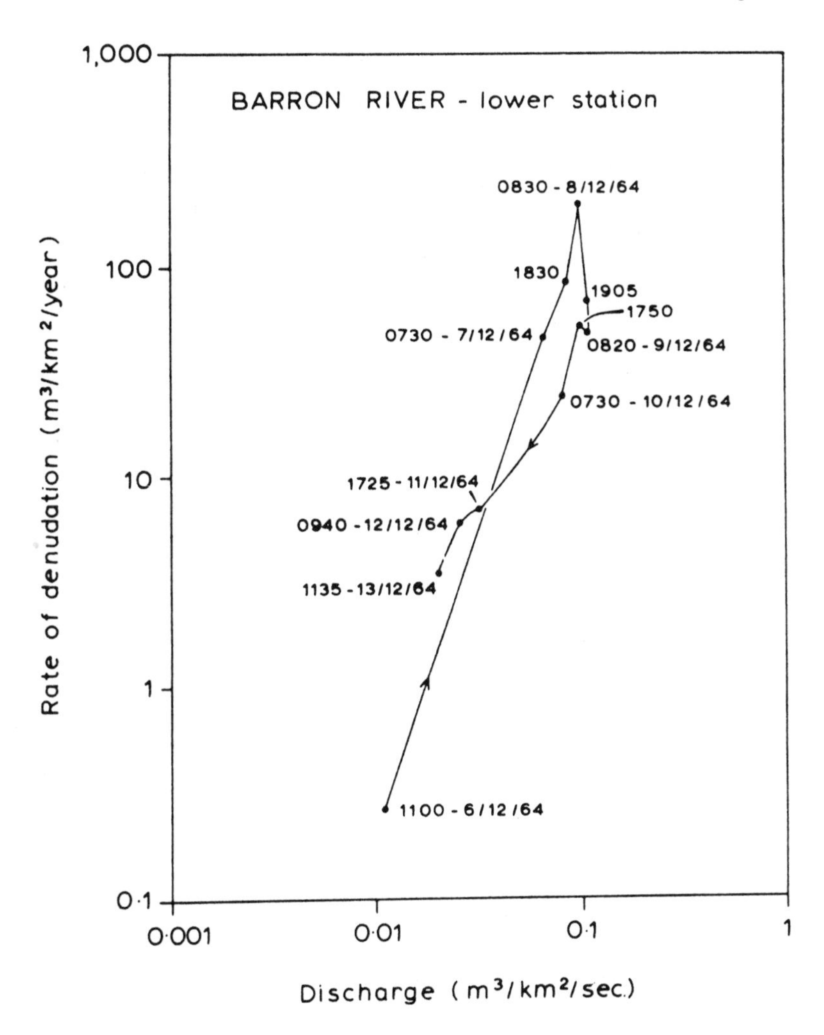

Fig. 6 — Relationship between suspended sediment load and discharge at the lower station on the Barron River during cyclone „Flora" December 1964.

than did the forest cover, causing therefore higher velocities in the streams which increase bank erosion.

In the cultivated area, despite the excellent contour methods employed by the Queensland Soil Conservation Branch, the cultivation of maize entails the exposure of large areas of bare soil at the beginning of the wet season, when the young maize plants start to grow. If, as happened in November 1964, heavy rainstorms occur unusually early, soil losses from maize-growing land may be very great. Unfortunately, not all farmers have accepted the advice of the Soil Conservation Branch, and some instances of rows of crops planted up and down slope can still be found. Soil erosion on these fields following storms when the intensity of rainfall is as much as 40 mm in 30 minutes is very great, and large accumulations of soil are found at the base of every slope planted in this way.

Erosion on the granite slopes where sclerophyll woodland occurs is greater than under natural rain forest, but not as great as in areas where man has exposed the basalt-derived soils. In the latter case, present-day erosion is removing part of the soil material accumulated by past weathering, whereas in the sclerophyll woodland, erosion today can only remove the debris produced by present weathering processes.

The Davies Creek Catchment

This creek is a tributary of the Barron, which enters that river well downstream of the lower station discussed in the previous paragraphs. Davies Creek rises in rugged, rain forest covered country of the Lamb Range, west of the town of Cairns. Deep weathering profiles are found under the rain forest on the granite, with typical unweathered granite core stones which eventually find their way into the stream bed. About 10 kilometres downstream from the head of the creek, as the western slopes of the Lamb Range steepen, and the rain shadow area in the lee side of the mountain from the south east trades is entered, soils become thinner and rain forest gives way to wet sclerophyll forest which, two or three hundred metres further downstream, grades into dry sclerophyll woodland, with rainforest species forming a gallery forest along the stream channel. The upper station studied here is situated just above the falls where Davies Creek begins to cascade down the western slopes of the range. Forestry activity is the only human activity in the catchment above the upper station.

Below Davies Creek Falls the creek reaches the level of the Barron-Mitchell plain, an old erosion surface, into which it is incised about 15 m through ancient podzolic soils on metamorphic rocks. This incision may represent the flood channel of flows with a ten to twenty year recurrence interval, as the road bridge 10 m above the creek bed is covered with flood waters about once a year. At the lower station Davies Creek has a minor channel only 10 m wide, but its flood channel is about 100 m wide and is largely covered with vegetation. Most of the catchment between the upper and lower stations has only had its sclerophyll woodland vegetation modified by grazing beef cattle. A few small areas are cultivated, mainly for tabacco.

The forestry activity in the upper part of the catchment involves much disturbance of the vegetation, not only along the permanent forestry roads, with their cuttings and embankments, but also along the snigging paths, along which logs are dragged by tracked vehicles. If a rainstorm strikes when road construction or logging is in progress, very large quantities of sediment can be washed into streams, as happened during a thunderstorm on January 8th 1964, when a sediment load of 20.7 $m^3/km^2/yr$ was recorded on a rain forest stream above the upper station. In the lower part of the catchment, little disturbance is required to initiate gully erosion in the ancient podzolic soils. Once the gully breaks through the hardpan about 1 metre below the soil suface, headward recession of the gully may be very rapid. Very high sediment concentrations occur in storm runoff waters from these gullies, 1,666 mg/l being recorded in a gully near the lower station, when the concentration in the main stream was only 12 mg /l.Unsurfaced roads and tracks may provide conditions under which gullies are initiated, particularly if they are not maintained. Tobacco crops are harvested before the wet season begins, and quite often the tobacco land is left freshly ploughed and bare in the wet season. Although the Barron-Mitchell land surface has but very gentle slopes, the impact of rainfalls as intense as 25 mm in 30 minutes, may involve considerable loss of soil, and even gully initiation in the cultivated land.

The Millstream Catchment

Rising on the rain forest covered western slopes of the Hugh Nelson Range, this river drains a large area of porphyritic rocks before flowing on to the Atherton Tableland basalts. The upper station is sited at the point where the river begins to flow back onto porphyry after crossing a narrow strip of basalt. This lithologic change is marked by a transition from rain forest to wet sclerophyll vegetation. As the rainfall decreases rapidly west of the Hugh Nelson Range, the rain forest is here near its ecological limits. Below the upper station, the Millstream enters drier country, where even the basalt-derived soils do not support rain forest. These gently sloping basalt soils have been cleared for cultivation, mainly of potatoes, and erosion of the exposed soil results. The scanty vegetation of the lower part of the catchment, and the relatively intense grazing by beef and dairy cattle, leave ample opportunities for erosion by surface runoff. The contrast between the upper and lower stations is well shown by the flood flow of 0.08 m^3/km^2sec from cyclone "Flora" in December 1964, when a suspended sediment load of 34 $m^3/km^2/yr$ was recorded at the upper station as against 95 $m^3/km^2/yr$ at the lower station.

The Sungei Gombak Catchment

This stream drains southwestwards from the main divide of West Malaysia into the Sungei Kelang at Kuala Lumpur. The upper part of the catchment is covered by natural rain forest, disturbed only by limited forestry activity, and by the main road running from Kuala Lumpur into Pahang. Running through the headwater area is a band of partially metamorphosed Palaeozoic sedimentary rocks, which offer a variety of lithologic conditions in contrast to the granite which prevails over the rest of the upper catchment. Both roadworks and forestry activity can provide masses of suspended sediment, as occurred in the course of the observations discussed here. A road widening scheme inevitably resulted in much soil slipping down the steep slopes of the valley, and producing the very high suspended sediment loads recorded at high flows at the upper station. The upper part of the Gombak catchment is so steep, that the river rises in less than an hour from a crystal-clear gentle stream to a raging muddy torrent. Under these conditions of rapid runoff, exposed soil particles are quickly swept into the stream.

On leaving the granite foothills of the main range, the Gombak flows over the tin bearing alluvium which covers the Palaeozoic rocks of the Kuala Lumpur district. This densely populated lowland area is used partly for rice cultivation and partly for large alluvial tin dredging and sluicing operations. The lower station is within the Kuala Lumpur urban area, where a whole series of pollutant effluent enters the river (Norris and Charlton, 1962). This intense human activity, particularly the tin mining, explains the high suspended sediment yields at low discharges at the lower station. The concentration of suspended sediment rarely falls below 100 mg/l, where as that at the upper station at low flows is between 1 and 5 mg/l. Much of the solid material entering the river cannot be evacuated by low flows, and therefore accumulates on the river bed. It enters into transport at high flows, and is carried some distance downstream, only to be deposited again when the river level falls. The net result is a gradual silting up of the river bed, decreasing the carrying capacity of the channel, so that floods occur at smaller discharges than previously. The removal of vegetation and increasing urbanisation further upstream have made runoff more rapid than before, so that flood peaks tend to be higher. Thus the flood plain dwellers and industrial users find they are faced with floods of increasing frequency and magnitude. Works on the Sungei Kelang in the central and southern parts of Kuala Lumpur have improved the effectiveness of the river channel in the lower parts of the town, and have prevented flooding in the central area, but in the Gombak district, the combined problem of flooding and sedimentation remains serious.

Consequences of man-made erosion

The story of the Sungei Gombak is all too often repeated in the humid tropics. Singapore draws most of its water supply from catchments in Johore, West Malaysia. Originally undisturbed rain forest covered catchments, such as those on Gunong Pulai, were used. However, these, and the three small catchmets on Singapore Island, were unable to meet the demand and supplies have had to be drawn from larger rivers such as the Scudai, Tebrau and Johore (Khong Kit Soon, 1963). These three rivers all suffer from high sediment loads caused by agricultural and other activity upstream of the water intakes. At the Scudai intake, a mechanical grab regularly removes large quantities of silt from above the intake weir. Suspended sediment concentrations range from 2 to 140 mg/l, values which are not exceptionally high, but when it is realised that the rainfall over the catchment is about 3,000 mm per year, and the dry weather flow is about 0.004 $m^3/km^2/sec$, these concentrations involve the movement of a great volume of silt each year. The multiple use of these catchment areas involves the Singapore Government in great expense for filter beds and flocculating tanks to remove the silt from the water supply.

In northeast Queensland, multiple use of the Barron waters involves several problems. Downstream from the lower station described previously, the Barron enters the Tinarro Reservoir which impounds water for the Mareeba-Dimbulah Irrigation Scheme. This reservoir has a capacity of 55,450,000 m^3. If the suspended sediment load of the Barron can be applied to the whole catchment area of the dam, the rate of siltation would be 7,730 m^3/yr, giving the dam a life of 7,200 years. In fact siltation will be more rapid than this, as the estimate calculated for the Barron is based on too short a period of observations to allow for hydrologic events of great magnitude, and bed load will increase the quantity of sediment by up to 50% of the suspended load.

Below where Davies Creek enters the Cholesy River, and thus the Barron, is the impounding weir of the Barron Falls Hydroelectric station. Much of the water draining from both the irrigation scheme and the maize growing area of the northern part of the Atherton Table land flows into the Barron carrying with it much sediment during storm runoff. This sediment causes severe problems at the impounding weir, which regularly has to be flushed clear of silt.

These and countless other example of the hydrologic changes wrought by man illustrate the scale of the problem of multiple use of catchment areas in the humid tropics. It is to be hoped that the combined knowledge of all concerned with the factors affecting river behaviour may be made available readily to those who have to make decisions on and those responsible for the management of catchment areas.

Acknowledgoments

The investigations in Australia were carried out during the tenure of a Postgraduate Scholarship at the Australian National University, and those in Malaysia and Singapore were sponsored by the University of Hull. The help of the Irrigation and Water Supply Commission, Queensland, the Commonwealth Bureau of Meteorology, Australia, the Drainage and Irrigation Department, Malaysia and the Water Department, Singapore in providing hydrologic information is most gratefully acknowledged. Laboratory facilities in Queensland were kindly made available by the Department of Primary Industries and the C.S.I.R.O. Tobacco Research Institute, and in Malaysia, by the Department of Geography, University of Malaya. Mr. Bah Tillah collected most of the water samples at the upper station on the Sungei Gombak and Mr. Soh Hoo Hong determined suspended sediment concentrations in Malaysia.

BIBLIOGRAPHY

KHONG KIT SOON, (1963). Formulation of a water resources development plan of the island of Singapore. *Water Resources Ser. Bangkok*, **23**, 149-155.
NORRIS, R. C. and CHARLTON, J. I., (1962). *A chemical and biological survey of the Sungei Gombak*, Kuala Lumpur, Govt. printer, 57 p.
NYE, P. H., (1961). Organic matter and nutrient cycles under moist tropical forest. *Plant and Soil*, **13**, 333-346.
NYE, P. H. and GREENLAND, D. J., (1960). The soil under shifting cultivation. *Tech. Commun. Commonw. Bur. Soils.*, **51**, 156 p.
SHALLOW, P. G. D., (1956). River Flow in the Cameron Highlands. *Hydro-electric Technical Memorandum No. 3 of the Central Electricity Board.*
VAN DIJK, J. W. and VOGELZANG, W. L. M., (1948). The influence of improper soil management on erosion velocity in the Tjiloetoeng Basin (Residency of Cheribon, West Java). *Meded, alg. Proefstn. Landb. Buitenz.*, **71**, 10 p.

43

Reprinted from *Trans. Amer. Geophys. Union,* 39(6), 1076–1084 (1958)

Yield of Sediment in Relation to Mean Annual Precipitation

W. B. LANGBEIN AND S. A. SCHUMM

Abstract—Effective mean annual precipitation is related to sediment yield from drainage basins throughout the climatic regions of the United States. Sediment yield is a maximum at about 10 to 14 inches of precipitation, decreasing sharply on both sides of this maximum, in one case owing to a deficiency of runoff and in the other to increased density of vegetation. Data are presented illustrating the increase in bulk density of vegetation with increased annual precipitation and the relation of relative erosion to vegetative density.

It is suggested that the effect of a climatic change on sediment yield depends not only upon direction of climate change, but also on the climate before the change. Sediment concentration in runoff is shown to increase with decreased annual precipitation, suggesting further that a decrease in precipitation will cause stream channel aggradation.

Introduction—The yield of sediment from a drainage basin is a complex process responding to all the variations that exist in precipitation, soils, vegetation, runoff, and land use. This study is aimed only toward a discernment of the gross variations in sediment yield that are associated with climate as defined by the annual precipitation. Such a study may contribute to an understanding of the effects of climatic change on erosion and of the regional variations in sediment yield. Data on sediment yields are now available in sufficient number for this kind of study, although still quite deficient in geographic coverage. Two major sources of sediment data exist. Records collected at about 170 gaging stations of the U. S. Geological Survey, where sediment transported by streams is measured, is one source of data; whereas, the other source of data is provided by the surveys of sediment trapped by reservoirs. Both kinds of data are used in this study.

Precipitation data—Precipitation is used as the dominant climatic factor in the study of sediment yield, because it affects vegetation and runoff. However, the effectiveness of a given amount of annual precipitation is not everywhere the same. Variations in temperature, rainfall intensity, number of storms, and seasonal and areal distribution of precipitation can also affect the yield of sediment. For example, *Leopold* [1951] in an analysis of rainfall variation in New Mexico, found that despite the absence of any trend in annual rainfall, changes in the number of storms produced a significant influence upon erosion. Although analyses of these effects are beyond the scope of this study, the effect of temperature, which controls the loss of water by evapotranspiration, can be readily taken into account. As is well known, the greater the temperature, the greater are the evapotranspiration demands upon soil moisture; hence, less moisture remains for runoff. More precipitation is required for a given amount of runoff in a warm climate than in a cool climate. Therefore, instead of using actual figures of annual precipitation, it is preferable to use figures of precipitation adjusted for the effect of annual temperature. However, in lieu of carrying out these extended computations, it appears possible to use the data on annual runoff which already reflect the influence of temperature. Annual runoff data are available for all the gaging station records and for most of the reservoir records. Because of the well-established relationships between annual precipitation and runoff, it is readily possible to estimate precipitation from the runoff figures.

We shall define effective precipitation as the amount of precipitation required to produce the known amount of runoff. Figure 1 shows a relationship between precipitation and runoff based on data given in Geological Survey Circular 52 [*Langbein*, 1949]. This graph has been used to convert known values of annual runoff to effective precipitation, based on a reference temperature of 50°F. In a warm climate, with temperature greater than 50°, the precipitation so estimated would be less than the actual amount of precipitation; in a cool climate, the effective precipitation so estimated would be more than the actual amount. This is the desired relationship.

Sediment-station data—In recent years a number of records of sediment yield, as measured at sediment-gaging stations, have become available. Annual loads were computed for about 100 stations giving preference to the smaller drainage areas in any region. All parts of the country,

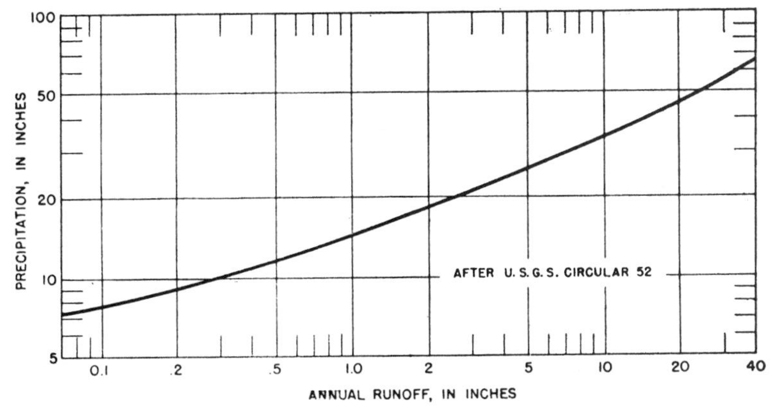

FIG. 1 – Relation between annual precipitation and runoff for a mean temperature of 50° F

where sediment records are collected, are represented.

The annual sediment loads were first arranged according to effective precipitation. They were next assembled into the class groups shown in Table 1, and the arithmetic averages were then computed for each group. Within any group the loads may vary tenfold, reflecting geologic and topographic factors not considered in this study. Each group mean is subject to a standard deviation of about 30 pct. The group averages are plotted in Figure 2. The curve shown was fitted to the data, subject to the condition (1) that it did not depart more than one standard deviation (30 pct) from any of the plotted group means, and (2) that it show zero yield for zero precipitation.

There is considerable opportunity for bias in the figures for load, because the relatively few records prohibit any high degree of selectivity. Few of the rivers drain areas in their primeval environment and moreover, land use can greatly affect the sediment yield. Farming, grazing, road construction, and channelization tend to increase sediment yield; reservoirs impound and, therefore, delay the movement of sediment. If these effects are uniform countrywide, then the overall results might be free of bias in the statistical sense, even though the absolute magnitudes may not be representative of primeval conditions. However, there is considerable variation, particularly with respect to intensity of agricultural operations, which perhaps are most intensive in the midcontinent region. The effects of various kinds of land use upon erosion vary with climate, physiography, soil type, and original vegetation. One surmises that the effect of cultivation is greater in the humid region, where effective precipitation is more than about 30 inches, because of the great contrast between original forest cover and tillage. The erosive reaction of some soil types to cultivation is evident for some small drainage basins in the humid region, which have sediment yields that approach or exceed those usual even in the arid country and are far above those to be expected within their particular range of annual precipitation. For example, sediment

TABLE 1 – *Group averages for data at sediment stations*

Range in effective precipitation	Number of records in each group	Average effective precipitation	Average yield
inch		inch	tons/sq mi
Less than 10	9	8	670
10 to 15	17	12.5	780
15 to 20	18	17.5	550
20 to 30	20	24	550
30 to 40	15	35	400
40 to 60	15	50	220

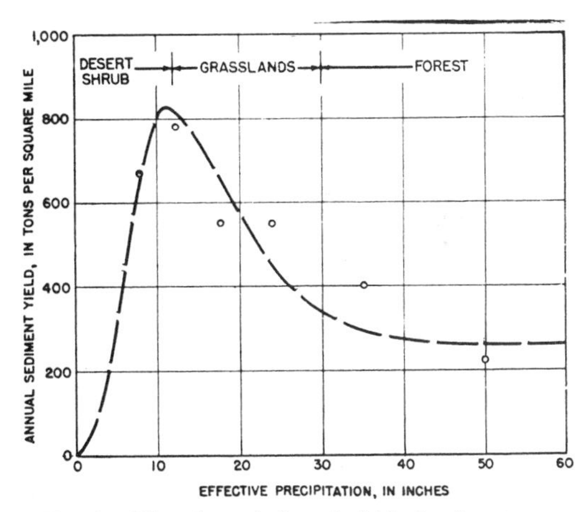

FIG. 2 – Climatic variation of yield of sediment as determined from records at sediment stations

yield from small drainage basins (0.1 to 1.0 sq mi) in the loess hills of Iowa and Nebraska is very high, largely because of poor conservation practices on wind deposited soils. These rates "are among the highest found anywhere in the country" [*Gottschalk* and *Brune*, 1950, p. 5].

Another source of bias is the relatively nonuniform distribution of sediment-gaging stations. Most are in the central part of the country, whereas virtually none is available in the Pacific Coast Region, in New England, or in the Gulf Region.

Reservoir sediment data—Although preference was given to the smaller drainage areas in using gaging-station records of suspended sediment, opportunities for choice in this regard were severely limited. Fortunately, surveys of sedimentation in reservoirs are more numerous, so that there was opportunity to be more selective in choosing reservoirs below small drainage areas, which on that account were presumed to be more indicative of sediment yield nearer the source. Data on reservoir sedimentation were compiled by the *Federal Inter-Agency River Basin Committee* [1953]. Rates of sedimentation were obtained from surveys of sediment accumulation, expressed as an annual rate in acre feet or tons per square mile of drainage area. For those reservoirs where the bulk density of the deposits was determined, the annual rates per square mile are given in terms of tons, otherwise the rates are given in terms of acre feet. In these cases, volumes in acre feet were converted into tons by assuming a density of deposit of 60 lb/cu ft, an average of reported densities. In selecting reservoirs, preference was given to those with capacities exceeding 50 ac ft/sq mi of drainage area, in order to select those which trap a large portion of the sediment that enters the reservoir.

Reservoirs with less than five square miles of drainage area appear to have highly variable rates of sedimentation. For very small areas, rates of sediment yield are greatly influenced by details of land use and local features of the terrain [*Brown*, 1950]. For this reason, reservoirs having drainage areas between 10 and 50 sq mi were used. Because no reservoirs in desert areas were listed in the Inter-Agency compilation, data for desert reservoirs were obtained from unpublished records collected by the U. S. Geological Survey. However, because these reservoirs were on drainage areas of ten square miles or less, rates of sediment yield for these desert reservoirs were adjusted downward to obtain equivalent rates from drainage areas of 30 sq mi, according to the 0.15 power rule explained below. The sediment data were arranged according to effective precipitation and grouped as shown in Table 2.

Group averages of the reservoir data are plotted in Figure 3. The general shape of the resulting curve is quite similar to the one obtained from the records of suspended sediment measured at river stations. The most evident difference is that the yields are about twice those indicated by the sediment-station records. There are significant differences between the two kinds of records. The sediment-station records do not include bed

TABLE 2. – *Group averages for reservoir data*

Range in effective precipitation	Number of reservoirs in each group	Average effective precipitation	Average yield	Remarks
inch		inch	tons/sq mi	
8–9	31	8.5	1400	15 reservoirs in San Rafael Swell, Utah, and 16 in Badger Wash, Colo.
10	38	10	1180	26 reservoirs in Twenty-mile Creek basin, Wyo., 7 in Cornfield Wash, N. Mex., and 5 general.
11	12	11	1500	General
14–25	18	19	1130	General
25–30	10	27.5	1430	General, including debris basins in Southern Calif. considered as one observation.
30–38	20	35.5	790	General
38–40	11	39	560	General
40–55	18	45	470	General
55–100	5	73	440	General

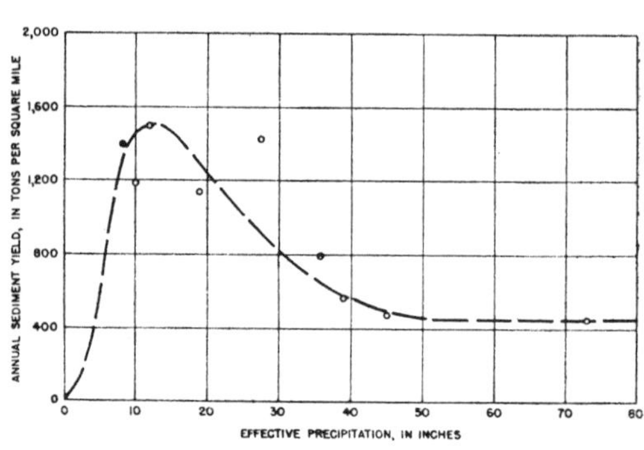

FIG. 3 – Climatic variation of yield of sediment as determined from reservoir surveys

load, which, being the coarser fraction of the load, is trapped by a reservoir. The effects are variable depending on relative amounts of bed and suspended loads at gaging stations and on the trap efficiency of the reservoirs. Moreover, reservoirs are generally built in terrain that offers favorable sites, which means drainage basins with steep slopes and hence higher rates of net erosion. However, in the comparison made here most of the difference is probably due to the effect of size of drainage area. Several studies have shown that sediment yields decrease with increased drainage area, reflecting the flatter gradients and the lesser probability that an intense storm will cover the entire drainage basin. Assuming that the graphs shown by *Brune* [1948] are correct for this effect, rates of sediment yield are inversely proportional to the 0.15 power of the drainage area. Noting that the drainage areas used for Figure 2 average about 1500 sq mi and those for Figure 3 about 30 sq mi, the sediment yields for reservoir data should average about $(1500/30)^{0.15}$, or 1.8 times that for the sediment-station data. This correction applied to the reservoir data would very nearly account for most of the difference between the curves of Figures 2 and 3.

Figures 2 and 3 appear to show a maximum sediment yield at about 12 inches annual effective precipitation, receding to a uniform yield from areas with more than 40 inches effective precipitation. The lack of data for climates with less than 5 inches of annual precipitation makes it difficult to determine the point of maximum yield with accuracy. Available data indicate, however, that it is at about 12 inches or less.

In a similar study of erosion rates and annual precipitation for large rivers of the world (Fig. 4), *Fournier* [1949] notes that the drainage basins are located on a parabolic curve in relation to their climatic character. For example, the upper limb of the parabola (greater than 43 inches) is formed by rivers typical of a monsoon climate: Ganges, Fleuve Rouge (Hung Ho), Yangtze, and some basins of southeastern United States; the middle segment of the curve between 24 and 43 inches of rainfall is formed by drainage systems located in regions with essentially equally distributed annual rainfall, as the Atlantic coastal rivers of the northeastern United States; the lower limb of the curve below 24 inches is formed by rivers draining regions of the more continental steppe or semiarid climates: Vaal, Indus, Rio Grande, Hwang-Ho, Tigris, and Colorado. Fournier concluded that the regions of maximum erosion are those in which monthly rainfall varies greatly, the monsoon and steppe climates.

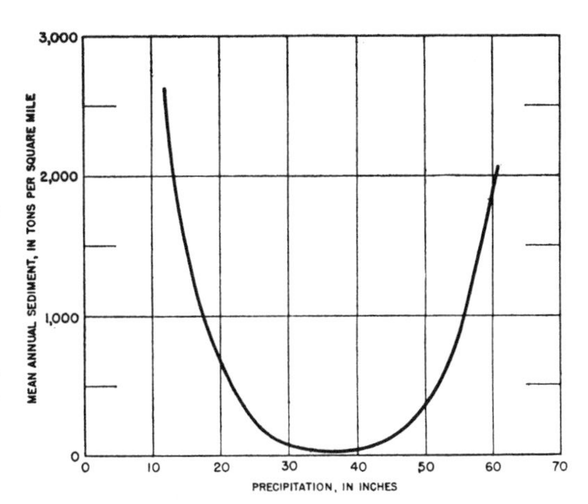

Fig. 4 – Relation of sediment yield to precipitation [after *Fournier*, 1949]

The midpart of his curve shows an annual yield of only five to seven tons per square mile which Fournier attributes to basins in which rainfall is uniform throughout the year such as those in the northeastern United States. These figures are much below that indicated by the few available records in that region which range from a minimum of about 40 tons/sq mi for Scantic River in Connecticut to 370 tons/sq mi for Lehigh River in Pennsylvania.

The lower limb of Fournier's curve terminates at a precipitation of about 12 inches, reaching an annual yield of about 2500 tons/sq mi. Although rates as high and even higher occur in many areas in the arid country, the figure of 2500 tons/year seems somewhat high as an average. The upward trend cannot continue if there is zero sediment yield (in rivers) for zero precipitation; the curve seemingly must reverse its upward trend and swing downward towards the origin.

The upper limb of Fournier's curve (above 43 inches precipitation) shows sharply increasing sediment yield with increasing precipitation, a trend that is not evident in Figures 2 and 3. However, it is possible that additional information in such areas of great rainfall as northern California and the Pacific Northwest may introduce an increasing trend in this part of those graphs.

Analysis of the climatic variation in sediment yield—The variation in sediment yield with climate can be explained by the operation of two factors each related to precipitation. The erosive influence of precipitation increases with its amount, through its direct impact in eroding soil and in generating runoff with further capacity for erosion and for transportation. Opposing this influence is the effect of vegetation, which increases in bulk with effective annual precipitation. In view of these precepts, it should be possible to analyze the curves shown in Figures 2 and 3 into their two components, the erosive effect of rainfall and the counteracting protective effect of vegetation associated with the rainfall.

These opposing actions can be represented by mathematical expressions of the following form

$$S = aP^m \frac{1}{1 + bP^n} \tag{1}$$

in which S is annual load in tons per square mile, P is effective annual precipitation, m and n are exponents, and a and b are coefficients. The factor aP^m in the above equation describes the erosive action of rainfall in the absence of vegetation. The die-away factor $1/(1 + bP^n)$ represents the protective action of vegetation. The factor aP^m increases continuously with increase in precipitation, P, whereas the factor $1/(1 + bP^n)$ is unity for zero precipitation, and decreases with increases in precipitation.

Eq. (1) can not be evaluated by the usual least-squares method. Hence it was evaluated by trial and error, graphical methods yielding the following approximate results.

$$S = \frac{10P^{2.3}}{1 + 0.0007P^{3.33}}$$

for sediment station data and

$$S = \frac{20P^{2.3}}{1 + 0.0007P^{3.33}}$$

for reservoir sediment data.

The factor $P^{2.3}$ describes the variation in sediment yield with constant cover. Analyses of measurements of rainfall, runoff, and soil loss made on small experimental plots operated by the Department of Agriculture [*Musgrave*, 1947], indicate that, other factors the same, erosion is proportional to the 1.75 power of the 30-minute rainfall intensity. However, it is rather difficult to draw a connection between the intensity of 30-minute precipitation and annual precipitation. Inspection of *Yarnell's* [1935] charts indicates one relationship exists in the eastern part of the country and another in the western areas. However, in both areas 30-minute intensities vary with annual rainfall to some power greater than unity. Hence, one can conclude that erosion will vary regionally to some power of the annual precipitation greater than 1.75.

The second factor, $1/(1 + 0.0007 P^{3.33})$, equals $S/aP^{2.3}$. This function, as graphed in Figure 5, purports to isolate the variation in sediment yield caused by differing degrees of vegetative cover. This function varies as shown in Table 3.

There is a good deal of information on the relation between different vegetal covers and rates of erosion in a given climate. However, most of this information deals with cultivated lands and

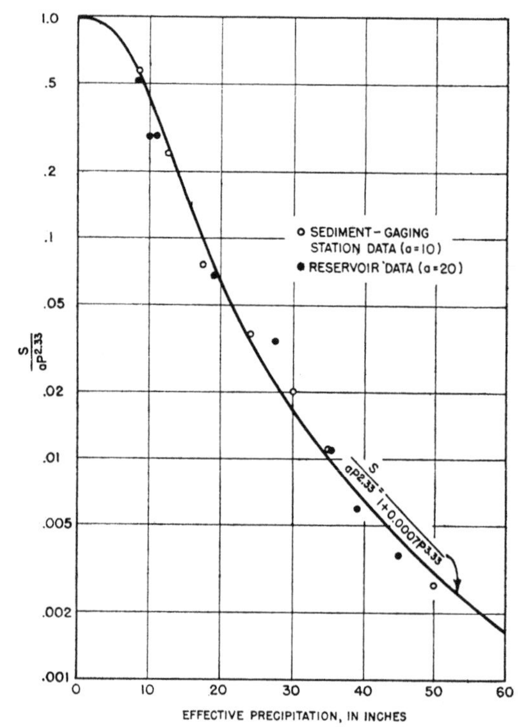

FIG. 5 – Decrease in relative sediment yield with increasing precipitation

TABLE 3 – *Variation in sediment yield associated with vegetative cover*

Effective precipitation	Vegetative cover[a]	$\frac{1}{1 + 0.0007 P^{3.33}}$
inch		
7	Desert shrub	0.69
13	Desert shrub	0.23
20	Grassland	0.06
30	Grassland	0.017
40	Forest	0.006
50	Forest	0.003

[a] Associated with effective precipitation.

few of the vegetal data are in quantitative terms. *Musgrave* [1947] attempted a quantitative evaluation of relative erosion based on data collected at experimental watersheds in the Pacific Northwest. The results agree quite well with the results given in Table 3.

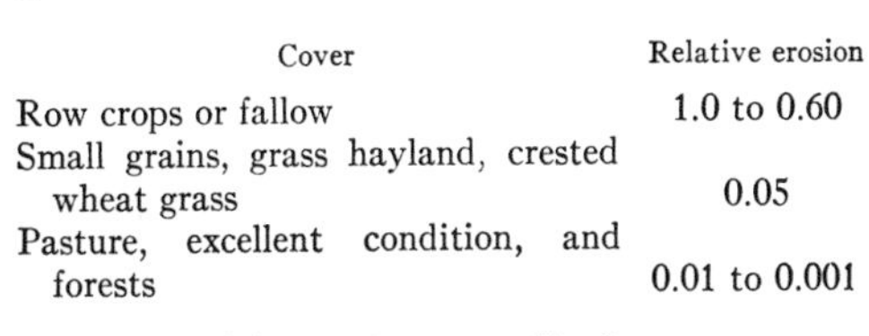

Cover	Relative erosion
Row crops or fallow	1.0 to 0.60
Small grains, grass hayland, crested wheat grass	0.05
Pasture, excellent condition, and forests	0.01 to 0.001

Formula (1) may be generalized as

$$S \propto R/V \tag{2}$$

where R is annual runoff, and V is mass density of vegetation. In any given region the two factors operate separately; thus, sediment loads may vary with runoff depending on land use and vegetal conditions. For example, *Brune* [1948] shows that, for a given land condition in the Midwest, sediment yield increases with runoff, and that, for a given runoff, sediment yield varies enormously with percent of tilled land. The present study, however, treats of the broad climatic variation in which both runoff and vegetation are each uniquely related to effective precipitation.

With increasing precipitation, sediment yield varies as shown on Figures 2 and 3, but runoff increases as shown on Figure 1. The ratio between sediment yield and runoff is a measure of the concentration. This quantity is generally reported in parts per million (ppm) by weight and may be computed by dividing sediment yield in tons per square mile by runoff computed in tons per square mile. Figure 6 shows results of this computation for the data in Table 2. The concentration decreases sharply with increasing precipitation.

Annual precipitation, as indicated by the annual runoff, is used as the sole climatic measure. We have considered differences in precipitation intensity and its seasonal distribution only so far as these influences are reflected in the amount of annual runoff. For example, low precipitation regimes are characteristically more variable than those of humid regions [*Conrad*, 1946], and, indeed, the short-period excesses in intensity show up in the runoff. However, we repeat, as we wrote in our introduction, that although climatic influences on sediment are more complex, a good deal can be learned from consideration of the annual precipitation.

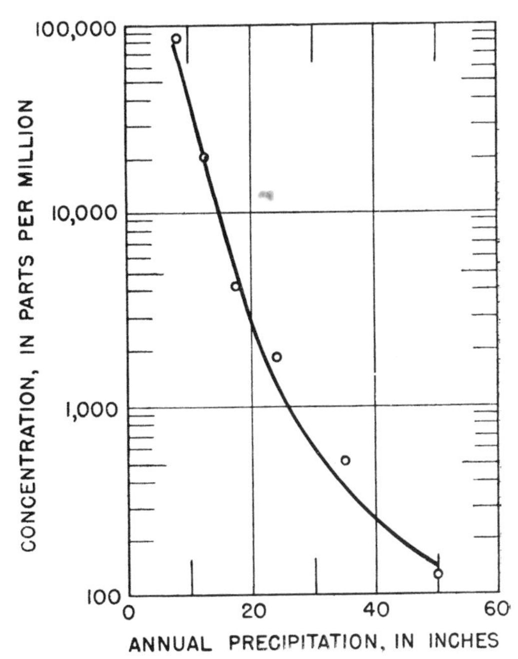

FIG. 6 – Variation of concentration with annual precipitation

Precipitation and vegetation—There can be no question of the highly significant effect of vegetation on erosion. For this reason, we have assembled information on climatic variation and vegetal bulk. The information on vegetal bulk contained in Table 4 was obtained mainly from published sources, ranging in reliability from carefully weighed quadrats to forest statistics and two estimates based on examination of photographs (for references, see Table 4). However, considering the more than 1000-fold variation in vegetal weight, as between desert shrubs to forests, great precision does not seem to be needed for the rough kind of study that seems possible at this time. Some of the data on vegetal weights were given directly in pounds per acre or equivalent. The forest data were obtained by dividing the reported cubic-foot volumes of saw-timber and pole-timber trees, given in millions of cubic feet for each state, by the respective forest area in acres. Unit weights of 45 lb/cu ft were used for hardwoods, 35 for soft woods, and 40 for mixed forests.

Table 4 also includes data on mean annual precipitation and temperature applicable to each case. The climatologic data were not usually given in the references cited and were obtained from U.S. Weather Bureau reports.

Figure 7 shows a plot of precipitation against

TABLE 4 – *Climatologic data and data on weight of vegetation*

Location	Type of vegetation	Mean annual precip.	Mean annual temp.	Weight of vegetation	Reference for vegetal bulk
		inch	°F	lb/ac	
Las Vegas, Nev.	Desert shrub	5	65	100	*McDougal* [1908, pl. 28]
Salt Lake Desert, Lakeside, Utah	Desert shrub	8	50	400	*McDougal* [1908, pl. 24]
Clark Co., Idaho	Sagebrush	12	40	891	*Blaisdell* [1953]
Fremont Co., Idaho	Sagebrush	12	40	1,273	*Blaisdell* [1953]
Coconino Wash, Ariz.	Grass	15	45	1,886	*Clements* [1922]
Burlington, Colo.	Grasses	17	52	2,251	*Weaver* [1923]
Phillipsburg, Kans.	Grasses	22	52	3,230	*Weaver* [1923]
Lincoln, Nebr.	Grasses	27	51	4,467	*Weaver* [1923]
Sandhills, Nebr.	Wheat grass	18	49	4,000	*Smith* [1895]
Lincoln, Nebr.	Grasses	27	51	6,224	*Kramer* and *Weaver* [1936]
Fraser forest, Colo.	Lodgepole pine	25	32	43,000	*Wilm* and *Dunford* [1948]
Rocky Mt. States	Conifers	28	38	54,000	*U. S. Forest Service* [1950]
Northeast Central States	Mixed forest	30	43	64,000	*U. S. Forest Service* [1950]
Northeast States	Hardwood forest	42	45	55,000	*U. S. Forest Service* [1950]
Southeast States	Mixed forest	51	60	48,000	*U. S. Forest Service* [1950]
Pacific Coast States	Conifers	64	47	150,000	*U. S. Forest Service* [1950]
Serro do Navio, Amapa Terr., Brazil	Hardwood forest	120		870,000	Field estimate by M. G. Wolman, 1956

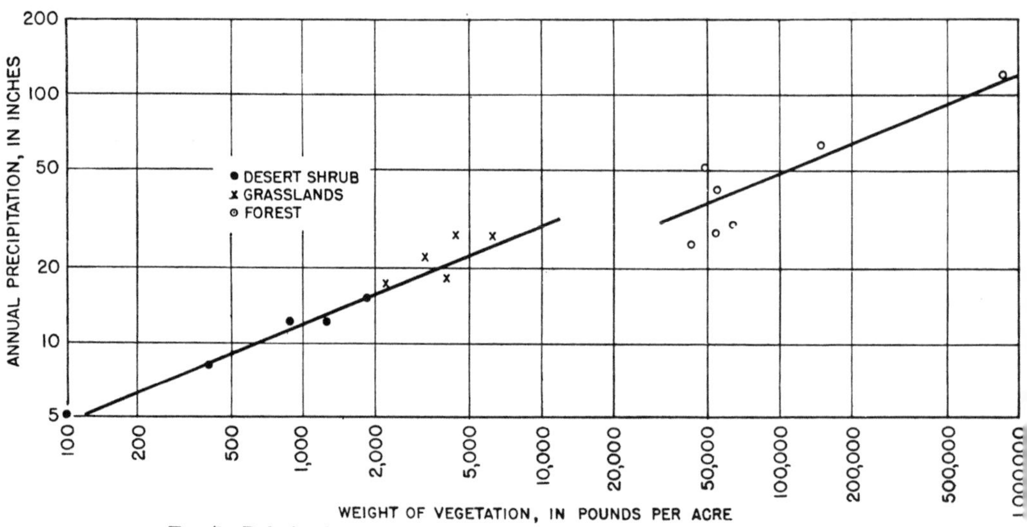

FIG. 7 - Relation between precipitation and weight of vegetation per unit area

vegetation weight. For the data available, the correlation seems quite high with a decided break between forest and nonforest types. The graph indicates that for equivalent rainfall, forests have about five times as much weight as grasses. With a longer life span, trees should understandably show greater total weight in place, although perhaps the annual growth (= annual decay for equilibrium) would be less than for grasses.

The seeming fact that desert shrubs are on the lower continuation of the line defined by the grasses seems anomalous. In arid and semi-arid regions the increase in vegetal bulk with rainfall rather simply reflects increasing opportunity for a greater number of plants, greater opportunity for each plant to reach maximum development for the species environment, and opportunity for growth of larger species. However, this does not explain the variation of forest bulk with rainfall greater than needed to satisfy optimum evapotranspiration demands for the climate. Forests are areas of water surplus in the climatic sense, yet vegetal density seems to vary with precipitation and temperature. Among the eastern states, for example, the vegetal bulk per unit area in Maine and North Carolina are about the same. The lesser annual precipitation in Maine appears to be compensated by a lesser temperature; whereas, in North Carolina higher annual precipitation is compensated by a higher temperature. The forested areas of Washington, Oregon, and California have about the same temperature; the forest densities seem to follow precipitation as follows:

State	Precipitation inch	Vegetation lb/ac
Washington	80	177,000
Oregon	60	158,000
California	53	120,000

The variation in unit weight shown in Figure 7 is made up of two components, one due to variation in weight among different communities of the same vegetation type, and that due to variation in weight among different types. The latter is very likely the dominant factor in the relationship on Figure 5. Beyond a certain limiting precipitation, say that for which precipitation is adequate to meet all evapotranspiration requirements, differences in vegetal bulk may reflect not so much growth factors as differences in plant types or associations. The heavy vegetal bulk in the Pacific Northwest, for example, may be the reflection of a difference in plant type rather than a direct effect of precipitation on growth.

We are considering here only gross relations, ignoring rather important variations that might be due to differences in species, topographic setting, or moisture conditions that might favor or discourage growth. For example, there are patches of timber in the valley bottoms, in the Great Plains, with weight densities far exceeding that of the grasslands. The data used to define this relationship are admittedly crude and subject to bias. The forest statistics, for example, generally exclude bark, leaves, flowers, fruit, and most branches. The ratio of these parts to the whole tree decreases with age. The existing data are for stands in various degrees of maturity, and most existing data exclude roots. The ratio of roots to aboveground growth is variable among different kinds of plants and may be a large source of error.

Then again, although one might conceive that the maximum amount of plant material should theoretically be correlative with climate, other factors such as aspect, depth, and nature of soil are of major influence. Ideally, vegetal den-

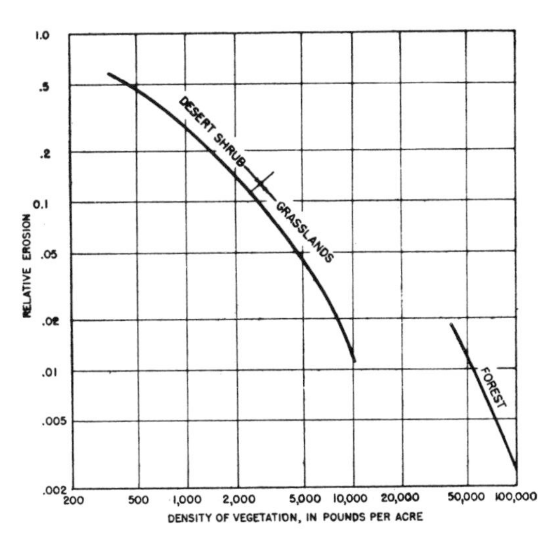

FIG. 8 – Relative erosion compared with density of vegetation

sities should be studied locally to arrive at a normal density for the regional climate. However, this kind of study would be beyond the scope of this discussion. Only the evident fact that vegetal densities are so variable over the range of climates experienced in this country makes it at all possible to use the existing data.

Interpreting the graph in Figure 5 as an indication of relative erosion associated with vegetation, as shown on Figure 7, the relationship shown on Figure 8 can be drawn. According to this graph, a relative change in vegetal density is effective on erosion throughout the climatic range, although the break between trees and grass suggests that per pound, grass is more effective in retarding erosion than trees.

Erosion and climate change—Examination of Figures 2 and 3 may be useful in visualizing not only variations in rates of net erosion between climatic zones in the United States but also the probable change in rates of erosion and stream activity during a climatic change.

Within the 0- to 12-inch precipitation zone an increase in annual rainfall would apparently be followed by an increase in erosion and vice versa; whereas, between about 12 to 45 inches of rainfall, erosion should decrease with increased precipitation. Above 45 inches of precipitation, erosion should remain about constant with increased precipitation, although Fournier's curve (Fig. 4) shows a marked increase in sediment yield above 43 inches of precipitation.

The direction of a change in sediment yield

with changing rainfall appears to be dependent on the amount of precipitation before the change. For example, in a drainage basin located in a region with mean precipitation ranging from about 10 to 15 inches, a change either to a wetter or drier climate might result in a decrease in erosion, in the one case owing to increased density of vegetation and in the other case owing to a decrease in runoff. The above discussion assumes unchanged temperature, but perhaps a change in mean annual precipitation would be accompanied by an inverse change in mean annual temperature, further enhancing its effects.

A change in stream character and activity, with climate change, can probably be understood best in relation to the changes in the ratio of sediment load to discharge as precipitation increases or decreases. Referring to Figure 6, it is apparent that as annual precipitation decreases, the concentration of sediment per unit of runoff increases. This suggests quite strongly that, other factors being the same, the increasing sediment loads associated with increasing dryness will cause aggradation, in an amount depending on the magnitude of the climatic change. *Mackin* [1948, pp. 493–495] has summarized changes to be expected in stream activity with changes in load and discharge. In every case, an increase in load or decrease in discharge with constant load results in aggradation and vice versa.

The decrease in annual runoff with decreased precipitation will necessitate an adjustment of stream gradient and shape according to established principles [*Leopold* and *Maddock*, 1953], such that the width and depth of the channel should decrease and gradient increase. These changes are consistent with aggradation. Of course, an increase in precipitation might be expected to result in degradation as sediment concentration decreases. The increased discharge will result in an increase in channel width and depth and a decrease in gradient. Numerous exceptions to the above generalizations can be cited, especially when glaciation, deforestation, cultivation, or a change in base level become important.

References

Blaisdell, J. P., Ecological effects of planned burning of sagebrush-grass range on the Upper Snake River Plains, *U. S. Dept. Agr. Tech. Bul. 1075*, 1953.

Brown, C. B., Effects of soil conservation, *Applie sedimentation*, P. D. Trask (ed), pp. 380–406, Joh Wiley and Sons, 707 pp., 1950.

Brune, Gunnar, Rates of sediment production i midwestern United States, *Soil Cons. Serv. TP-6* 40 pp., 1948.

Clements, F. E., Destruction of range by prairie dog *Carnegie Inst. Washington Yearbook 21*, 1922.

Conrad, V. A., *Methods in climatology*, Harvard Univ Press, 1946.

Federal Inter-Agency River Basin Comm., *Summary of reservoir sedimentation surveys for the Unite States through 1950*, Subcommittee on Sedimentatior Sedimentation Bul., 31 pp., 1953.

Fournier, M. F., Les facteurs climatiques de l'erosio du sol, *Bul. Assn. Geogr. Francais 203* (Seance d 11 Juin 1949), 97–103, 1949.

Gottschalk, L. C., and G. M. Brune, Sediment design criteria for the Missouri Basin loess hills, *Soi Cons. Serv. Tech. Pub. 97*, 21 pp., 1950.

Kramer, J., and J. E. Weaver, Relative efficiency c roots and tops of plants in protecting the soil fron erosion, *Nebr. Univ. Cons. & Surv. Div. Bul. 12*, p 94, 1936.

Langbein, W. B., Annual runoff in the United States *U. S. Geol. Surv. Circ. 52*, 11 pp., 1949.

Leopold, L. B., Rainfall frequency: An aspect o climatic variation, *Trans. Amer. Geophy. Union*, **32** 347–357, 1951.

Leopold, L. B., and T. Maddock, Jr., The hydrauli geometry of stream channels and some physiographi implications, *U. S. Geol. Surv. Prof. Paper 252*, 57 pp 1953.

Mackin, J. H., Concept of the graded river, *Bul. Geol Soc. Amer.*, **59**, 463–512, 1948.

McDougal, D. T., Botanical features of North American deserts, *Carnegie Inst. Washington Pub. 99*, 1908

Musgrave, G. W., The quantitative evaluation o water erosion—A first approximation, *J. Soil Wate Cons.*, **2**, 133–138, 1947.

Smith, T. G., Forage conditions of the prairie region *U. S. Dept. of Agr., Yearbook*, 1895.

Weaver, J. E., Plant production as a measure o environment, *Carnegie Inst. Washington Yearbook 22* 1923.

Wilm, H. G., and E. G. Dunford, Effect of timbe cutting on water available for streamflow, *U. S. Dept of Agr. Tech. Bull. 968*, 12–14, Nov. 1948.

Yarnell, D. L., Rainfall-intensity frequency data *U. S. Dept. Agr. Misc. Pub. 204*, 67 pp., 1935.

U. S. Forest Service, Basic forest statistics of the U. S as of Jan. 1945, Sept. 1950.

U. S. Geological Survey, Washington, D. C. (W.B.L. and U. S. Geological Survey, Denver Federal Center Denver, Colorado (S.A.S.)

(Communicated manuscript received April 25, 1958.)

44

Reprinted from *Permafrost in Canada. Its Influence on Northern Development,*
University of Toronto Press, Toronto, 1970, pp. 204–207

PERMAFROST IN CANADA. ITS INFLUENCE ON NORTHERN DEVELOPMENT

A. J. E. Brown

[*Editor's Note:* In the original, material precedes this excerpt.]

CONCLUSION

An examination of the ways in which permafrost affects the various spheres of man's existence in northern Canada shows that it exerts considerable influence on his activities there. Whether the permafrost occurs in scattered islands as near the southern limit of the discontinuous zone, or whether it is widespread and hundreds of feet thick as in the continuous zone, it is a factor meriting serious consideration in the development of an area.

It is possible to divide the permafrost region into subregions of varying types of permafrost problems and the means of coping with them. For example, one well-established division is discontinuous and continuous zones. Areas of thawed ground are found in the former zone in contrast to the latter zone where permafrost is found everywhere beneath the ground surface. Therefore, it is sometimes possible to avoid permafrost in the discontinuous zone and employ methods used in temperate regions whereas it is impossible to avoid permafrost in the continuous zone. Construction and other activities in the discontinuous zone are complicated, however, by the erratic and often unpredictable occurrence of permafrost and the proximity of its temperature to 32° F.

In the construction of buildings, for example, permafrost may be avoided in the discontinuous zone and the structure designed for the properties of the soil in the thawed rather than the frozen state. Frozen ground can often be thawed and prevented from reforming by stripping the vegetation and allowing the heat from the building to maintain the soil in the thawed state. Problems can arise, however, where the permafrost, which is close to 32° F, may thaw slowly during the life span of the structure. In the continuous zone, it is impossible to avoid permafrost and buildings must be designed in most cases for the properties of the soil in the frozen state.

Roads and railroads are slightly different because they extend many miles over which the permafrost may change in extent and thickness. In the discontinuous zone, it is possible to avoid permafrost in some areas and design for the properties of the soil in its thawed state. In the continuous zone, the soil is perennially frozen everywhere and construction techniques must be modified accordingly.

Mining presents a somewhat different situation because the site of this activity is determined by the location of the orebodies. In the discontinuous zone, an orebody may be either unfrozen, partly frozen, or completely frozen depending on the extent and thickness of permafrost in the vicinity. In the continuous zone, an orebody may lie in or beneath the permafrost depending on the thickness of the frozen ground.

The relation of agricultural activities to permafrost distribution is similar to that of buildings. In the discontinuous zone, it is frequently possible to avoid permafrost whereas in the continuous zone permafrost is encountered everywhere near the ground surface.

Although there are broad differences between the discontinuous and continuous zones, it is difficult to subdivide these zones. This is because human activities are concentrated at scattered points separated by vast tracts of uninhabited land and the development at any one spot is conditioned by local factors. The method employed at any point of activity is to evaluate the existing natural conditions and use the best possible techniques of construction and exploitation keeping in mind transportation costs, availability of local materials, labour and other cost factors. It will probably be many years before human activities are sufficiently widespread across the permafrost region to enable the subdivision of each permafrost zone into areas based on categories of permafrost problems and their solutions.

Because permafrost occurs widely in Alaska and the USSR some details of their experience have been included. The physical properties and effects of permafrost are similar in the three countries and the methods of coping are generally similar. In each country it is recognized that kept frozen, soils having high ice contents provide good bearing capacity for structures, but that allowed to thaw, they lose this bearing strength. Structures founded on such soils will be damaged or even destroyed. The location of transportation routes in permafrost regions is governed by the same principles in each country: to locate the route on well-drained soils with low ice contents and to prevent thawing of the base course – which causes severe settlement of the grade and makes the route unusable. Mines in permafrost regions of the three countries face such problems as water seepage into shafts and drifts resulting in massive ice accumulation, and frozen ore which resists blasting and other conventional extraction methods used in temperate areas. Agriculture is hampered by such problems as the presence (because of the proximity of the permafrost table) of bodies of ice in the ground; these thaw when the land is cleared, causing such large depressions that farm machinery cannot be used for cultivation. The above are but a few of the major problems which are an indication of the many-sided influence of permafrost.

Economic factors and the development of certain practices mean that these countries do differ in their approach to the problem of permafrost. For example, in some settlements in Canada and Alaska, water and sewage lines are carried in utilidors above ground. In the USSR, on the other hand, where utilidors are rarely used, one method is to excavate a trench several feet deep, bury the pipe in this trench, backfill with coarse-grained soils which are not frost-susceptible, and heat the water or sewage to a sufficiently high temperature to prevent freezing. Another method is to place the pipes in a precast concrete conduit which is half buried in the ground above the permafrost table. In Canada buildings are mostly one or two storeys high and frequently of wood frame construction on wood pile foundations, in contrast to the USSR where many buildings up to ten storeys are constructed of concrete blocks or precast concrete panels on precast reinforced concrete piles. The difference here appears to be that in the USSR large cities with large populations have arisen in the permafrost region as a result of widespread exploitation of natural resources; and the need for large structures to accommodate the various functions of these cities has been recognized even though the cost of construction of such buildings on permafrost is high and greater than in temperate regions. On the other hand, settlements in Canada's permafrost region are small and few in number. The need for large multi-storey buildings is not pressing and smaller wood frame buildings are adequate and perform well. Thus although it would be possible to construct large masonry buildings in Canada's North they are not required.

Permafrost is only one of many factors which hamper developments in northern Canada. The severe climate and brief summer season drastically shorten the length of the construction and agriculture seasons. Long winters raise annual heating costs. The long period of continual darkness in winter is both a physical and psychological deterrent to any operation. Land transportation routes into the North are limited and water routes are usable for only a few months in the summer. Because of the remoteness of the area from southern regions and the difficulty of marketing exploited resources so as to compete economically with similar ones in temperate regions, resource exploration and exploitation is limited. It is not lack of technical knowledge so much as logistical problems and economic considerations which have caused the development of the permafrost region of Canada to be slow.

Permafrost is only one of a number of factors influencing activities but it is a phenomenon common to the entire north of Canada. Its generally adverse effects result in increased costs of construction and operation thus impeding the possibilities for settlement and the prospects of transportation, mining and agriculture.

In spite of these special problems not encountered in temperate regions, development of the permafrost region will continue at an accelerated rate – the ability to cope has already been demonstrated, and improvements in technology, combined with careful site selection and evaluation of the permafrost problem, indicate

45

Reprinted from *Land Evaluations,* G. A. Stewart, ed., Papers of a CSIRO Symp.
UNESCO, 1968, pp. 12-13, 15, 21-22, 25-28

REVIEW AND CONCEPTS OF LAND CLASSIFICATION

J. A. Mabbutt

[*Editor's Note:* In the original, material precedes this excerpt.]

THE GENETIC APPROACH

Introduction

Attempts to arrive at distinctive land units by repeated subdivision on the basis of causal environmental factors may be grouped as the genetic approach. This has its origins in the development of physical geography in the nineteenth century under the influence of botanists and geologists concerned with the genetic groupings of natural phenomena and the environmental controls of their associations and distributions.

Reconciliation of this outlook with a primary aim of the geographer, namely the recognition of unity in diversity as shown in the distinctive character of portions of the earth's surface, led to the concept of genetically based, 'natural' regions.

Theoretical arguments in support of such a genetic scheme are:

1. It is a logical breakdown, and similarities between widely separated areas should be predictable where the basic controls are similar.
2. It offers a rational hierachy and should allow further investigation and subdivision within the one framework.
3. It has the promise of universality.

'Natural' and Morphologic Divisions

In the English literature, perhaps the best-known of such schemes is that of Herbertson (1905), which may be regarded as typical. Herbertson's types of natural regions are based on climate, 'since it not merely affects the physical features, but also because it best summarizes the various influences acting on the surface.' The limiting criteria of temperature and amount and distribution of rainfall actually employed have been chosen (like those of later world climatic classifications) for their significance for natural vegetation, this being regarded as an expressive synthesis of the whole complex of climate.

Herbertson claimed that configuration entered in at a lower level in his scheme of subdivision, but although his regional types are defined in climatic terms, closer study of his maps shows that morphology is the overriding criterion used in the actual delimitation of regions, and that climatic boundaries have been adjusted where necessary to avoid conflict. Linton (1951) has pointed out that most so-called natural regions are in fact morphologically delimited. Climatic boundaries are zonal, after all, nor can realistic land units be delimited on criteria which are invisible or hypothetical (as 'natural vegetation' is for an increasingly large part of the earth).

The Davisian interpretation of land forms as a function of structure, process, and stage enabled morphologic subdivision also to proceed on genetic lines, and this is reflected in the division of the United States by Fenneman (1916), in which the physiographic provinces are areas characterized by 'unity or similarity of physiographic history'.

[*Editor's Note:* Material has been omitted at this point.]

THE LANDSCAPE APPROACH

Introduction

Tired of fitting boundaries which did not exist around areas which did not matter, regional geographers abandoned the search for the elusive 'natural region' and sought real objects of study in distinctive parts of the observed environment. As a system of regional geography, the method is particularly associated with Unstead (1933), who aimed to show how the small regions so identified could be combined into areas of successively higher order . . . 'to a system of regions developed in this way the term "synthetic" may be applied . . . the process is one, not of division and subdivision of areas, but of combination of the smaller regions in order to arrive at the larger ones. By this means more accuracy is assured . . .'

Unstead's primary units had to be small enough for detailed scientific investigation and at the same time be distinctive geographical entities. He proposed for them the term 'stow', identified by unity of relief, though with characteristic structure, hydrology, plant cover, and land use. Nevertheless he recognized that the stow was made up of yet smaller components or 'features' of the status of minor land forms. His second order unit, the 'tract', consisted of a grouping of types of stow, and its unity might derive from one or all of relief, structure, and soils. Above the tract were levels of regional grouping expressive of characteristic relief type or climate.

At about this time, people concerned with land assessment for such practical purposes as forestry and economic planning also began to analyze terrain patterns at the scale of visual experience. I shall use 'landscape approach' for the path to land classification via the areal identity of land. Methods employed by various groups, although developed somewhat independently of each other, show striking similarity of working concept and of descriptive structure.

[*Editor's Note:* Material has been omitted at this point.]

THE PARAMETRIC APPROACH

Introduction

One may define the parametric approach as the division and classification of land on the basis of selected attribute values. For instance, a hypsometric map demonstrates a classification of land based on elevation, with class limits at chosen contours. Employment of the parametric approach ranges from general-purpose surveys considering many attributes, to classification for special purposes and on a narrower basis, and also includes the stiffening of more qualitative systems through the infusion of parametric ingredients.

The tenor of claims made for the parametric approach is that it achieves a more precise definition of land and that it avoids the subjectivity of the landscape method; being quantitative, it allows comparison between and affords greater consistency within land evaluation projects; and it is in terms suited to automatic scanners and computers.

Increasing advocacy of the parametric approach stems partly from dissatisfaction with the results of the landscape method of analysis, with its physiographic and morphogenetic bias, its qualitative framework, and its elements of subjectivity and inference. Not only, it is claimed, is the level of reproducibility, and hence of reliability, low, but the way in which land is defined makes it difficult to measure variance, to formulate rational sampling, or to express probability limits for the findings. But there are more positive reasons also, among them the introduction of new sensors enabling direct scanning of attributes which had formerly to be inferred from associated features. With the advent of these there disappears an important relative advantage of the landscape approach, namely the easy recognition of its land components. Another innovation favouring the parametric approach is electronic data handling, which sets quantitative criteria at a premium and which allows the incorporation of a much greater range of defining attributes. To some degree, also, recent moves towards parametric land classification have resulted from the influence of a new class of practitioners, notably engineers, whose rather specific demands require answers in quantitative form.

Problems inherent in the approach include the choice of attributes and the delimitation of attribute classes; the variance of attributes in space and time, and the translation into terms of land of the areas of occurrence and association of selected parameters.

[*Editor's Note:* Material has been omitted at this point.]

COMPARISONS AND COMBINATIONS

In summary, it is useful to revert to the fundamental requirements of land classification discussed at the outset.

The problem of defining land character along orderly principles so as to categorize and compare land types is answered in three ways. Under the genetic approach a theoretical solution is sought and an attempt made to particularize by degrees, using environmental controls as successive criteria, until a sufficiently narrow definition, in genetic terms, has been reached; however, it appears that no combination of criteria of this sort can yield the precision required for land-use planning. The landscape approach offers a largely empirical solution. The character of land is sought through its appearance as in an aerial view, and it is claimed that with an understanding of the genesis or dynamics of the land we may read the inherent character from the external forms, whereby types of land so defined will be found to be consistent in attributes discovered after closer analysis and experimentation. It is claimed that this ready means of assessing land character as a whole over large areas offers a precision and reliability adequate for moderately intensive land use. Dissatisfied with this last claim, adherents of the parametric approach are prepared to sacrifice comprehensiveness and ease of recognition for the reliability and quantitative output of a definition based on measured properties and note that in any case the picture they offer is becoming increasingly complete and readily obtained with new techniques of scanning and computing.

Since, clearly, what cannot be defined cannot be depicted, the genetic approach fails to deal with the problem of delimiting land types on a scale realistic for land use. The remaining approaches offer different solutions. The landscape approach distinguishes units of land which are invested with an overall character, whilst the parametric approach ascertains patterns of occurrence of selected attributes which may be combined into areal complexes. The landscape units are of realistic size for land-use planning and their delimitation presents little problem; however, a minimal probability that apparent similarity means consistent attributes could require a sampling intensity which might offset this ease of mapping. Areas defined under the parametric approach will be of high reliability, but their mapping may present problems unless scanning of attributes is possible. The more that is known about probable demands on the land, the less are the disadvantages of its narrowly defined land units. An important advantage of the parametric approach is the greater flexibility of spatial concept, with its possibilities of factor analyses of various types.

The problem of depicting the association of land units is answered directly under the landscape approach in that the recognition and demarcation of units is commonly at the level of component groupings which are regarded as being functionally or genetically linked. Somewhat similarly, the object of the genetic approach is to show each unit as grouped in a region of higher order in terms of a genetic factor operative at the larger scale. The isarithmic patterns which form the output of the parametric approach offer an alternative answer in that they indicate the working of areal-process factors at a range of scales, for a number of scanned attributes, and without commitment to any regional framework.

It has been shown that, at the scale of detail required for practical purposes, land can be mapped into landscape units or land-attribute occurrences, and that both methods, in their different ways, answer the requirements of land classification. The two approaches are really not alternatives and can in fact be combined with profit. The relative advantages and hence the degree of emphasis to be placed on one approach or the other are not fixed, but will vary with the circumstances under which they are applied and the technical resources that can be drawn upon. With what is presently regarded as standard equipment, the landscape approach offers the possibility of more rapid survey at low cost. The reliability, although not outstanding, is consistent with reconnaissance investigation as at present conceived, and can be raised, within limits and at a price, by more intensive sampling. Ultimately, there comes a level of investigation at which the greater precision and reliability of the parametric approach are needed and to the extent that improvements in scanning render the method more comprehensive, its inherent advantages of reliability will be exploited, even at the reconnaissance level of investigation.

In practice, the methods are combined to reinforce each other. For instance, past survey procedure of the CSIRO Division of Land Research would be defined as the landscape approach to land classification, but the planning of the operation, including the selection of sampling points, may be guided by a preliminary breakdown of the area on broad physiographic lines, comparable with the genetic approach (Christian, Jennings, and Twidale 1957). Similarly, the test sites of the Waterways Experiment Station may be classified by parametric methods, but for the identification of land in other little-known areas in analogous terms it may be necessary to identify these forms with physiographic components and patterns as under the landscape approach. One can envisage many cases in which the definition and reproducibility of landscape units could be improved by the testing of associated parameters; on the other hand, it may be some time before the parametric approach can provide a sufficiently comprehensive and mappable definition of land for general-purpose classification without some additional support from landscape analysis. It seems inevitable that, with increasing auto-scanning and data handling, as our knowledge of land increases, and as the demands of the land user become more specialized, that parametric analysis will become general practice. But perhaps the ultimate advantage of the approach will be the flexibility of spatial concept that it allows. There seems to be no reason why, at more advanced stages of investigation, the regional concept of convention may not be abandoned completely in favour of measures of functional interdependence such as connectivity, for with progressive development it is the links rather than the breaks in the landscape with which we become increasingly concerned.

REFERENCES

CHRISTIAN, C. S., JENNINGS, J. N., and TWIDALE, C. R. (1957). Geomorphology. Ch. 5. In *Guide Book to Research Data for Arid Zone Development.* (UNESCO: Paris.)

FENNEMAN, M. N. (1916). Physiographic divisions of the United States. *Ann. Ass. Am. Geogr.* **6**, 19–98.

HERBERTSON, A. J. (1905). The major natural regions: an assay in systematic geography. *Geogrl J.* **20**, 300–12.

LINTON, D. L. (1951). The delimitation of morphological regions. Ch. 11. In *London Essays in Geography.* (Longmans Green: London.)

UNSTEAD, J. F. (1933). A system of regional geography. *Geog.* **18**, 185–7.

46

Reprinted from *Z. Geomorph.*, 16(4), 367–373 (1972)

PRINCIPLES IN A GEOMORPHOLOGICAL APPROACH TO LAND CLASSIFICATION

R. L. Wright

[*Editor's Note:* In the original, material precedes this excerpt.]

Geomorphological surveys

In considering the potential role of geomorphology in land classification, it is necessary to examine other, more specialized geomorphological approaches to regional description. In recent years these have been of two main types: landform surveys in which different parts of an area are "classed" quite separately, each in terms of its salient geomorphological characteristics, measured or inferred; and "special-feature" investigations in which all parts of an area are differentiated on the basis of measured values of a selected geomorphological property or combination of properties. The first type forms the basis of "detailed geomorphological mapping". It provides a marked contrast to the special-feature method and has been applied much more widely, especially in Europe, with KLIMASZEWSKI of Poland and TRICART of France in the forefront. The objectives of detailed geomorphological mapping are to obtain as complete an inventory of the landforms of an area, and as full an understanding of them, as possible. Based on "a precise, detailed field survey", to use TRICART's (1965) terms, it aims at an "explanatory description of all of the elements of the relief", and "every geomorphological feature must appear". Similarly, KLIMASZEWSKI (1961) explicitly states that the work should incorporate "a record of all landforms as observed in the field, together with particulars of their dimensions, origin and age", for a geomorphological map "must not only present a full picture of the relief but must be constructed so that the evolution of the relief can also be clearly interpreted".

To give the full explanatory description that is intended, leaders in this field stress that four kinds of data must be considered: morphographic, morphometric, morphogenetic and morphochronologic. In practice, however, emphasis is placed upon genetic and chronologic considerations. Thus the Polish map legend (KLIMASZEWSKI 1963) and that proposed by TRICART (1965) comprise detailed genetic-chronologic classifications of landforms. In both cases the main groups –for example, "fluvial" (KLIMASZEWSKI) or "work of running water" (TRICART)– contain erosional and depositional subclasses. Individual landforms are represented by symbols, colour being used by KLIMASZEWSKI to indicate both origin and age, and by TRICART to represent age and lithology. Most workers in this field have adopted the same general principles as KLIMASZEWSKI, TRICART and their colleagues, though with some differences in detail.[7]

An attempt has been made to formulate a standardized scheme since the first conference of the I.G.U. sub-commission on geomorphological mapping held at Cracow in 1962. In the light of resolutions made there, a "unified key to the detailed geomorphological map" has been devised. This was first presented in preliminary form (BASHENINA et al. 1966) but has been recently finalized (BASHENINA et al. 1968). In most regards, however, the unified key is closely similar to the Polish map legend, though it includes in its finalized version six slope classes to supplement contour data–a notable departure from the preliminary version and the Polish prototype. These classes (including 0–2°, 3–5°, 6–15°, for example) are selected "on the basis of morphogenetic as well as technical and economic criteria" (BASHENINA et al. 1968).

Detailed geomorphological maps as conceived by KLIMASZEWSKI and TRICART are understandably limited in number because intensive field work is required. But the specialist nature of those published is most apparent. Thus TRICART's (1965) type example is an explanatory representation of the Vaunage area, with a particularly detailed and impressive genetic-chronologic breakdown for the lowland parts. However, information about the actual character of landforms is restricted: including only contours and rock type for much of the upland–though this is clearly diverse terrain as evidenced by the contour pattern–and only contours and broad textural groupings of surficial materials for much of the lowlands. As with integrated resource surveys, this kind of approach provides an inventory rather than a classification of landforms, with various parts of the same area separated in quite different terms–according to age, or genesis, or bedrock, or morphometry, for example. The classificatory limitations of the system itself are evident in TRICART's (1965) scheme, in that "classes" within a particular morphogenetic group are differentiated on a varied basis: according to chemical composition of materials in some cases and texture in others; according to form dimensions in some cases and vegetative cover in others. Such an approach often fails to achieve any kind of regional synthesis, however, a consequence that was aptly put by KELLOGG (1940): "What appears ... to be a single map ... may have thousands of different kinds of symbols, different kinds of areas, in which it is unconsciously assumed that each selected characteristic has exactly the same significance under all conditions."

Furthermore, the applied value of this method for non-geomorphologists is curtailed by the kinds of data that are emphasized and the terms used in their presentation. Although it is contended that a wide range of factual as well as "explanatory" information should be represented in these maps, only a minute fraction of the geomorphological complexity of any area can be considered.

[7] See, for example, *Geographical Studies* No. 46, Warsaw (1963).

TRICART's (1965) standpoint is quite clear: morphogenetic information must be the "focal point" of the map; morphometric and other data being suppressed to avoid overloading the map and because "specific geomorphological data" must have priority. In practice such maps reflect to a marked degree the training, interests and abilities of the observers. Moreover, all are couched in technical terms, mainly unintelligible to other than a trained geomorphologist. In addition, with their "explanatory" theme there is a notable lack of quantitative information, both for form and materials, in many of these maps. This is a striking omission if they are to be of practical value. However, the maps are not regarded as "ends in themselves", and Polish workers especially urge that they should be adaptable as tools in other fields. STARKEL's "geomorphological-improvement" map of the upper San basin (BRYDAK & PLASKACZ 1963) illustrates this, and, with its emphasis on morphological representation within a framework of slope classes, represents a significant departure from the standard genetic-chronologic types.

At the same time as the "explanatory" system of geomorphological mapping was evolving in Europe, an empirical approach to the problem was conceived in the United Kingdom. Developed mainly by LINTON, SAVIGEAR & WATERS, the approach is based on precise mapping of the shape of the ground in terms of "flats" and "slopes"–LINTON's "ultimate units of relief". The technique requires detailed survey of the slope discontinuities which delimit these "facets" (WATERS 1958), and their "classification" according to curvature or angularity (SAVIGEAR 1956, 1965). The facets thus defined can be given generalized expression in terms of structure, process and stage, and have considerable pedological and ecological implications (LINTON 1951). Theoretically, therefore, they provide a framework for the construction of any number of types of geomorphological maps according to the nature of the terrain and the interests of the observer. However, in contrast with the morphogenetic-chronologic bias of the European approach, the maps produced by these British workers have been chiefly limited to morphologic-morphometric types.

Of attempts at combining the two approaches–"explanatory" and empirical–those of Belgian geomorphologists are especially noteworthy. Thus, aiming to produce maps which are at the same time descriptive and interpretative, workers at the "Centre National de Recherches Geomorphologique" first carried out morphological surveys, then on this basis proceeded to an interpretation of the genesis and evolution of the landforms (MACAR et al. 1960; GULLENTOPS 1963). In maps produced by the "Centre", morphological facts are presented in the British style and with slope units further defined according to various systems of gradient classes. Morphogenetic-chronologic information is either superimposed or given separately. While not achieving the level of morphological detail of SAVIGEAR (1965), for instance, nor the full interpretative statements of the best European examples, the Belgian attempts incorporate a successful blend of both methods.

The second type of more specialized geomorphological approach to regional differentiation is the "special-feature" method, of which the "terrain analogs" prepared for the United States Army (VAN LOPIK & KOLB 1959) provide an example. In this American work, areas are classified in terms of land-surface

27*

characteristics, or "terrain factors", important in military evaluation, and including various geomorphological properties. Separate maps of an area are constructed in terms of each of these "factors" using essentially arbitrarily defined classes but incorporating where possible military and "naturalistic" considerations. Small-scale maps of extensive desert areas in the northern hemisphere are produced, based on existing maps and extrapolated information from analogous terrain in the United States. Thus, while the terrain classes represent fine distinctions in most cases, actual mapping units are necessarily generalized.

A further example of the special-feature approach is the land-surface form map of the United States, produced by HAMMOND (1954) primarily to meet the requirements of geographers. Mapping units are expressed in terms of a combination of three properties–amplitude of relief, per cent of near-level land, and "profile type", or the vertical arrangement of near-level land. Classes for each of these are somewhat arbitrary, values being chosen which would separate "distinctive kinds" of terrain found in North America. On this basis various "terrain types", each comprising a particular combination of the preconceived classes for the three selected properties, are recognized and mapped at a broad scale from existing topographic maps. Similarly, DEITCHMAN (1959) devised a method of terrain classification in order to define transport system requirements, and applied it to produce a very generalized representation of an area extending from Western Europe through South Asia to Korea. Various environmental "characteristics" are considered, but with the geomorphological attributes of slope and amplitude of relief of prime concern. In terms of these and annual rainfall, and in a manner somewhat analogous to that of HAMMOND, the area is differentiated according to a preconceived set of terrain types. More intensive studies of this kind include investigations into the problem of cross-country movement of men and vehicles (STRAHLER & KOONS 1960; STONE & DUGUNDJI 1965). A guiding principle in the special-feature approach is that surveys must be rigidly planned in accordance with the use to which information is to be put. This is especially evident in these more detailed examples which are based on collecting precise, numerical data for those properties actually determining resistance to overland movement, including ground slope and surface "roughness", or microrelief, for instance.

In contrast with the explanatory method of detailed geomorphological mapping, a "parametric" approach (MABBUTT 1968)–as employed in special-feature studies–is especially relevant in land classification. In this way, terrain character can be expressed precisely and objectively according to measured ranges of values for specific properties. As MABBUTT makes clear, the approach permits a more rigorously scientific evaluation of land, and is in terms suited to modern techniques of data collection, analysis and storage. One difficulty, however, is the spatial translation of samplingpoint data, for, as noted above, accurate regionalization on this basis requires an extremely close sampling network. This question receives little attention in the special-feature studies referred to, and, in work of this kind, spatial patterns are commonly differentiated in terms of preconceived terrain types and based on interpretation of contour maps and aerial photographs. For example, STRAHLER & KOONS (1960) indicate that types "composed of the same terrain elements" might be identified from aerial photographs mainly, in order to provide a regional framework to be subsequently tested and finalized in the field. However, MABBUTT suggests the problem may be resolved by isarithmic mapping, for, depending on the purpose of the study, this could provide a suitable alternative to regional identification. Where land classification and regionalization are required, the special-feature method is best suited to intensive studies of small areas for clearly specified purposes. This is not to imply that general-purpose classifications of larger areas cannot be based on measured values of terrain properties. But in this case the special-feature method needs reinforcing by some system of land units compounded from areal "individuals" rather than simply point-data, and built-up on the ground rather than according to preconceived terrain types. This would permit information to be collected, compared and extrapolated methodically, especially where it is impractical to sample terrain conditions using either a truly random or a regular basis. It would also enable regional contrasts to be distinguished with greater precision than would otherwise be possible.

Literatur

BASHENINA, N. V., J. GELLERT, F. JOLY, M. KLIMASZEWSKI & E. SCHOLZ (1966): The unified key to the detailed geomorphological map. – I.G.U. Subcomm. on Geomorph. Mapping. Mimeograph.

– (1968): The unified key to the detailed geomorphological map of the world. – Folia Geographica, Ser. Geographica-Physica. 2, II, Krakow.

BRYDAK, K. & J. PLASKACZ (1963): Eine kurze Mitteilung über die Bedeutung der geomorphologischen Karte bzw. Bonitierungskarte für die Bearbeitung des Raumbewirtschaftungsplanes. – Geogr. Stud. (Warsaw). 46: 27–31.

DEITCHMAN, S. J. (1959): Classification and quantitative description of large geographic areas to define transport system requirements. – In: Symposium on Quantitative Terrain Studies. Am. Assn. Adv. Sci. Chicago.

GULLENTOPS, F. B. (1963): La cartographie géomorphologique en Belgique. – Geogr. Stud. (Warsaw). 46: 57–58.

HAMMOND, E. H. (1954): Small-scale continental landform maps. – Ann. Assn. Am. Geogrs. 44: 33–42.

KELLOGG, C. E. (1940): The theory of land classification. 1. The contributions of soil science and agronomy to rural land classification. – Missouri Agric. Expt. Sta. Bull. 421: 164–173.

KLIMASZEWSKI, M. (1961): The problems of the geomorphological and hydrographic map on the example of the Upper Silesian industrial district. – Geogr. Stud. (Warsaw). 25: 73–81.

– (1963): Landform list and signs used in the detailed geomorphological map. – Geogr. Stud. (Warsaw). 46: 139–177.

LINTON, D. L. (1951): The delimitation of morphological regions. – In: London Essays in Geography. Longmans, London.

MABBUTT, J. A. (1968): Review of concepts of land classification. – In: Land Evaluation (Ed. G. A. STEWART). Macmillan, Melbourne.

MACAR, P., P. DE BETHUNE, J. MAMMERICKX & G. SERET (1960): Travaux preparatoires à l'elaboration d'une carte géomorphologique de Belgique. – Ann. Soc. Géol. de Belg. 84: 179–198.

SAVIGEAR, R. A. G. (1956): Technique and terminology in the investigation of slope forms – Premier Rapp. de la Comm. pour l'Étude des Versants, Union Géogr. Internat. 66–76.

– (1965): A technique of morphological mapping. – Ann. Assn. Am. Geogrs. 55: 514–538.

STONE, R. O. & J. DUGUNDJI (1965): A study of microrelief – its mapping, classification and quantification by means of Fourier analysis. – Eng. Geol. 1: 89–187.

STRAHLER, A. N. & D. KOONS (1960): Objective and quantitive field methods of terrain analysis. – Final Rep. Project NR 387-021. Dept. of Geol., Columbia Univ.

TRICART, J. (1965): Principes et Methodes de la Géomorphologie. – Masson, Paris.

VAN LOPIK, J. R. & C. R. KOLB (1959): A technique for preparing desert terrain analogs. – U.S. Army Corps Engrs. Tech. Rep. 3–506. Vicksburg, Mississippi.

WATERS, R. S. (1958): Morphological mapping. – Geography. 43: 10–17.

AUTHOR CITATION INDEX

SUBJECT INDEX

About the Editor

CUCHLAINE A. M. KING is Professor of Physical Geography and at present Head of Department of the Geography Department, Nottingham University, England. She teaches courses in geomorphology, physical geography, and oceanography.

Professor King received the degrees of B.A. (1943), M.A. (1946), Ph.D. (1949), and Sc.D. (1973) from the University of Cambridge, England, in geography and geomorphology. In 1972–1973 she held a Senior Visiting Fellowship of the National Science Foundation at the State University of New York, Binghamton. She is author and co-author of 12 books, including second editions, on various aspects of geography and geomorphology, with special emphasis on coastal and glacial geomorphology.